THE AIRLINER

CABIN
ENVIRONMENT

AND THE

HEALTH

OF

PASSENGERS

AND

CREW

Committee on Air Quality in Passenger Cabins of
Commercial Aircraft

Board on Environmental Studies and Toxicology

Division on Earth and Life Studies

National Research Council

NATIONAL ACADEMY PRESS
Washington, D.C.

MAY 0 6 2002

NATIONAL ACADEMY PRESS 2101 Constitution Ave., N.W. Washington, D.C. 20418

NOTICE: The project that is the subject of this report was approved by the Governing Board of the National Research Council, whose members are drawn from the councils of the National Academy of Sciences, the National Academy of Engineering, and the Institute of Medicine. The members of the committee responsible for the report were chosen for their special competences and with regard for appropriate balance.

This project was supported by Award No. DTFA0100P100P10285 between the National Academy of Sciences and the U.S. Department of Transportation. Any opinions, findings, conclusions, or recommendations expressed in this publication are those of the author(s) and do not necessarily reflect the view of the organizations or agencies that provided support for this project.

Library of Congress Control Number 2001099122

International Standard Book Number 0-309-08289-7

Cover photograph by Steve Cole, Photodisc

Additional copies of this report are available from:

National Academy Press
2101 Constitution Ave., NW
Box 285
Washington, DC 20055

800-624-6242
202-334-3313 (in the Washington metropolitan area)
http://www.nap.edu

THE NATIONAL ACADEMIES
National Academy of Sciences
National Academy of Engineering
Institute of Medicine
National Research Council

The **National Academy of Sciences** is a private, nonprofit, self-perpetuating society of distinguished scholars engaged in scientific and engineering research, dedicated to the furtherance of science and technology and to their use for the general welfare. Upon the authority of the charter granted to it by the Congress in 1863, the Academy has a mandate that requires it to advise the federal government on scientific and technical matters. Dr. Bruce M. Alberts is president of the National Academy of Sciences.

The **National Academy of Engineering** was established in 1964, under the charter of the National Academy of Sciences, as a parallel organization of outstanding engineers. It is autonomous in its administration and in the selection of its members, sharing with the National Academy of Sciences the responsibility for advising the federal government. The National Academy of Engineering also sponsors engineering programs aimed at meeting national needs, encourages education and research, and recognizes the superior achievements of engineers. Dr. Wm. A. Wulf is president of the National Academy of Engineering.

The **Institute of Medicine** was established in 1970 by the National Academy of Sciences to secure the services of eminent members of appropriate professions in the examination of policy matters pertaining to the health of the public. The Institute acts under the responsibility given to the National Academy of Sciences by its congressional charter to be an adviser to the federal government and, upon its own initiative, to identify issues of medical care, research, and education. Dr. Kenneth I. Shine is president of the Institute of Medicine.

The **National Research Council** was organized by the National Academy of Sciences in 1916 to associate the broad community of science and technology with the Academy's purposes of furthering knowledge and advising the federal government. Functioning in accordance with general policies determined by the Academy, the Council has become the principal operating agency of both the National Academy of Sciences and the National Academy of Engineering in providing services to the government, the public, and the scientific and engineering communities. The Council is administered jointly by both Academies and the Institute of Medicine. Dr. Bruce M. Alberts and Dr. Wm. A. Wulf are chairman and vice chairman, respectively, of the National Research Council.

Rethinking the Ozone Problem in Urban and Regional Air Pollution (1991)
Decline of the Sea Turtles (1990)

Copies of these reports may be ordered from the National Academy Press
(800) 624-6242 or (202) 334-3313
www.nap.edu

Preface

In 1986, a committee of the National Research Council (NRC), the principal operating arm of the National Academy of Sciences and the National Academy of Engineering, produced a report requested by Congress titled *The Airliner Cabin Environment: Air Quality and Safety*. That report recommended the elimination of smoking on most domestic airline flights and a number of other actions to address health and safety problems and to obtain better data on cabin air quality. In response to that report, the Federal Aviation Administration (FAA) took several actions, including the banning of smoking on all domestic flights. However, the health complaints of passengers and cabin crew continue. Their complaints tend to be broad and nonspecific and to have multiple possible causes, including air contaminants, so it is difficult to define or discern a precise illness or syndrome.

As a result of continued concerns about aircraft cabin air quality and health issues raised by passengers and cabin crew, Congress directed FAA in the Wendell H. Ford Aviation Investment and Reform Act of the 21st Century, enacted in 2000, to request that the NRC perform another independent study to examine cabin air quality.

In this report, the Committee on Air Quality in Passenger Cabins of Commercial Aircraft reviews what is known about air quality in passenger cabins, emphasizing studies conducted since the 1986 report. The committee specifically examined the aircraft environmental control systems, the sources of contaminants in aircraft cabins, and the toxicity and health effects associated with these contaminants; it provides a number of recommendations for potential approaches for improving cabin air quality.

Preface

This report has been reviewed in draft form by persons chosen for their diverse perspectives and technical expertise, in accordance with procedures approved by the NRC's Report Review Committee. The purposes of this independent review were to provide candid and critical comments to assist the institution in making its published report as sound as possible and to ensure that the report meets institutional standards of objectivity, evidence, and responsiveness to the study charge. The review comments and draft manuscript remain confidential to protect the integrity of the deliberative process. We wish to thank the following for their review of this report: Charles E. Becker (emeritus), University of California, San Francisco School of Medicine, Snow Mass Village, Colorado; Franklin D. Farrington, Boeing Company, Long Beach, California; Ashok Gadgil, Lawrence Berkeley National Laboratory, Berkeley, California; R. Richard Heppe (retired), California Lockheed, Solvang, California; Donald F. Hornig (emeritus), Brown University, Little Compton, Rhode Island; Nadia S. Juzych, Michigan Public Health Institute, Birmingham, Michigan; Roger O. McClellan (emeritus), Chemical Industry Institute of Toxicology, Albuquerque, New Mexico; James M. Melius, New York State Laborers' Health and Safety Trust Fund, Albany, New York; Shelly Miller, University of Colorado, Boulder, Colorado; Niren L. Nagda, Energen Consulting, Inc., Germantown, Maryland; P. Barry Ryan, Emory University, Atlanta, Georgia; John C. Sagebiel, Desert Research Institute, Reno, Nevada; Calvin C. Willhite, California Environmental Protection Agency, Berkeley, California.

Although the reviewers listed above have provided many constructive comments and suggestions, they were not asked to endorse the conclusions or recommendations, nor did they see the final draft of the report before its release. The review of the report was overseen by John C. Bailar, III (emeritus), University of Chicago, Chicago, Illinois, and Edward C. Bishop, Parsons Engineering Science, Inc., Fairfax, Virginia. Appointed by the NRC, they were responsible for making certain that an independent examination of this report was carried out in accordance with institutional procedures and that all review comments were carefully considered. Responsibility for the final content of the report rests entirely with the author committee and the institution.

The committee gratefully acknowledges the following for making presentations to the committee: Charles Ruehle and Thomas Nagle, FAA; Martha Waters, Elizabeth Whelan, and Kevin Dunn, National Institute for Occupational Safety and Health; Christopher Witkowski and Judith Murawski, Association of Flight Attendants; Olney Anthony, International Association of Machinists Union and Aerospace Workers (IAM); David Space and Richard Johnson, Boeing Corporation; Martin Dechow, Airbus Corporation; Richard Fox and George Rusch, Honeywell Corporation; Raynard Fenster, Information Overload Corporation; and Jolanda Janczewski, Consolidated Safety Services, Inc. The committee also wishes to thank the following who provided further background information: Gene Kirkendahl and Stephen Happenny, FAA; Ron Shepard, IAM; Jim McClendon, Alaska Airlines; John Downey, BAE Systems; Mac Cookson, Steve Ramdeen, and Kilisi Vailu'u, United Airlines; Sarah Knife, General Electric Aircraft Engines; Keith Morgan, Pratt & Whitney; Wayne Daughtrey, ExxonMobil Corporation; Vincent Johnston, Boeing Corporation; and Robert Wright, U.S. Air Force. The committee gives special thanks to staff at United Airlines who provided site visits of its major maintenance facilities in Oakland, California, and Indianapolis, Indiana, and provided us with additional background information: Clayton Satterlee, Yvonne Daverin, John Upchurch, Anita Davis, Roger Rube, Robert Patterson, Steve Lewis, and Rick Ransom.

The committee is thankful for the useful input of Charles Schumann in the early deliberations of this study. The committee is also grateful for the assistance of the NRC staff in preparing this report. Staff members who contributed to this effort are Eileen Abt, project director; Roberta Wedge, senior program officer; Ellen Mantus, program officer; Norman Grossblatt, editor; Ruth Crossgrove, managing editor; Lucy Fusco, senior project assistant; Mirsada Karalic-Loncarevic, research assistant; and Bryan Shipley, project assistant. I would also like to thank all the members of the committee for their dedicated efforts throughout the development of this report.

Finally, the committee extends its heartfelt condolences to those who lost family, friends, and colleagues in the events of September 11, 2001. These events will undoubtably have extensive repercussions on all aspects of air transportation. Although safety is always the overriding priority for air transportation, air quality in the aircraft cabin will also continue to be an important factor affecting the health of passengers and crew. The committee hopes that this report will make a long-lasting contribution to the goal of ensuring the health of all who fly aboard commercial aircraft.

Morton Lippmann, *Chair*
Committee on Air Quality in Passenger
Cabins of Commercial Aircraft

Contents

Contents

The Airliner Cabin Environment
and the Health of
Passengers and Crew

Summary

The number of people traveling by commercial aircraft in recent years is unprecedented. Over the last 30 years, the number of air passengers worldwide has nearly quadrupled, from 383 million in 1970 to 1,462 million in 1998. More older and younger people are flying, including adults with medical conditions (e.g., cardiovascular and pulmonary disease), children, and infants.

The aircraft cabin is similar to other indoor environments, such as homes and offices, in that people are exposed to a mixture of outside and recirculated air. However, the cabin environment is different in many respects—for example, the high occupant density, the inability of occupants to leave at will, and the need for pressurization. In flight, people encounter a combination of environmental factors that includes low humidity, reduced air pressure, and potential exposure to air contaminants, such as ozone (O_3), carbon monoxide (CO), various organic chemicals, and biological agents.

Over the years, passengers and cabin crew (flight attendants) have repeatedly raised questions regarding air quality in the aircraft cabin. In 1986, a committee of the National Research Council (NRC), the principal operating arm of the National Academy of Sciences and the National Academy of Engineering, produced a report requested by Congress titled *The Airliner Cabin Environment: Air Quality and Safety*. That report recommended the elimination of smoking on most domestic airline flights and other actions to address health and safety problems and to obtain better data on cabin air quality. In response, the Federal Aviation Administration (FAA) took several actions, including a ban on smoking on all domestic flights. However, 15 years later, many of the other issues about aircraft cabin air quality have yet to be ade-

1

quately addressed by FAA and the airline industry, and new health questions have been raised by the public and cabin crew.

In response to the unresolved issues, Congress—in the Wendell H. Ford Aviation Investment and Reform Act of the 21st Century, enacted in 2000—directed the FAA to ask the NRC to perform another independent study to assess airborne contaminants in commercial aircraft, to evaluate their toxicity and associated health effects, and to recommend approaches to improve cabin air quality.

THE CHARGE TO THE COMMITTEE

The NRC convened a new committee, the Committee on Air Quality in Passenger Cabins of Commercial Aircraft, which prepared this report. The committee's members were selected for expertise in industrial hygiene, exposure assessment, toxicology, occupational and aerospace medicine, epidemiology, microbiology, aerospace and environmental engineering, air monitoring, ventilation and airflow modeling, and environmental chemistry. The committee was charged to address the following topics:

1. Contaminants of concern, including pathogens and substances that are used in the maintenance, operation, or treatment of aircraft, including seasonal fuels and deicing fluids.
2. The systems of passenger cabin air supply on aircraft and ways in which contaminants might enter such systems.
3. The toxic effects of the contaminants of concern, their byproducts, the products of their degradation, and other factors, such as temperature and relative humidity, that might influence health effects.
4. Measurements of the contaminants of concern in the air of passenger cabins during domestic and foreign air transportation and comparison with measurements in public buildings, including airports.
5. Potential approaches to improve cabin air quality, including the introduction of an alternative supply of air for the aircraft passengers and crew to replace bleed air.

The committee was not asked or constituted to address the possible effects of ionizing and nonionizing radiation. Furthermore, the committee did not, nor was it asked to, evaluate the potential costs of implementing any of its recommendations.

THE COMMITTEE'S APPROACH TO ITS CHARGE

The committee heard, in public session, presentations from FAA, the Association of Flight Attendants (AFA), the National Institute for Occupational Safety and Health, the International Association of Machinists and Aerospace Workers (IAM), manufacturers of aircraft and aircraft equipment (Boeing, Airbus, and Honeywell), and consulting firms (Consolidated Safety Services, Inc., and Information Overload Corporation). The committee evaluated the body of literature on air quality in commercial aircraft, emphasizing studies conducted since the 1986 NRC report. It also solicited information from FAA, airlines, aircraft and engine manufacturers, AFA, IAM, and manufacturers of engine lubricating oils and hydraulic fluids. Committee members visited United Airlines heavy maintenance facilities in Oakland, California, and Indianapolis, Indiana.

In addressing its charge, the committee focused on air quality in aircraft regulated by the FAA, on national and international flights. It did not focus on specific types or models of aircraft, but rather examined air quality in commercial aircraft in general. Various aircraft models might differ in cabin air quality, but the committee considered the exposure and health-related issues to be applicable to most commercial aircraft systems.

THE COMMITTEE'S EVALUATION

Aircraft Systems

Commercial jet aircraft operate in an external environment that varies widely in temperature, air pressure, and relative humidity as they move from taxiing and takeoff through cruise to descent and landing. To transport passengers and crew through environmental extremes, an aircraft is equipped with an environmental control system (ECS) designed to maintain a safe, healthful, and comfortable environment for the passengers and crew. The air provided to the passengers and crew on jet aircraft is typically a combination of outside air brought in through the engines and air that is taken from the cabin, filtered, and recirculated. The ECS is designed to minimize the introduction of harmful contaminants into the cabin and to control cabin pressure, ventilation, temperature, and humidity.

To promote safe and healthful air aboard commercial aircraft, FAA has established design and operational specifications—Federal Aviation Regulations (FARs) in 14 CFR 21, 14 CFR 25, 14 CFR 121, and 14 CFR 125—for O_3, CO, carbon dioxide (CO_2), ventilation, and cabin pressure.

After reviewing the role and function of the ECS on most aircraft, the committee concluded that the ECS, when operated as specified by the manufacturer, should provide an ample supply of air to pressurize the cabin, meet general comfort conditions, and dilute or otherwise reduce normally occurring odors, heat, and contaminants. The committee noted, however, that the current design standard of a minimum of 0.55 lb of outside air per minute per occupant (FAR 25.831) is less than one-half to two-thirds the ventilation rate recommended in American Society of Heating, Refrigerating, and Air-Conditioning Engineers (ASHRAE) Standard 62-1999, which was developed for building environments. Whether the building ventilation standard is appropriate for the aircraft cabin environment has not been established.

Exposures on Aircraft

Although the ECS is designed to minimize the concentrations of contaminants in the cabin, contaminant exposures do occur. They can originate outside the aircraft, inside the aircraft, and in the ECS itself. There are two distinct types of contaminant exposures: those which occur under routine operating conditions and those which occur under abnormal operating conditions. Contaminant exposures that occur under routine conditions include odors and gases emitted by passengers, O_3 that enters with ventilation air during high-altitude cruise, organic compounds emitted from residual cleaning materials and other materials in the cabin, and infectious agents, allergens, irritants, and toxicants. During nonroutine events, contaminant exposures result from the intake of chemical contaminants (e.g., engine lubricating oils, hydraulic fluids, deicing fluids, and their degradation products) into the ECS and then into the cabin.

A number of studies have attempted to collect data on occupant exposures to air contaminants in aircraft cabins under routine conditions. The data represent only a small number of flights, and the studies have varied considerably in their sampling strategies, the environmental factors monitored, and the measurement methods used. Consequently, cabin air quality under routine conditions has not been well characterized. Furthermore, no published studies describe quantitative measurements of air quality under abnormal operating conditions.

Exposures to contaminants that originate outside the aircraft differ, depending on whether the aircraft is on the ground, in ascent, in cruise, or in descent. When the aircraft is on the ground, exposures to outdoor air pollutants (e.g., O_3, CO, and particulate matter) are determined primarily by the ambient concentrations at the airport. During cruise at high altitudes, O_3 concentrations are elevated in ambient air. Although the FAA requires that O_3 concentrations in the aircraft cabin be maintained within specified limits, studies indicate that cabin O_3 concentrations on some flights may exceed the FAA regulatory standards and the Environmental Protection Agency national ambient (outdoor) air standards.

Passengers and crew themselves are the sources of several contaminants (e.g., bioeffluents, viruses, bacteria, allergens, and fungal spores) that originate in the cabin. Furthermore, structural components of the aircraft, luggage, personal articles, animals brought on board, food, and sanitation fluids can be sources of vapors or particles. Cabin surfaces can be sources of residues of cleaning compounds, pesticides, and dust. Passengers and crew have raised questions about exposure to pesticides (e.g., *d*-phenothrin and permethrin), because they are sprayed on selected international flights to limit the spread of insect pests, but no quantitative data are available on passenger or crew exposures to these compounds.

The ECS can be a source of contamination. Problems arise when engine lubricating oils, hydraulic fluids, or deicing fluids unintentionally enter the cabin through the air-supply system from the engines in what is called bleed air. Many cabin crews and passengers have reported incidents of smoke or odors in the cabin. No exposure data are available to identify the contaminants in cabin air during air-quality incidents, but laboratory studies suggest that many compounds are released when the fluids mentioned above are heated to the high temperatures that occur in the bleed-air system.

Health Considerations

Available exposure information suggests that environmental factors, including air contaminants, can be responsible for some of the numerous complaints of acute and chronic health effects in cabin crew and passengers. The complaints tend to be so broad and nonspecific and can have so many causes that it is difficult to define or discern a precise illness or syndrome. The current data collection systems administered by the National Air and Space Administration and AFA, designed to report health complaints of cabin crew and passengers, do not have standardized, systematic methods to collect and record these reports. Furthermore, FAA does not collect health-effects data. Therefore, establishing a causal relationship between cabin air quality and the health complaints of cabin crew and passengers is extremely difficult.

Among the possible causes of the symptoms reported by passengers and cabin crew are the cabin environment itself (e.g., cabin pressure and relative humidity), contaminants (e.g., O_3, pesticides, biological agents, and constituents and degradation products of engine lubricating oils and hydraulic fluids), physiological stressors (e.g., fatigue, cramped seats, and jet lag), and exacerbation of pre-existing conditions in sensitive groups.

Regarding the flight environment, the committee identified two cabin air-quality characteristics that should be given high priority for further investigation: reduced oxygen partial pressure and elevated O_3 concentrations. Although reduced oxygen partial pressure in the aircraft cabin at cruise altitude should not affect healthy people adversely, health-compromised people, particularly those with cardiopulmonary disease, might experience a variety of symptoms. Infants could also be adversely affected because of their greater oxygen requirements. Elevated O_3 concentrations have been associated with airway irritation, decreased lung function, exacerbation of asthma, and impairments of the immune system.

The presence of some biological agents in cabin air, primarily airborne allergens, has also raised questions. Exposures to allergens (e.g., cat dander) have been reported to cause health effects, but have not been definitively documented in aircraft. Transmission of infectious agents from person to person has been documented to occur in aircraft, but the most important transmission factors appear to be high occupant density and the proximity of passengers. Transmission does not appear to be facilitated by aircraft ventilation systems.

Other cabin air contaminants or characteristics during routine operations are generally not expected to cause adverse health effects. One possible exception is the low relative humidity that occurs on nearly all flights. Low relative humidity might cause some temporary discomfort (e.g., drying of the eyes, nose, and skin) in cabin occupants; its role in causing or exacerbating short- or long-term health effects has not been established. Another possible exception is exposure to pesticides that are applied on some international flights; these chemicals can cause skin irritation and are reported to be neurotoxic, although of low toxicity in humans.

During abnormal operating conditions, exposure to several contaminants might occur. The engine lubricating oils and hydraulic fluids used in commercial aircraft are composed of a variety of organic constituents, including tricresyl phosphate, a known neurotoxicant. If the oils and fluids and their potential degradation products (e.g., CO and formaldehyde) enter the aircraft cabin, they will adversely affect cabin air quality. No data have definitively linked exposure to these compounds with reported health effects in cabin occupants.

Findings

The primary air-quality characteristics evaluated by the committee are summarized in Table S-1, which presents information on potential health effects, frequency of occurrence, and quality of the available data. The committee ranked the characteristics as of low, moderate, or high concern, on the basis of the likelihood of exposure and the potential severity of their effect. For example, hydraulic fluids or engine lubricating oils were ranked as of moderate concern because the potential severity of their effects is high but the likelihood of exposure to them at high concentrations is believed to be low. An important point to note is that the ability to evaluate a characteristic is limited in most cases because of a lack of data on exposure or health effects.

RECOMMENDATIONS

1. FAA should rigorously demonstrate in public reports the adequacy of current and proposed FARs related to cabin air quality and should provide quantitative evidence and rationales to support sections of the FARs that establish air-quality-related design and operational standards for aircraft (standards for CO, CO_2, O_3, ventilation, and cabin pressure). If a specific standard is found to be inadequate to protect the health and ensure the comfort of passengers and crew, FAA should revise it. For ventilation, the committee recommends that an operational standard consistent with the design standard be established.

2. FAA should take effective measures to ensure that the current FAR for O_3 (average concentrations not to exceed 0.1 ppm above 27,000 ft, and peak concentrations not to exceed 0.25 ppm above 32,000 ft) is met on all flights, regardless of altitude. These measures should include a requirement that either O_3 converters be installed, used, and maintained on all aircraft

TABLE S-1 Air-Quality Characteristics: Potential Health Impacts and Likelihood of Exposure

Characteristic[a]	Potential Health Impacts	Frequency of Exposure	Availability of Information
High Concern			
Cabin pressure	Serious health effects may occur in some people (e.g., infants and those with cardiorespiratory diseases) due to decreased oxygen pressure. Temporary pain or discomfort due to gas expansion (e.g., middle ear or sinuses) may occur.	Reduced cabin pressure occurs on nearly all flights.	Reliable measurements are available; health effects in some sensitive groups are uncertain.
Ozone	Health effects (e.g., airway irritation and reduced lung function) may occur at concentrations as low as 0.1 ppm with increasing severity at higher concentrations, exposure durations, and respiratory rates.	Elevated concentrations are expected primarily on aircraft without O_3 converters that fly at high altitudes; substantial uncertainty exists as to frequency and duration of elevated concentrations on these flights.	Few systematic measurements made since the 1986 NRC report.
Moderate Concern			
Airborne allergens	Inhalation can result in irritated eyes and nose, sinusitis, acute exacerbations of asthma, or anaphylaxis.	Frequency and intensity of exposure sufficient to cause sensitization or symptoms is not known.	Few exposure data are available; only self-reported information on hypersensitivity responses is available.
Carbon monoxide	Headaches and lightheadedness occur at low concentrations; more severe health effects result from higher concentrations and longer durations.	High concentrations could occur during air-quality incidents. Frequency of incidents is highly uncertain, but believed to be low.	Reliable measurements are available for normal operating conditions; no data are available for incidents.

Hydraulic fluids or engine oils (constituents or degradation products)	Mild to severe health effects can result from exposure to these fluids or their degradation products.	Frequency of incidents in which these fluids or degradation products enter the cabin is very uncertain, but is expected to be relatively low.	No quantitative exposure data are available. Little information is available on health effects related to smoke, mists, or odors in aircraft cabin.
Infectious agents	Exposure may have no effect or cause an infection with or without symptoms.	Presence of some infectious agent is likely, but the frequency of exposures that result in infection is not known.	Little information is available on the transmission of infectious agents on aircraft.
Pesticides	Health effects (e.g., skin rashes) can result from dermal or inhalation exposure.	Exposure is likely on selected aircraft used for international flights.	No exposure data are available; only self-reported information on health effects is available.
Low Concern			
Carbon dioxide	Indicator of ventilation adequacy. Elevated concentrations are associated with increased perceptions of poor air quality.	Concentrations are generally below FAA regulatory limits.	Reliable measurements are available only for normal operating conditions.
Deicing fluids	Health effects can result from inhalation of high concentrations.	Frequency is expected to be very low.	No information is available on incidences of fluids entering aircraft.
Nuisance odors	Annoyance and mucosal irritation can occur.	Can be present on any flight.	Reliable information is available from surveys of cabin occupants.
Relative humidity	Temporary drying of skin, eyes, and mucous membranes can occur at low relative humidity (10 to 20%).	Low relative humidity occurs on most flights.	Reliable and accurate measurements in aircraft are available.

[a] Listing in each concern group is alphabetical; the committee did not rank the characteristics within a group.

capable of flying at or above those altitudes, or strict operating limits be set with regard to altitudes and routes for aircraft without converters to ensure that the O_3 concentrations are not exceeded in reasonable worst-case scenarios. To ensure compliance with the O_3 requirements, FAA should conduct monitoring to verify that the O_3 controls are operating properly (see also recommendation 8).

3. FAA should investigate and publicly report on the need for and feasibility of installing air-cleaning equipment for removing particles and vapors from the air supplied by the ECS on all aircraft to prevent or minimize the introduction of contaminants into the passenger cabin during ground operation, normal flight, and air-quality incidents.

4. FAA should require a CO monitor in the air supply ducts to passenger cabins and establish standard operating procedures for responding to elevated CO concentrations.

5. Because of the potential for serious health effects related to exposures of sensitive people to allergens, the need to prohibit transport of small animals in aircraft cabins should be investigated, and cabin crews should be trained to recognize and respond to severe, potentially life-threatening responses (e.g., anaphylaxis, severe asthma attacks) that hypersensitive people might experience because of exposure to airborne allergens.

6. Increased efforts should be made to provide cabin crew, passengers, and health professionals with information on health issues related to air travel. To that end, FAA and the airlines should work with such organizations as the American Medical Association and the Aerospace Medical Association to improve health professionals' awareness of the need to advise patients on the potential risks of flying, including risks associated with decreased cabin pressure, flying with active infections, increased susceptibility to infection, or hypersensitivity.

7. The committee reiterates the recommendation of the 1986 NRC report that a regulation be established to require removal of passengers from an aircraft within 30 minutes after a ventilation failure or shutdown on the ground and to ensure the maintenance of full ventilation whenever on-board or ground-based air conditioning is available.

8. To be consistent with FAA's mission to promote aviation safety, an air-quality and health-surveillance program should be established. The objectives and approaches of this program are summarized in Table S-2. The health and air-quality components should be coordinated so that the data are collected in a manner that allows analysis of the suggested relationship between health effects or complaints and cabin air quality.

9. To answer specific questions about cabin air quality, a research program should be established (see Table S-2). The committee considers the following research questions to be of high priority:

- O_3. How is the O_3 concentration in the cabin environment affected by various factors (e.g., ambient concentrations, reaction with surfaces, the presence and effectiveness of catalytic converters), and what is the relationship between cabin O_3 concentrations and health effects on cabin occupants?
- *Cabin pressure and oxygen partial pressure.* What is the effect of cabin pressure altitude on susceptible cabin occupants, including infants, pregnant women, and people with cardiovascular disease?
- *Outside-air ventilation.* Does the ECS provide sufficient quantity and distribution of outside air to meet the FAA regulatory requirements (FAR 25.831), and to what extent is cabin ventilation associated with complaints from passengers and cabin crew? Can it be verified that infectious-disease agents are transmitted primarily between people in close proximity? Does recirculation of cabin air increase cabin occupants' risk of exposure?
- *Air-quality incidents.* What is the toxicity of the constituents or degradation products of engine lubricating oils, hydraulic fluids, and deicing fluids, and is there a relationship between exposures to them and reported health effects on cabin crew? How are these oils, fluids, and degradation products distributed from the engines into the ECS and throughout the cabin environment?
- *Pesticide exposure.* What are the magnitudes of exposures to pesticides in aircraft cabins, and what is the relationship between the exposures and reported symptoms?
- *Relative humidity.* What is the contribution of low relative humidity to the perception of dryness, and do other factors cause or contribute to the irritation associated with the dry cabin environment during flight?

10. The committee recommends that Congress designate a lead federal agency and provide sufficient funds to conduct or direct the research program proposed in recommendation 9, which is aimed at filling major knowledge gaps identified in this report. An independent advisory committee with appropriate scientific, medical, and engineering expertise should be formed to oversee the research program to ensure that its objectives are met and the results publicly disseminated.

12

TABLE S-2 Surveillance and Research Programs

Surveillance Program	Research Program
Objectives	
• To determine aircraft compliance with existing FARs for air quality	• To investigate possible association between specific air-quality characteristics and health effects or complaints
• To characterize accurately air quality and establish temporal trends of air-quality characteristics in a broad sample of representative aircraft	• To evaluate the physical and chemical factors affecting specific air-quality characteristics in aircraft cabins
• To estimate the frequency of nonroutine operations in which serious degradation of cabin air quality occurs	• To determine whether FARs for air quality are adequate to protect health and ensure comfort of passengers and crew
• To document systematically health effects or complaints of passengers and crew related to routine conditions of flight or air-quality incidents; to be effective, this effort must be conducted and coordinated in conjunction with air-quality monitoring	• To determine exposure to selected contaminants (e.g., constituents of engine oils and hydraulic fluids, their degradation products, and pesticides) and establish their potential toxicity more fully
Approach	
• Continuously monitor and record O_3, CO, CO_2, fine particles, cabin pressure, temperature, and relative humidity	• Use continuous monitoring data from surveillance program when possible
• Sample a representative number of flights over a period of 1-2 years	• Monitor additional air-quality characteristics on selected flights as necessary (e.g., integrated particulate-matter sampling to assess exposure to selected contaminants)

- Continue to monitor flights to ensure accurate characterization of air quality as new aircraft come online and aircraft equipment ages or is upgraded

- Conduct a program for the systematic collection, analysis, and reporting of health data with the cabin crew as the primary study group

- Identify and monitor "problem" aircraft and review maintenance and repair records to evaluate issues associated with air-quality incidents

- Collect selected health data (e.g., pulse-oximetry data to assess arterial O_2 saturation of passengers and crew)

- Conduct laboratory and other ground-based studies to characterize air distribution and circulation and contaminant generation, transport, and degradation in the cabin and the ECS

1

Introduction

The current volume, speed, and reach of air travel are unprecedented; technology has made it easy and readily available (Wilson 1995; WHO 1998). Over the last 20 years, the world's population is estimated to have grown at about 2% per year, but the traveling population has grown at 6% per year (Weiss 2001). From 1970 to 1998, the number of aircraft passengers worldwide almost quadrupled, from 383 million to 1,462 million. There has also been an increase in the number of older people flying, including those with health conditions (e.g., cardiovascular and pulmonary diseases) that may make them more susceptible to the effects of flight. In addition, the number of flights and the fraction of seats occupied (load factor) have increased, seats are more densely packed (especially in economy class), delay times are longer, and a greater number of miles are traveled. Between 1986 and 1999, the load factor for U.S. carriers serving domestic and foreign locations increased by about 13% and 21%, respectively. And from 1986 to 1998, the average U.S. domestic trip length increased from 767 miles to 813 miles, and the average foreign trip length increased from 2,570 miles to 3,074 miles (AIA 2000).

The aircraft cabin is similar to other indoor environments, such as homes and offices, in that people are exposed to a mixture of outside and recirculated air. (The outside air in the aircraft is usually supplied by a compressor on the engine and is also called bleed air.) But the cabin environment is different in many respects—for example, the high occupant density, the inability of occupants to leave at will, and the need for pressurization. In flight, people encounter a combination of environmental factors that includes low humidity, low air pressure, and sometimes exposure to air contaminants, such as ozone (O_3), carbon monoxide (CO), various organic chemicals, and biological agents. Passengers and cabin crew have long complained about the air quality in commercial aircraft. These complaints include fatigue, dizziness, headaches, sinus and ear problems, dry eyes, sore throat, and occasionally more serious effects, such as nervous system disorders and incapacitation. Data on the overall percentages of passengers and cabin crew who report complaints about cabin air have not been systematically collected, a few small surveys provide illustrative examples (see Tables 6-3 through 6-6 in Chapter 6).

Aircraft passengers are sedentary most of the time on any flight, but that is not true of the cabin crew, who are responsible for the safety and comfort of the passengers. Cabin crew, who number over 105,000 in the United States, are 20-80 years old, with the majority being between 30 and 55 years. Flight attendants work at a higher energy level than passengers and are exposed to cabin air for longer durations. They are typically in flight 50-80 h per month, and their maximal flight hours range from 75 to 105 h per month (AFA 2001).

In response to concerns raised by the flying public and the Association of Flight Attendants (AFA) regarding the air quality aboard commercial aircraft, Congress directed the Federal Aviation Administration (FAA), in the Wendell H. Ford Aviation Investment and Reform Act of the 21st Century (passed in April 2000), to request that the National Research Council (NRC) perform an independent study to assess the contaminants of concern in commercial aircraft and their toxicological and health effects, and provide recommendations for approaches to improving cabin air quality. In response, the NRC convened the Committee on Air Quality in Passenger Cabins of Commercial Aircraft, whose members include experts in industrial hygiene, exposure assessment, toxicology, occupational and aerospace medicine, epidemiology, microbiology, aerospace and environmental engineering, air-quality monitoring, ventilation and airflow modeling, and environmental chemistry. The committee was charged with the following specific issues:

1. Contaminants of concern, as determined by the committee, including pathogens and substances used in the maintenance, operation, or treatment of aircraft, including those that may result from seasonal changes in fuels and from the use of deicing fluids.

2. The systems of passenger cabin air supply on aircraft and the means by which contaminants may enter such systems.

3. The toxic effects of the contaminants of concern, their byproducts, and the products of their degradation and other factors, such as temperature and relative humidity, that may influence health effects.

4. Measurements of the contaminants of concern in the air of passenger cabins during domestic and international flights, foreign air transportation, and comparisons of these measurements with those taken in public buildings, including airports.

5. Potential approaches to improving cabin air quality, including the replacement of engine and auxiliary power unit bleed air with an alternative supply of air for passengers and crew.

In addressing its task, the committee sought to assess air quality in aircraft in general, not in specific types or models of aircraft, because it considered that concerns about exposures and health effects are potentially applicable to all commercial aircraft systems. However, the descriptions of systems (see Chapter 2) apply principally to large aircraft (more than 100 passengers) and the information on these systems was provided primarily by the major aircraft manufacturers. Thus, the committee's assessments might not be directly applicable to aircraft with smaller seating capacities. Because this report focuses on air quality in aircraft regulated by FAA, both flights within the United States and flights to or from other countries are considered.

This report is intended for a wide audience, including FAA, members of Congress, cabin crew, aircraft manufacturers, airline companies, and the general public.

EXPOSURES ON AIRCRAFT

In the aircraft cabin, both passengers and cabin crew may be exposed to numerous air contaminants, including CO from engine exhaust, O_3 that enters with outside air, organic compounds generated by emissions from materials in the cabin and the human body, and infectious agents, allergens, irritants, and other contaminants of biological origin. Air quality incidents[1] have been re-

---

[1]The committee defines air-quality incidents as events that result in the intake of potential contaminants, including engine oils and hydraulic fluids, through the environmental control system into the cabin.

ported during which passengers and cabin crew were exposed to other contaminants. Such incidents can result from inadvertent releases of engine oils, hydraulic fluids, deicing fluids, and their decomposition products into the air that enters the cabin. In general, the frequency of such incidents is not known because many of the data are considered proprietary by the airlines and were not made available to the committee and because there are no comprehensive, systematic methods for the collection of exposure and health effects information. AFA has reported a frequency of 7.6 incidents per 10,000 flights of a single airline on the basis of review of many sources of information, including reports filed by flight attendants, insurance companies, the Occupational Safety and Health Administration (OSHA), and medical professionals (Witkowski 1997).

In addition to passenger and cabin crew exposures to airborne contaminants, physiological stressors are inherent in flight, may contribute to complaints about cabin air quality, and may exacerbate underlying health problems. For instance, aircraft cabins are pressurized to an equivalent altitude of 5,000-8,000 ft, relative humidity is typically below 20%, seating is often cramped, and people may experience jet lag.

Many guidelines and standards established for air quality in aircraft cabins are applicable to routine exposures in other indoor and outdoor environments. Table 1-1 lists contaminants that may be encountered under routine conditions in aircraft and the exposures recommended or legally established by various organizations. The organizations include FAA, the American Society of Heating, Refrigerating and Air-Conditioning Engineers (ASHRAE), the Environmental Protection Agency (EPA), OSHA, and the American Conference of Governmental Industrial Hygienists (ACGIH). FAA is the only organization with regulatory authority to establish standards for the aircraft cabin environment. ASHRAE has committees that provide guidelines on exposure in indoor environments, including those of aircraft (ASHRAE 2001). EPA promulgates national ambient air-quality standards (NAAQSs) for outdoor air. OSHA establishes permissible occupational exposure limits (PELs), and ACGIH recommends threshold limit values (TLVs) to protect worker health. Because the limits established by OSHA and ACGIH are intended for application in workplaces populated by healthy adults of working age, they are not intended to apply in situations where infants, children, the elderly, or those with medical conditions might be exposed; these subpopulations, included among aircraft passengers, are addressed by EPA's NAAQSs.

TABLE 1-1 Limits on Contaminants That May Be Found in Aircraft Cabin Air

Contaminants	FAA	ASHRAE[a]	EPA NAAQS[b]	OSHA PEL[c]	ACGIH TLV[d]
Ozone[e,f]	0.1 ppm 0.25 ppm	0.05 ppm	0.12 ppm (1 h) 0.08 ppm (8 h)	0.1 ppm	0.05 ppm (TWA) (heavy work) 0.08 ppm (moderate work) 0.1 ppm (light work)
Carbon dioxide	5000 ppm	700 ppm above ambient[g]	na	5000 ppm	5000 ppm (TWA), 30,000 ppm (STEL)
Carbon monoxide	50 ppm	9 ppm (8 h) 35 ppm (1 h)	35 ppm (1 h) 9 ppm (8 h)	50 ppm	25 ppm (TWA)
Nitrogen dioxide	na	0.055 ppm (annual average)	0.05 ppm (annual average)	5 ppm	3 ppm (TWA), 5 ppm (STEL)
PM$_{10}$[h]	na	na	150 µg/m^3 (24 h)	na	na
PM$_{2.5}$[h]	na	na	65 µg/m^3 (24 h)	na	na
Formaldehyde	na	na	na	0.75 ppm (TWA) 2 ppm (STEL)	0.3 ppm (ceiling)
Acetic acid	na	na	na	10 ppm	10 ppm (TWA) 15 ppm (STEL)
Acetone	na	na	na	na	500 ppm (TWA), 750 ppm (STEL)
Acetylaldehyde	na	na	na	200 ppm (TWA)	25 ppm (ceiling)
Acrolein	na	na	na	0.1 ppm	0.1 ppm (ceiling)
Benzene	na	na	na	1 ppm	0.5 ppm (TWA) 2.5 ppm (STEL)

(Continued)

TABLE 1-1 Continued

Contaminants	FAA	ASHRAE[a]	EPA NAAQS[b]	OSHA PEL[c]	ACGIH TLV[d]
Ethanol	na	na	na	1000 ppm	1000 ppm (TWA)
Ethylene glycol	na	na	na	50 ppm (ceiling)	39.4 ppm (ceiling)
Toluene	na	na	na	200 ppm	50 ppm (TWA)
Xylene	na	na	na	100 ppm	100 ppm (TWA)
					150 ppm (STEL)
Bacteria	na	na	na	na	na
Fungi	na	na	na	na	na
Pyrethrum	na	na	na	5 mg/m³	5 mg/m³

[a] ASHRAE 62-1999.

[b] EPA NAAQS, 40 CFR 50.

[c] PEL= OSHA permissible exposure limit.

[d] TWA = time-weighted average concentration in a normal 8-h workday and a 40-h workweek, to which nearly all workers may be repeatedly exposed, day after day, without adverse effect (ACGIH 1999). STEL = short-term exposure level is a 15-min TWA exposure that should not be exceeded at any time during the workday (ACGIH 1999).

[e] FAA airworthiness standards (14 CFR 25) for ozone: "0.25 parts per million by volume, sea level equivalent, at any time above 32,000 ft; and 0.1 parts per million by volume, sea level equivalent, time-weighted average during any 3-h interval."

[f] National Institute for Occupational Safety and Health (NIOSH) recommended exposure limit (REL) not to be exceeded at any time for O_3 is 0.10 ppm (NIOSH 1997); California Air Resources Board California ambient air-quality standard (CAAQS) for O_3 is 0.09 ppm for 1-h exposure (CARB 1999); and World Health Organization guideline for O_3 is 0.06 ppm for 8-h exposure (WHO 2000).

[g] Applies to use of carbon dioxide as a proxy for odors from bioeffluents; not a limit on exposure to carbon dioxide.

[h] PM_{10} = particulate matter less than 10 microns in diameter; $PM_{2.5}$ = particulate matter less than 2.5 microns in diameter.

REGULATORY ASPECTS OF CABIN AIR QUALITY

FAA has regulatory authority over the operation of civil aircraft, including aviation safety, stemming from the congressional passage of the Federal Aviation Act in 1958 (Public Law 85-726). In 1970, Congress passed the Occupational Safety and Health (OSH) Act which was intended to ensure safe and healthful working conditions (Public Law 91-516). Under the OSH Act (Section 4[b][1]), federal agencies were granted the right to exercise jurisdiction over their own workers. In 1975, FAA asserted its jurisdiction over the safety and health of cockpit and cabin crew (40 FR 29114, DOT 1975). Specifically, FAA stated in 40 FR 29114:

> Every factor affecting the safety and healthy working conditions of aircraft crew members involves matters inseparably related to the FAA's occupational safety and health responsibilities under the [Federal Aviation] act. With respect to civil aircraft in operation, the overall FAA regulatory program, outlined in part above, fully occupies and exhausts the field of aircraft crew member safety and health.

It is important to note that FAA regulatory authority over occupational safety and health applies when an aircraft is "in operation." *In operation* is defined as the time starting when the aircraft is first boarded by a crew member, preparatory to a flight, to when the last crew member leaves the aircraft after completion of the flight, including stops on the ground during which at least one crew member remains on the aircraft even if the engines are shut down (40 FR 29114, July 10, 1975).

In addition to its regulatory authority over cockpit and cabin crew, FAA is authorized to protect the health and safety of passengers, as expressed in 49 USC 40101D and 49 USC 44701A, which provide FAA with broad authority to maintain the safety and security of air commerce.

As a result of that regulatory authority over safety and health, FAA has promulgated specifications for air quality in commercial aircraft in Federal Aviation Regulations (FARs): 14 CFR 21, 14 CFR 25, 14 CFR 121, and 14 CFR 125). Those regulations address O_3, CO, carbon dioxide (CO_2), ventilation, and cabin pressure. Regulations in 14 CFR 25 are airworthiness standards for commercial aircraft; they are intended as design specifictions for aircraft that are subject to certification under 14 CFR 21.[2] In contrast, 14 CFR 121 is intended as an operational standard and applies to domestic, flag (foreign), and supplemental air carriers. (Appendix C contains the FARs that are relevant to cabin air quality.) Regulations similar to the U.S. regulations established by FAA are applied to European aircraft by the European Joint Airworthiness Authority (JAA) and are termed Joint Aviation Regulations.

The air-quality design specifications in 14 CFR 25 are for ventilation, O_3, CO, and CO_2. The ventilation standard (Section 25.831) requires that the ventilation system be designed to provide enough uncontaminated air to enable crew members to perform their duties without undue discomfort or fatigue and to provide reasonable passenger comfort. Specifically, for normal operating conditions, the ventilation system must be designed to provide each occupant with an airflow containing at least 0.55 lb of "fresh" air per minute, equivalent to 10 ft³/per min (cfm) at 8,000-ft cabin altitude. The ventilation standard was revised in June 1996 to include cabin occupants. Before June 1996, Section 25.831 had specified only that for the crew compartment (the cockpit) a minimum of 10 cfm of fresh air per crew member (pilots and flight engineers) was required (61 FR 28683).[3] This ventilation standard was revised in June 1996 to include cabin occupants. FAA determined that the change in the standard was needed because cabin crew members, who are active during flights, must be able to perform their duties in the cabin without discomfort and fatigue. In addition, FAA concluded that fresh airflow in the aircraft is necessary to provide adequate smoke clearance in the event of smoke accumulation due to a system failure.

The ventilation standard also specifies that the air of the cockpit and cabin must be free of harmful or hazardous concentrations of gases or vapors (14 CFR 25, Section 831). According to the standard, CO concentrations in excess of 1 part in 20,000 parts of air (50 ppm) are considered hazardous, and CO_2 concentrations during flight may not exceed 0.5% by volume (sea-level

---

[2]Certification is the process by which the FAA ensures that the design of aircraft complies with statutes and that these regulations and standards are met by manufacturers and air carriers in the course of designing, producing, operating, and maintaining aircraft.

[3]It is important to note that this amended standard does not apply to existing aircraft types—whether produced in the past or the future. Rather, it applies to aircraft types (designed and certified after June 5, 1996) and derivatives for which an application for certification was filed on or after this date of the regulation.

equivalent[4]) (5,000 ppm) in compartments normally occupied by passengers or crew members. In 1996, the CO_2 regulation was reduced from 3% by volume (sea-level equivalent) (30,000 ppm) on the basis of the 1986 NRC committee's recommendation that 30,000 ppm was much higher than was recommended for other indoor environments (61 FR 63952).

The FAR (Section 25.832) states that aircraft cabin O_3 concentrations during flight must be shown not to exceed 0.25 ppm by volume (sea-level equivalent) at any time above flight level 320 (that is, 32,000 ft or 10.7 km) or to exceed 0.1 ppm by volume (sea-level equivalent) for a time-weighted average (TWA) during any 3-h interval above flight level 270 (27,000 ft or 9 km). The regulation is expressed in more detail as an operational standard in 14 CFR 121. They are based on complaints of crew members and passengers about discomfort due to high O_3 concentrations at high altitudes (DOT 1980)[5].

In addition to design standards for ventilation, CO, CO_2, and O_3, FAA requires a cabin-pressure altitude of not more than 8,000 ft at the maximal operating altitude of the aircraft under normal conditions (Section 25.841). The standard was published in the FARs in 1964 (29 FR 18291, December 24, 1964), but no rationale was ever provided.

14 CFR 121, unlike 14 CFR 25, is an operational standard that not only describes the appropriate O_3 concentrations in the cabin at particular altitudes, but also specifies ventilation requirements (Section 121.219). Specifically, Section 121.219 states that each passenger or crew compartment must be "suitably" ventilated. In addition, Section 121.219 states that CO concentrations may not be more than 1 part in 20,000 parts of air (50 ppm), and fuel fumes may not be present.

HISTORY OF PREVIOUS CABIN AIR-QUALITY STUDIES

The NRC committee that produced *The Airliner Cabin Environment* (NRC 1986) addressed some of the same issues as the current committee. That committee was tasked with determining whether characteristics of cabin

---

[4]Sea-level equivalent refers to conditions of 25°C (77°F) and 760 mm Hg (15 psi) (FAR 25, Section 832).

[5]This regulation specifies that FAA will conduct spot checks to ensure the effectiveness of O_3 control devices (14 CFR 25 and 121. Airplane cabin O_3 contamination. Fed. Regist.45(14):3880-3885. January 27, 1980).

air could be responsible for health problems of passengers and cabin crew. The cabin air characteristics looked at included the quantity of outside air, the quality of onboard air, the extent of pressurization, the characteristics of humidification, and the presence of contaminants, such as bacteria, fungi, environmental tobacco smoke, CO, CO_2, and O_3. In response to its tasks, the 1986 committee concluded that "empirical evidence is lacking in quality and quantity for a scientific evaluation of the quality of airliner cabin air or of the probability of health effects of short or long exposure to it" (NRC 1986). The committee proposed numerous conclusions and recommendations to FAA that addressed several air-quality issues, including environmental tobacco smoke (ETS), CO_2, O_3, ventilation, and the need for exposure and health monitoring. Those most relevant to the current committee's task are discussed below.

- *Ventilation.* The 1986 report concluded that if the current equipment were used under full passenger loads, ventilation would be at the minimum for acceptable indoor air quality when smoking was not permitted and other contaminant sources were not present. The 1986 report recommended that because ventilation rate was one of the controlling factors for cabin air quality and because air-quality data were insufficient, FAA should implement a data-collection program that measures airflow and contamination in aircraft cabins.
- *Carbon dioxide.* The 1986 report recommended a review of the CO_2 standard of 30,000 ppm, which was much higher than standards for other indoor environments, including workplaces.
- *Humidity.* The 1986 report found no conclusive evidence of extensive or serious adverse health effects of low relative humidity. Therefore, it did not recommend supplemental humidification of cabin air.
- *Ozone.* The 1986 report concluded that O_3 in aircraft could reach concentrations above those in the FARs. The committee recommended that FAA conduct a review to ensure that cabin O_3 concentrations comply with the regulations.
- *Environmental tobacco smoke.* The 1986 report recommended a ban on smoking in all commercial domestic flights.
- *Bioaerosols.* Because of the lack of data on bioaerosols in aircraft cabins, the committee could not conclude whether exposures posed a health hazard. Because of concern regarding the potential transmission of infectious agents, particularly while an aircraft is on the ground and the ventilation system is not operated at full capacity, the 1986 report recommended a regulation that requires removal of passengers from an aircraft within 30 minutes after a

ventilation failure or shutdown on the ground and maintenance of full ventilation whenever onboard or ground air conditioning is available. In addition, the 1986 report recommended that maximal airflow be used with full passenger loads to decrease the potential for microbial exposure and that recirculated air be filtered.

• *Volatile organic compounds.* The 1986 committee found no studies on the concentrations of volatile organic compounds (VOCs) or substances that might be emitted from disinfectants or cleaning materials.

• *Pressurization.* The 1986 report concluded that current pressurization criteria and regulations are adequate to protect the traveling public. However, it also concluded that the medical profession should use a more efficient system to warn those with medical conditions who might be at greater risk because of reduced pressure.

• *Data collection.* The 1986 report concluded that there was a lack of data for a scientific evaluation of aircraft cabin air quality and associated health effects. It recommended that FAA establish programs for the systematic measurement (by unbiased groups) of CO, respirable particles (RSPs), biological agents, O_3, ventilation rates, and cabin pressure. It also recommended that FAA establish a program to monitor health effects of cabin crew.

After the 1986 report, FAA adopted several of that committee's recommendations. In 1988, Congress passed Public Law 100-202 banning smoking on commercial flights with durations less than 2 h. In 1989, legislation banned smoking on nearly all domestic flights with durations of less than 6 h (Public Law 101-164). In 1996, FAA reduced the CO_2 standard from 30,000 to 5,000 ppm, on the basis of recommendations of the 1986 committee (61 FR 63952). In response to the committee's recommendation that FAA establish a program for the measurement of exposure variables, the Department of Transportation sponsored a study by Nagda et al. (1989) to evaluate health risks posed by exposures to ETS—including nicotine, RSPs, and CO—and other contaminants (O_3, bacteria, fungi, and CO_2) on randomly selected smoking and nonsmoking flights. FAA interpreted the 1986 committee's use of the term *program* to mean a one-time study (DOT 1987). The current committee finds it regrettable that FAA interpreted the term that way, since the 1986 committee's clear intent was to establish continuing monitoring and surveillance.

Since the 1986 report and FAA's response to its conclusions and recommendations (DOT 1987), a number of studies have made air-quality measurements on aircraft during commercial flights. Table 1-2 presents a summary of

TABLE 1-2 Contaminant Concentrations Reported in Published Studies

Contaminants or Characteristic		Nagda et al. 1989[a]	CSS 1994	Dechow 1996	Spengler et al. 1997	ASHRAE/CSS 1999	Haghighat et al. 1999	Lee et al. 1999[a]	Waters et al. 2001	Nagda et al. 2001[b]
No. of flights		92	35	x	6	8	43	16	37	10
Ozone, ppb	mean	22 ± 23		x	x	51 ± 15		x	200 ± 180	
	min, max	x, 78			2, 10	<20, 122		0, 90	<50, 1,000	
Carbon dioxide, ppm	mean	1,756 ± 660	1162		1400	1,469 ± 225	386-1,091[c]	683-1,557[c]	1,387 ± 351	1,380
	min, max	765, 3,157	x, x		1,200, 1,800	942, 1,959	293, 2,013	423, 2,900	664, 4,238	x, 1,755
Carbon monoxide, ppm	mean	0.6			0.7	x		1.9-2.39[c]	0.87 ± 0.65	0.2
	min, max	x, 1.3			0.8, 1.3	<0.1, 7		1.0, 4.0	<0.2, 9.4	x, 0.8
Nitrogen oxides, ppb	mean				36			4.5-49.6[f]	580 ± 700	
	min, max				23, 60			x, x	<200, 3,100	
Particulate matter, μg/m³	mean	37 (PM$_{3.5}$)	176 (PM$_{10}$)		x (total particles)			1-17[c,d]	x (PM$_{10}$)	<10 (PM$_{2.5}$ & PM$_{10}$)
	min, max	x, 199[e]	140, 200		3, 10			nd, 1980	30, 380	
VOC, μg/m³, with ethanol	mean		x	x	3171[e]	900 ± 450				
	min, max		x, 2,200 (ppb)	x, 2,200[e] (ppb)	608, 1,805[e]	380, 1,500				

Parameter	Statistic							
Formaldehyde, ppb	mean		7		2.9 ± 1.7		x	7.2 (µg/m³)
	min, max		3, 26		<0.6, 4.9		0, <0.07	x, 13 (µg/m³)
Bacteria, CFU/m³	mean	131.1 ± 123.4	x	201[f]	x		x	
	min, max	x, 642	0, 360	20, 1,700	x, x	39, 244	44, 93	
Fungi, CFU/m³	mean	9.0 ± 12.7	x		x		x	
	min, max	x, 61	0, 110		<1, 37		17, 107	
Temperature, °C	mean	24.1 ± 1.6	24.4	23.0	23 ± 1.7	20.3-23.8[c]	21.3-25.3[c]	
	min, max	21.0, 27.2	x, x	22.2, 25.6	17.8, 26.1	19, 27	17.8, 26.3	23, 26
Relative humidity, %	mean	21.5 ± 5.1	16.8	18	14 ± 3.2	x	10.0-42.6[c]	10.5
	min, max	9.9, 30.8	x, x	17, 19	8.8, 27.8	1.8, x	4.9, 55.5	x, 34.3
Cabin-pressure altitude, ft	mean	4,344	x		x			
	min, max	2,415, 7,212	5,500, 6,900	x, 6,950				5,500, 8,000[g]

x, data not provided; CFU, colony-forming units.

[a] Data from nonsmoking flights.

[b] Values represent those in cabin during cruise.

[c] Range of means.

[d] Particle size range measured was not specified.

[e] Values from Space et al. (2000).

[f] Geometric mean.

[g] Range varied depending on aircraft type. For B767 and B747, cabin-pressure altitude was 5,500-6,500 ft during cruise; for B737, approximately 8,000 ft.

some of the studies, showing the principal contaminants measured and reported concentrations (CSS 1994; ASHRAE/CSS 1999; Dechow 1996; Dechow et al.1997; Haghighat et al. 1999; Lee et al. 1999; Nagda et al. 1989; Nagda et al. 1992; Nagda et al. 2001; Spengler et al. 1997; Waters et al. 2001). Collectively, the studies presented in Table 1-2 measured a large number of contaminants and characteristics, including O_3, CO_2, CO, nitrogen oxides (NO_x), particulate matter (PM), VOCs, formaldehyde, bacteria, fungi, temperature, humidity, and cabin pressure altitude. Nagda et al. (1989) was the only study to measure air exchange rates, the rate at which cabin air is replaced with outdoor air. The type of aircraft, the number of flights on which measurements were made, and how flights were selected for monitoring differed among studies. In addition, measurement techniques varied considerably. Most of the studies showed that relative humidity was below 20 to 30% minimum levels specified by ASHRAE for comfort (ASHRAE Standard 55-92; CSS 1994; Haghighat et al. 1999). Most of the studies showed that CO_2 concentrations were higher than ASHRAE standard 62, which would result in concentrations of about 1,100 ppm or less (ASHRAE standard 62; Haghighat et al. 1999; Nagda et al. 1989). (It is important to note that ASHRAE Standards 55-92 and 62 are not explicit for aircraft, but rather are recommended levels for indoor buildings.) Table 1-2 indicates that average relative humidity ranged from 10.0%-42.6%, and average CO_2 concentrations from 386 to1,756 ppm. Spengler et al. (1997), the only study that made concomitant measurements in other forms of transportation—including trains, interstate buses, and subways—determined that concentrations of contaminants that were measured in aircraft cabin environments (CO_2, CO, particles, VOCs, O_3, and nitrogen dioxide, bacteria, and fungi) were similar to those observed in other forms of public transportation.

In addition to the studies presented in Table 1-2, two independent inquiries into cabin air quality were recently conducted, by the British House of Lords (House of Lords 2000) and by the Australian Senate Rural and Regional Affairs and Transport References Committee (Parliament of the Commonwealth of Australia 2000). Those inquiries were motivated by complaints and growing public concerns regarding cabin air quality and associated health effects. The House of Lords inquiry focused on air travel in general; the Australian Senate committee examined principally the British Aerospace 146 (BAe 146) aircraft and concerns about exposures to engine oils. The House of Lords inquiry concluded that there was no significant impact of air travel on health for the majority of travelers, but it also noted the lack of knowledge on issues regarding the healthfulness of cabin air, and it recommended that the government commission further research and that passengers be notified of health concerns regarding flying. The Australian Senate committee offered much stronger conclusions, including that cabin and cockpit crew flying BAe 146 aircraft suffered occupational health effects of exposures to constituents of engine oils that took a long time to recognize. The committee's recommendations were therefore more extensive and included the redesign of the BAe 146 air circulation system, the introduction of regulations specifying air-quality monitoring and compulsory reporting guidelines for all passenger aircraft, and the development of a research program to study the effect of aircraft cabin air on cabin crew and passengers.

Although more data on cabin air quality have been collected since the 1986 NRC report, they have not been collected in a systematic manner that would conclusively address many of the questions that were raised in the 1986 report. For that reason and because of concerns about other potential exposures in aircraft that were not addressed in the 1986 report (including leaks of engine oil, hydraulic fluid, and deicing fluid, and exposure to pesticides), another NRC study was commissioned.

ORGANIZATION OF THE REPORT

The remainder of this report is organized into seven chapters. Chapter 2 presents information on the purpose and operation of the aircraft environmental control system (ECS). Chapter 3 addresses sources of chemical contaminants in aircraft; it focuses on the different types of exposures (including exposure sources inside and outside the aircraft) and how the ECS can act as a contaminant source. Chapter 4 reviews exposure and health data on biological agents; biological agents associated with hypersensitivity diseases are discussed first, followed by a discussion of agents that cause infectious diseases. Chapter 5 presents health effects of exposure to chemical contaminants; its emphasis is on the toxicology of chemical contaminants that are of greatest concern. Chapter 6 reviews the current database of health-surveillance and epidemiology studies. Chapter 7 addresses air-quality measurement techniques and applications that are available for gathering the necessary data. Chapter 8 explores the approaches needed to address the outstanding questions regarding aircraft cabin air quality; specifically, it lays out a surveillance and research program that integrates health-surveillance and air-quality monitoring techniques.

REFERENCES

ACGIH (American Conference of Governmental Industrial Hygienists). 1999. Documentation of the Threshold Limit Values and Biological Exposure Indices, Supplements to the 6th Ed. American Conference of Governmental Industrial Hygienists, Cincinnati, OH.

AFA (Association of Flight Attendants). 2001. Information on flight attendant demographics. Memo from C. Witkowski, Director of Safety and Health, Association of Flight Attendants, Washington, DC, to Eileen Abt, National Research Council. May 15, 2001.

AIA (Aerospace Industries Association). 2000. Aerospace Facts and Figures, 1999-2000. Aerospace Industries Association. [Online]. Available: www.aia-aerospace.org. [June 21, 2001].

ASHRAE (American Society of Heating Refrigerating and Air-Conditioning Engineers). 1999. ASHRAE Standard—Ventilation for Acceptable Indoor Air Quality. ANSI/ASHRAE 62-1999. American Society of Heating Refrigerating and Air-Conditioning Engineers, Atlanta, GA.

ASHRAE/CSS (American Society of Heating Refrigerating and Air-Conditioning Engineers and /Consolidated Safety Services). 1999. Relate Air Quality and Other Factors to Symptoms Reported by Passengers and Crew on Commercial Transport Category Aircraft. Final Report. ASHRAE Research Project 957-RP. Results of Cooperative Research Between the American Society of Heating, Refrigerating and Air-Conditioning Engineers, Inc., and Consolidated Services, Inc. February 1999.

ASHRAE (American Society of Heating Refrigerating and Air-Conditioning Engineers). 2001. ASHRAE Standard 161. Air Quality Within Commercial Aircraft - Committee Review Draft. American Society of Heating Refrigerating and Air-Conditioning Engineers, Atlanta, GA. January 2, 2001.

CARB (California Air Resources Board). 1999. California Ambient Air Quality Standards. Sacramento, CA: California Air Resources Board. [Online]. Available: http://arbis.arb.ca.gov/aqs/aqs.htm (October 18, 2001).

CSS (Consolidated Safety Services). 1994. Airline Cabin Air Quality Study. Prepared for Air Transport Association of America, Washington, DC. April 1994.

Dechow, M. 1996. Airbus Cabin Air Quality - Only the Best! FAST Airbus Technical Digest 20(Dec.):24-29.

Dechow, M., H. Sohn, and J. Steinhanses. 1997. Concentrations of selected contaminants in cabin air of airbus aircrafts. Chemosphere. 35(1):21-31.

DOT (Department of Transportation). 1987. Airline Cabin Air Quality. Report to Congress. Washington, DC: Department of Transportation, Federal Aviation Administration. February.

Haghighat, F., F. Allar, A.C. Megril, P. Blondeau, and R. Shimotakahara. 1999. Measurement of thermal comfort and indoor air quality aboard 43 flights on commercial airlines. Indoor Built Environ. 8(1):58-66.

House of Lords. 2000. Air Travel and Health. House of Lords, Session 1999-2000, 5th Report/ HL Paper 121. Select Committee on Science and Technology, House of Lords, Parliament, Great Britain.

Lee, S.-C., C.-S. Poon, X.-D. Li, and F. Luk. 1999. Indoor air quality investigation on commercial aircraft. Indoor Air 9(3):180-187.

Nagda, N.L., M.D. Fortmann, M.D. Koontz, S.R. Baker, and M.E. Ginevan. 1989. Airliner Cabin Environment: Contaminant Measurements, Health Risks, and Mitigation Options. DOT-P-15-89-5. NTIS/PB91-159384. Prepared by GEOMET Technologies, Germantown, MD, for the U.S. Department of Transportation, Washington DC.

Nagda, N.L., M.D. Koontz, A.G. Konheim, and S.K. Hammond. 1992. Measurement of cabin air quality aboard commercial airliners. Atmos. Environ. Part A Gen. Top. 26(12):2203-2210.

Nagda, N.L., H.E. Rector, Z. Li, and E.H. Hunt. 2001. Determine Aircraft Supply Air Contaminants in the Engine Bleed Air Supply System on Commercial Aircraft. ENERGEN Report AS20151. Prepared for American Society of Heating, Refrigerating, and Air-Conditioning Engineers, Atlanta, GA, by ENERGEN Consulting, Inc., Germantown, MD. March 2001.

NIOSH (National Institute for Occupational Safety and Health). 1997. Ozone. Pp. 238-239 in Pocket Guide to Chemical Hazards. DHHS(NIOSH) 97-140. Cincinnati, OH: National Institute for Occupational Safety and Health.

NRC (National Research Council). 1986. The Airliner Cabin Environment: Air Quality and Safety. Washington, DC: National Academy Press.

Parliament of the Commonwealth of Australia. 2000. Air Safety and Cabin Air Quality in the BAe 146 Aircraft. Report by the Senate Rural and Regional Affairs and Transport References Committee, Parliament House, Canberra. October 2000.

Space, D.R., R.A. Johnson, W.L. Rankin, and N.L. Nagda. 2000. The airplane cabin environment: past, present and future research. Pp. 189-214 in Air Quality and Comfort in Airliner Cabins, N.L. Nagda, ed.. ASTM STP 1393. West Conshohocken: American Society for Testing Materials.

Spengler, J., H. Burge, T. Dumyahn, M. Muilenberg, and D. Forester. 1997. Environmental Survey on Aircraft and Ground-Based Commercial Transportation Vehicles. Prepared by Department of Environmental Health, Harvard University School of Public Health, Boston, MA, for Commercial Airplane Group, The Boeing Company, Seattle, WA. May 31, 1997.

Waters, M.,T. Bloom, and B. Grajewski. 2001. Cabin Air Quality Exposure Assessment. National Institute for Occupational Safety and Health, Cincinnati, OH, Federal Aviation Administration Civil Aeromedical Institute. Presented to the NRC Committee on Air Quality in Passenger Cabins of Commercial Aircraft, January 3, 2001. National Academy of Science, Washington, DC.

Weiss, E.L. 2001. Epidemiologic alert at international airports. Pp. 530-533 in Textbook of Travel Medicine and Health, 2nd Ed., H.L. DuPont, and R. Steffen, eds. Hamilton, Ontario: BC Decker.

WHO (World Health Organization). 1998. Tuberculosis and Air Travel: Guidelines for Prevention and Control. Geneva, Switzerland: World Health Organization.

WHO (World Health Organization). 2000. Guidelines for Air Quality. Geneva: World Health Organization, Cluster of Sustainable Development and Healthy Environment, Dept. of Protection of the Human Environment, Occupational and Environmental Health Programme.

Wilson, M.E. 1995. Travel and the emergence of infectious diseases. Emerg Infect. Dis. 1(2):39-46.

Witkowski, C. 1997. Review of Air Quality Incidents at "Airliner B". Association of Flight Attendants, AFL-CIO, Washington, DC. [Public presentation to the NRC Committee on Air Quality in Passenger Cabins of Commercial Aircraft, by J. Murawski January 3, 2001. National Academy of Science, Washington, DC].

2

Environmental Control

Commercial jet aircraft are designed to carry passengers safely and comfortably from one point to another. The external environments of the aircraft include taxiing, takeoff, cruise, and descent; outside temperature from below −55°C (−65°F) to over 50°C (122°F); ambient pressure from about 10.1 kPa (1.5 psi) to 101 kPa (15 psi); and water content from virtually dry to greater than saturation. For aircraft to transport people in those extremes of external environment, they are equipped with environmental control systems (ECSs) that provide a suitable indoor environment.

A number of aircraft systems are involved in meeting the environmental needs, including the propulsion system (engines), which is a source of pressurized air; the pneumatic system, which processes and distributes the pressurized air; and the ECS, which conditions the pressurized air and supplies it to the cabin. For the purposes of this report, each component or subsystem that is integral in providing the necessary environmental conditions in the aircraft cabin is considered to be part of the ECS, even if it is technically part of another aircraft system.

This chapter first describes the important functions of the ECS, including background information on principles of ventilation, temperature control, and humidity control. It then describes the equipment and subsystems that make up the ECS; the descriptions of aircraft systems in this chapter apply principally to large aircraft (more than 100 passengers) and might not be applicable to all aircraft. Finally, standards that are potentially related to aircraft cabin environments and aircraft ECSs are examined.

ENVIRONMENTAL CONDITIONS

During flight, the aircraft cabin is a ventilated, enclosed environment whose occupants are totally dependent on the air provided by the ECS. The ECS is designed to provide a healthy and comfortable environment for the aircraft occupants from the time crew members and passengers first board for a flight until all passengers and crew members deplane after a flight. The ECS must pressurize the aircraft cabin and maintain its temperature within tolerable limits. Most other functions are subordinate to those requirements at cruise altitudes.

The aircraft ECS is different from ECSs used in most other applications, such as buildings and surface vehicles, in that it must be able to operate in extremes of temperature, ambient air quality, and air pressure. The primary role of an aircraft ECS is to protect the occupants of the aircraft from those extreme conditions. Commercial aircraft operate over a broad range of temperatures from $-55°C$ to $50°C$ ($-65°F$ to $122°F$) at ground level and as low as $-80°C$ ($-112°F$) at an altitude of 12,000 m (39,400 ft). As shown in Figure 2-1, at a typical cruise altitude of 11,000 m (36,000 ft), the air temperature is usually about $-55°C$ ($-65°F$) but can range from about $-70°C$ to $-30°C$ ($-92°F$ to $-20°F$) (ASHRAE 1999a).

More critically, at a typical cruise altitude of 11,000 m (36,000 ft), the atmospheric pressure is only about one-fifth that at sea level (Figure 2-2). Although the relative concentration of oxygen at that altitude is nearly the same as at sea level, the partial pressure of the oxygen (PO_2) is only about 4.7 kPa(0.69 psi) compared with 21 kPa (3.1 psi) at sea level and is far below what is necessary to sustain human life. Furthermore, the ambient air quality on the ground and at low altitudes can range from pristine to extremely polluted in urban environments. The ECS meets those needs through integrated subsystems that pressurize the cabin when in flight, control thermal conditions in the cabin, and ventilate the cabin with outside air to prevent a buildup of contaminants that might cause discomfort or present a health hazard.

Pressure

In flight, the ECS maintains the cabin pressure and therefore the oxygen partial pressure at acceptable levels by compressing the low-pressure outside air and supplying it to the cabin. The air pressure in aircraft cabins is com-

35

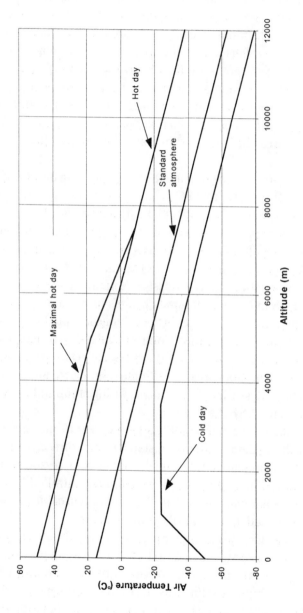

FIGURE 2-1 Typical temperature design conditions for aircraft. Source: Adapted from ASHRAE 1999a.

monly expressed as a "pressure altitude" equivalent. Cabin pressure altitude is the distance above sea level at which the atmosphere exerts the same pressure as the actual pressure in the aircraft cabin. The relationship between atmospheric pressure and pressure altitude is based on the corresponding curve in Figure 2-2. The minimal cabin pressure is set by Federal Aviation Regulation (FAR) 25, which requires the pressurization system to "provide a cabin pressure altitude of not more than 8000 ft [2,440 m]" under normal operating conditions. This limit of 2,440 m (8,000 ft) corresponds to a cabin pressure of 75 kPa (10.9 psi) as shown in Figure 2-2. Thus, the cabin pressure can range from a maximum of 101 kPa (14.7 psi) on the ground at sea level to a minimum of 75 kPa (10.9 psi) in flight regardless of the altitude at which the aircraft flies.

The primary purpose of the pressurization is to maintain the PO_2 at acceptable levels. Figure 2-2 shows that PO_2 values at sea level and at a pressure altitude of 2,440 m (8,000 ft) are 21 kPa (3.1 psi) and 16 kPa (2.3 psi), respectively. Thus, the minimal PO_2 allowed in the aircraft cabin at the maximal allowed cabin pressure altitude of 2,440 m (8,000 ft) is 74% of the sea level value (Federal Aviation Regulations (FAR) Section 25.841).

In addition to generating enough pressure to maintain the necessary PO_2 in the cabin, the ECS must prevent rapid changes in cabin pressure. Rapid changes in pressure can cause changes in the volume occupied by gases in the body cavities and result in discomfort. Controlling the rate of change in pressure is particularly important during ascent and descent. During normal operation, the rate of change in cabin pressure altitude is limited to not more than 5 m/s (about 1,000 ft/min), sea-level equivalent, during climb and 2.3 m/s (450 ft/min) during descent (ASHRAE 1999a).

The aircraft skin and pressure bulkhead at the rear of the cabin form a pressure hull that allows the aircraft to withstand the pressurization necessary during flight. In flight, pressurized air from the engine compressors is supplied continuously to the cabin, and the cabin pressure is controlled by outflow valves; the valves are automatically controlled to maintain cabin pressure but can be manually overridden by controls in the cockpit.

For structural reasons, the difference between internal and external pressures is not allowed to exceed about 55-62 kPa (8-9 psi), depending on the aircraft. On some aircraft, the cabin pressure altitude is controlled to the lowest possible value (highest cabin pressure) and would not reach 2,440 m (8,000 ft) until the aircraft reaches its maximum operational altitude (e.g., 14,300 m [47,000 ft]). On other aircraft, the cabin pressure altitude is con-

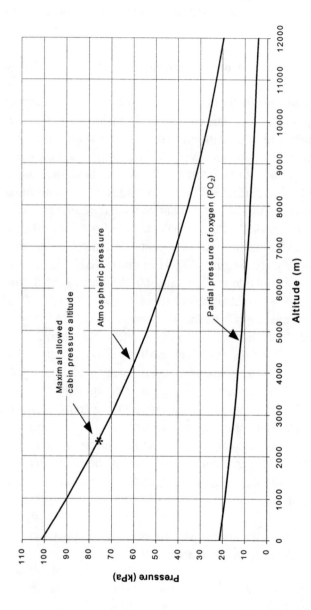

FIGURE 2-2 Effect of altitude on atmospheric pressure.

trolled to the highest allowed value (lowest pressure) to minimize structural loads from pressurization. Because of the interrelationships between flight altitude, cabin pressure, and structural load on the aircraft, any change in pressure altitude requirements will affect the altitudes at which many aircraft can operate.

Contamination

Contaminants generated in the aircraft cabin air are eliminated by ventilating the cabin with outside air. The compressed outside air that is used for pressurization in the cabin is the same air that is used for ventilation. Pressurization and ventilation, however, serve very different purposes. For ventilation, outside air is used to dilute contaminants in the air and flush them out of the cabin. As described below, the rate of flow of outside air has a substantial and direct impact on the concentration of contaminants in the cabin air. The flow rate has a negligible effect on the PO_2, in that only a tiny portion of the oxygen in this air is consumed by the aircraft occupants. A typical sedentary adult consumes oxygen at about 0.44 g/min (0.001 lb/min) (Nishi 1981). With the FAR 25 minimal design outside-air flow rate of 0.25 kg/min (0.55 lb/min) per cabin occupant, oxygen is brought into the cabin at 0.058 kg/min (0.127 lb/min) per person. Oxygen consumption by the occupants reduces the PO_2 levels by about 0.8% in this case, compared with a PO_2 reduction of up to 25% due to the reduced cabin pressure, as explained earlier. Thus, adequate oxygen concentrations in the cabin are maintained, even at ventilation rates far below those specified in FAR 25, as long as the cabin is adequately pressurized.

Contaminants can originate in the cabin itself or in sources outside the cabin. Furthermore, the concentrations of contaminants in the cabin are subject to change as a result of fluctuations in the source emission and ventilation rates. Some contaminants degrade or react with other chemicals in the cabin. The following sections discuss the generation, distribution, and elimination of contaminants in cabin air.

Contaminants Originating in the Cabin

The basic steady-state ventilation equation for a particular contaminant "i" may be expressed as follows (derived from ASHRAE 1997a):

$$D_{c,i} = D_{o,i} + S_i / V_o, \qquad (2\text{-}1)$$

where

D_c is contaminant density in cabin air, kg/m^3 (lb/ft^3),

D_o is density of contaminant in outside air used for ventilation, kg/m^3 (lb/ft^3),

S is strength of contaminant source, kg/s (lb/s), and

V_o is ventilation rate of outside air, m^3/s (ft^3/s).

To be accurate, both V_o and $D_{o,i}$ should be evaluated at cabin temperature and pressure (see Box 2-1). For gaseous contaminants, it is easier to work in terms of concentrations rather than densities, and the above equation can be expressed as follows:

$$C_{c,i} = C_{o,i} + (S_i MW_a)/(m_o MW_i), \qquad (2\text{-}2)$$

where

C_c is volume fraction of contaminant in cabin air,

C_o is volume fraction of contaminant in outside air used for ventilation,

S is strength of contaminant source, kg/s (lb/s),

m_o is ventilation rate of outside air, kg/s (lb/s),

MW_a is molecular weight of air (28.96), and

MW is molecular weight of contaminant.

An example of the application of Equation 2-2 for carbon dioxide (CO_2) is as follows. The CO_2 concentration in the cabin air may be related to the rate at which outside air is supplied to the cabin by the ventilation system. A typical sedentary person will generate CO_2 at about 7.7×10^{-6} kg/s (ASHRAE 1999b). The concentration of CO_2 in clean outdoor air is about 0.037%. The molecular weight of CO_2 is 44.01 g/mol. If the occupants are the only source of CO_2 in the cabin, Equation 2-2 becomes

$$C_{c,CO2} = 0.00037 + N(7.7 \times 10^{-6})(0.658/m_o), \qquad (2\text{-}3)$$

where N is the number of occupants and 0.658 is the ratio of the molecular weights of air and CO_2. Equation 2-3 can be used to relate ventilation rates to measured values of CO_2 concentrations as long as respiration is the dominant source of CO_2 in the cabin and the outside CO_2 concentration is not above typical values. The CO_2 concentration with the FAR 25 minimal design ventilation rate for aircraft of 0.0042 kg/s per person (0.25 kg/min) can be estimated as

$$C_{c,CO2} = 0.00037 + (7.7 \times 10^{-6})(0.658/0.0042) = 0.00158 = 1{,}580 \text{ ppm.}$$

Other ventilation rates will result in higher or lower CO_2 concentrations according to Equation 2-3. It should be pointed out that CO_2, at the concentrations present in this example is not noticeable by the occupants, nor is it considered hazardous. However, occupant-generated CO_2 is produced roughly in proportion to other occupant-generated bioeffluents that can affect perceived air quality. The concentration of CO_2 is sometimes used as an indicator of the concentration of other contaminants (ASHRAE 1999b).

The ventilation requirements in the FAR are given in terms of mass flow. However, it is common to state ventilation flows in volumetric terms such as liters per second or cubic feet per minute; this practice can lead to confusion in that the relationship between mass flows and volumetric flows depends on the ambient pressure and temperature (see Box 2-1).

Contaminants Originating Outside the Cabin

The preceding discussion dealt with contaminants that are generated in the cabin and that can be effectively controlled by ventilation. However, other contaminants can be in the outside air, such as ozone (O_3) or can be picked up in the air supply system, such as leaking oil. Obviously, it is not possible to control or eliminate those contaminants through an increased ventilation flow rate. If the source of the contaminant exists for only a short time (e.g., during deicing), effective control can be achieved by turning off the flow of outside air while the source is present. That control measure is not an option in flight, because of the requirements for pressurization; nor is it an option when the source is present for more than a short time (e.g., 15 min). Some reduction in concentrations of such cabin air contaminants can be achieved by using the minimal practical flow of outside air and increasing the flow of recirculated air if the recirculation filters are effective at removing the contaminants in question (see recirculation section later in this chapter).

At high altitudes, especially at high latitudes, O_3 concentrations in the outside air can be high enough for their introduction into the cabin to result in O_3 concentrations that exceed the FAR 25 limit of 0.25 ppm by volume at any time above 32,000 ft (9,800 m) or above a time-weighted average of 0.1 ppm during any 3-h flight above 27,000 ft (8,200 m). Therefore, catalytic destruction of the O_3 in the incoming air is used on some aircraft ECS to meet the FAR requirement.

BOX 2-1 Units for Expressing Ventilation

It is common to express ventilation flow rates in volumetric terms—liters per second (L/s) or cubic feet per minute (cfm). However, this practice leads to ambiguity unless the pressure and temperature are also stated. Air density is proportional to atmospheric pressure and inversely proportional to absolute temperature. Thus, air that will occupy 1 m³ (35 ft³) at sea level will expand to over 3 m³ (106 ft³) at pressures and temperatures typical of outside air at cruise altitudes. The importance of this effect can be demonstrated by examining the ventilation requirement of 0.25 kg/min (0.55 lb/min) in FAR 25. At sea level and the standard atmospheric temperature of 15°C (59°F), the corresponding volumetric flow rate is 3.4 L/s (7.2 cfm); at the maximal allowed cabin pressure altitude of 2,440 m (8,000 ft) and a typical cabin temperature of 22°C (72°F), it is 4.7 L/s (9.9 cfm). At ambient atmospheric pressure at an altitude of 12,000 m (39,300 ft) and standard atmospheric temperature of -63°C (-81°F), the flow rate is 13.0 L/s (27.6 cfm).

For ventilation purposes, the mass flow rate, not the volume flow rate, is most important. Thus, FAR 25 is correctly stated in this regard. Unfortunately, the data on most ventilation systems, including aircraft ventilation systems, are expressed in terms of volumetric flow. Temperature and pressure information generally is not included with those data, so there can be uncertainty as to the amount of air flowing. For aircraft, it can be assumed that the flow data do not correspond to outside ambient conditions at cruise altitudes. However, it is not always clear whether they correspond to cabin conditions, standard conditions, or some other conditions.

With the exception of O_3, the outside air at cruise altitudes is generally quite pure and requires no additional cleaning. The outside air at or near ground level, however, can contain a wide variety of contaminants from industrial and urban sources. In addition to outside air contaminants, leaking hydraulic fluid, spilled fuel, or deicing fluid can be entrained in the air supply systems; few, if any, aircraft have cleaning systems to remove any of these contaminants.

Transient Response of Cabin Environmental Conditions

Equations 2-1 through 2-3 describe contaminant concentrations under steady-state conditions. Contaminant concentrations in the cabin do not change immediately when the controlling characteristics of ventilation flow rate and contaminant source strength are changed. It takes time for contaminant concentrations to build up to steady-state conditions after introduction of a source and to decline after the source is removed or ventilation begins. The time it takes for contaminant concentrations to approach steady-state conditions in aircraft is short, typically around 5-15 min, and is proportional to the quantity derived by dividing the volume of the space being ventilated by the ventilation rate. Such a rapid response means that there is only a short lag in the buildup of contaminants once they are introduced. It also means that the contaminants are flushed from the cabin quickly once the source is eliminated. In that respect, aircraft differ from buildings, in which it can take several hours to reach a steady state when ventilation rates are those recommended in American Society of Heating, Refrigeration, and Air-conditioning Engineers (ASHRAE) Standard 62 (1999b).

Because contaminants can concentrate so quickly in an aircraft cabin, it is important that the ECS not be shut down for an extended period when the aircraft is occupied (except in an emergency). When the ECS is not operating, contaminant concentrations can become excessive and temperature uncomfortably high rapidly—in less than 15 min for a fully loaded aircraft in a hot environment.

Reactive Contaminants

Some contaminants react with other substances or decompose after they enter the cabin environment. Whether the contaminants decompose or combine with other chemicals once they are in the cabin can have an important effect on the contaminant concentrations in the cabin (Weschler and Shields 2000). The residence time of a contaminant in the cabin (the average length of time from introduction of the contaminant until it is flushed from the cabin by ventilation) has an important influence on the concentration of reactive contaminants in that it determines how long a contaminant has to decompose or react. As with transient responses discussed previously, residence time is proportional to the quantity derived by dividing the volume of the space by the ventilation flow rate, so residence time in aircraft is typically much shorter than in buildings. Residence time is particularly important for O_3 and its byproducts (see Chapter 3 for additional discussion).

Temperature Control

Temperature control in the aircraft cabin is critical for safety at high altitudes and is important for occupant comfort at all altitudes. Comfortable conditions are maintained in the cabin by supplying cool or warm air to the cabin as needed. Because of the high occupant density, cooling of the cabin is required in most circumstances, particularly on the ground and at low altitudes in warm climates. Supplying air to the cabin at an appropriate temperature is a key function of the ECS.

The temperature of the air that must be provided to the cabin can be determined from the steady-state heat balance for an aircraft cabin, which in equation form is analogous to Equation 2-2:

$$T_c = T_s + (Q/m_s)(1/c_p), \qquad (2\text{-}4)$$

where

T_c is the temperature of the air in the cabin, °C (°F),

T_s is the temperature of the air supplied to the cabin, °C (°F),

Q is the amount of heat that must be removed from the cabin, W (Btu/s),

m_s is the flow rate of conditioned air supplied to the cabin, kg/s (lb/s), and

c_p is the specific heat of the air, 1000 J/kg·°C, (0.25 Btu/lb·°F).

The ECS designer must determine the combination of m_s and T_s that will result in the desired T_c for a given Q.

The air conditioning systems can provide air at a wide range of temperatures. However, there is a lower limit at which air can be supplied to the cabin without creating uncomfortable cold drafts near the inlets, typically about 10°C (50°F), and even at that temperature only with good air circulation in the cabin and good diffuser design (ASHRAE 1997a). Consequently, the system must be designed with an air flow rate that is adequate to meet the largest heat load with this temperature of supply air. An example will demonstrate this principle.

The average heat generation by a comfortable, sedentary person, excluding heat loss due to evaporation of moisture, is about 70 W (ASHRAE 1997b). The total heat load in an aircraft cabin will include heat loads from electronics and heat gain through the aircraft skin. For the purpose of this example, the total heat load will be taken as twice the occupant-generated amount, or

140 W/person. If the cabin temperature is kept in the middle of the ASHRAE "comfort envelope" (ASHRAE 1992) at 23°C and the air is supplied to the cabin at 10°C, Equation 2-4 can be used to determine the required rate of flow of conditioned air to the cabin:

$$23°C = 10°C + (140/m_s)(1/1,000) => m_s = 0.0108 \text{ kg/s·person} = 0.646 \text{ kg/min·person.}$$

That is, adequate temperature control in the cabin requires that conditioned air be supplied to the cabin at about 0.65 kg/min (1.4 lb/min) per person to maintain a comfortable temperature. This requirement is more than twice the FAR 25 requirement of 0.25 kg/min per person for outside air and provides one rationale for recirculating cabin air (see section on recirculation).

Air Distribution and Circulation

The ventilation system does more than supply outside air to the cabin. It also distributes and circulates air in the cabin. Moving outside air into the cabin at one or a few locations will not provide adequate contaminant removal and acceptable thermal conditions throughout the cabin. On the contrary, parts of the cabin would get very cold and other parts hot; similarly, parts of the cabin would have clean, fresh air and other parts stagnant, stale, and unpleasant air. An important function of the ECS is to distribute fresh air throughout the cabin by providing good air circulation for uniform temperature conditions; another is to flush out contaminated air.

Figure 2-3 shows an example of air circulation in an aircraft cabin. Typically, air is supplied and exhausted along the whole length of the cabin. Although the air mixes locally in the cabin, the air supply and air exhaust flow rates are matched along the length of the cabin as much as possible to minimize net flows along the length of the aircraft. Distribution of the air to the cabin occurs through diffusers located in the center of the ceiling of the aisles, above the windows, or along the overhead baggage compartments. Generally, it is supplied to the cabin from one or more linear diffusers that run the length of the cabin. Wide-body aircraft, in particular, use multiple diffusers.

The velocities and circulation patterns in the cabin are determined by the location, configuration, and flow rates of the inlet diffusers. The incoming air must be balanced with exhaust of an equal amount of air from the cabin.

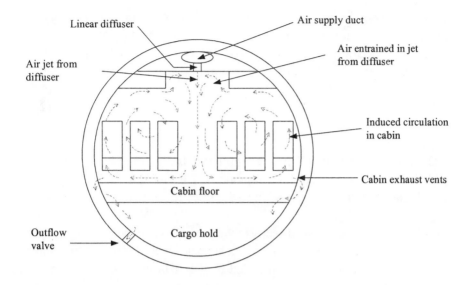

FIGURE 2-3 Cross-section of aircraft with air circulation paths.

Exhausted air generally is removed from the cabin at floor level and at the side walls and sometimes through the ceiling. Some of the exhausted air is recirculated to the cabin after passing through a filter, the balance of the exhausted air may pass around or through the cargo hold before being dumped overboard through an outflow valve. Exhaust fans extract air from the lavatories and galley areas, and it is ducted directly to the outflow valves to avoid contaminating air in other parts of the cabin with odors or other contaminants from these areas.

Ideally, the spread of contaminants along the length of the cabin is avoided by balancing inflow and outflow at all locations. That is not 100% effective even if the inlet and exhaust flows are perfectly matched at all locations, inasmuch as the circulation generated by the ECS transports contaminants from one part of the cabin to other parts and a substantial random component of the air motion transports airborne contaminants in all directions. Random velocity components of about 0.1-0.2 m/s (20-40 fpm) may exist throughout much of the occupied zone of a typical aircraft cabin (Jones et al. 2001).

The need for circulation to provide thermal uniformity generally sets the requirement for the total flow rate of air that must be supplied to the cabin. The circulation is induced by the air flowing out of the diffusers into the cabin;

with good diffuser design, the circulation will be greater than the incoming flow by a factor of 5 or more (ASHRAE 1997a). If ΔT_{al} is the maximal acceptable temperature variation from one spot to another in the cabin, the minimal acceptable temperature of the inlet air is given, to a first approximation, by

$$T_i = T_c - K \, \Delta T_{al}, \tag{2-5}$$

where

  T_i is the temperature of the air supplied to the cabin, °C (°F),
  T_c is the temperature of the cabin air, °C (°F), and
  K is the ratio of the circulation to the inlet air flow, dimensionless.

If conditioned air is supplied directly to the cabin, T_i is the same as T_s in Equation 2-4, and the flow rate of air, m_s, must be set accordingly. If the conditioned air is first mixed with recirculated air, T_i will be greater than T_s, and the total flow rate of air will be greater than the flow rate of conditioned air required according to Equation 2-4.

Recirculation

Aircraft designers have four important factors that must be considered when determining flow rate requirements for the ECS: the flow rate of *outside air* required to remove contaminants from the cabin, the flow rate of *conditioned air* required to remove heat from the cabin, the *total flow rate of air* required to provide adequate circulation in the cabin, and the flow rate of *outside air* required to pressurize the aircraft. In older aircraft (see Table 2-1), all the air supplied to the cabin comes from outside air, as shown in Figure 2-4. The flow rate of outside air must be large enough to meet the maximums of all four requirements even though it results in outside air flow rates larger than specified in FAR 25. This practice is inefficient in that supplying compressed air to the cabin in flight has a substantial energy cost.

The FAR 25 requirement for outside air for ventilation purposes is generally more than adequate for pressurization purposes as well; in the context of this discussion, pressurization is not a controlling factor for determining flow rates. The flow requirements needed to address the three remaining factors do not necessarily coincide. It was seen in a previous example that the required flow of conditioned air required for temperature control could be more

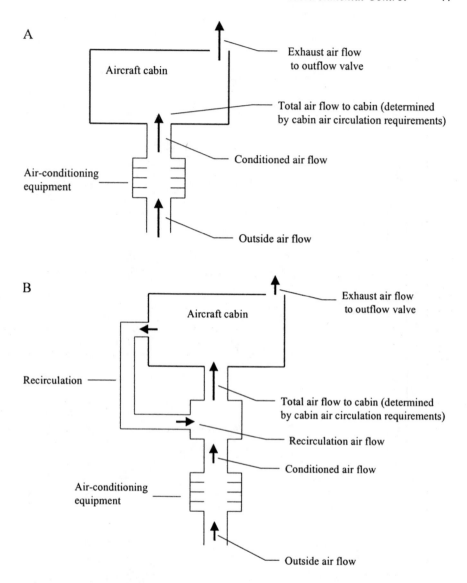

FIGURE 2-4 Cabin air supply without recirculation and with recirculation: (A) without recirculation—all flow of air to cabin comes from outside air, and flow rate of outside air must be large enough to meet greatest needs for cabin air circulation, temperature control, and contaminant control; (B) with recirculation—total flow of air to cabin is outside air flow and recirculation air flow combined. Flow rate of outside air is matched to needs for contaminant and temperature control and extra flow needed for cabin circulation is met with recirculation air.

TABLE 2-1 Basic Characteristics of Aircraft Ventilation Systems

Aircraft Manufacturer	Model No.	Air Source		Air Recirculation, %[a]		No. Air Exchanges per Hour[a,b]	
		Ram Air	Bleed Air	Cabin	Cockpit[c]	Cabin	Cockpit[c]
Boeing	707, 720						
	Turbojet	✓		0	0	NA	NA
	Turbo fan		✓	0	0	NA	NA
	727-100	Backup	✓	0	0	21.6-24.8	31.7-36.2
	727-200	Backup		0	0	23.6-29.2	42.2-56.8
	737-100		✓	0	0	12.0-26.1	17.5-43.5
	737-200		✓	0	0	11.0-23.9	17.5-43.5
	737-300		✓	37-47	0	21.8-24.8	35.0-40.5
	747-100		✓	23-27	0	19.2-22.6	29.9-37.1
	747-200		✓	23-27	0	19.2-22.6	29.9-37.1
	747-300		✓	23-27	0	18.0-21.3	29.9-37.1
	757		✓	48-55	0	29.7-34.3	57.6-61.9
	767-200		✓	45-53	49-55	20.1-24.6	61.3-68.4
Douglas	DC-8-10	✓	Backup	0	0	20.3	22.8
	DC-8-50	✓	Backup	0	0	20.3	22.8
	DC-8-62	✓	Backup	0	0	19.2	22.8
	DC-8-71	✓	Backup	34-49	0	26.6-31.5	25.0-35.7
	DC-8-72	✓	Backup	34-49	0	20.9-24.7	25.0-35.7
	DC-8-73	✓	Backup	34-49	0	26.6-31.5	25.0-35.7

	DC-9-10	✓	0	0	20.6-27.4	39.2-52.5
	DC-9-20	✓	0	0	20.6-27.4	39.2-52.5
	DC-9-30	✓	0	0	13.3-27.3	32.7-66.8
	DC-9-40	✓	0	0	12.2-25.0	32.7-66.8
	DC-9-50	✓	0	0	11.1-22.9	32.7-66.8
	MD-81	✓	0-44	0	16.7-25.9	35.3-67.2
	MD-82	✓	0-44	0	16.7-25.9	35.3-67.2
	DC-10	✓	0 (originally)	0	21.6-23.2	77.5-83.8
	DC-10	✓	38	0	21.6-23.2	77.5-83.8
Airbus	A-300-100	✓	0	0	61.0-19.0	66.4-78.7
	A-300-200	✓	0	0	61.0-19.0	66.4-78.7
	A-300-300	✓	0	0	61.0-19.0	66.4-78.7
	A-300B4-600	✓	41-53	0	21.0-25.3	46.2-58.6
	A-310-200	✓	37-53	0	20.6-25.4	49.1-64.1
Lockheed	L-1011-1	✓	0	0	14.0-29.7	36.3-77.4
	L-1011-100	✓	0	0	14.0-29.7	36.3-77.4
	L-1011-200	✓	0	0	14.0-29.7	36.3-77.4
	L-1011-500	✓	0	0	15.2-32.3	36.3-77.4

[a] Based on reported full-flow schedule. Air exchange rates reported include data on takeoff, climb, cruise, and descent.
[b] Includes outside air and recirculated air when applicable.
[c] Data provided for comparison with recirculation and air exchange rates in passenger cabin.

Source: Adapted from Lorengo and Porter (1986).

than twice the FAR-specified value for outside air for contaminant control. The practice of recirculation was introduced to address this issue. As shown in Figure 2-4B, recirculation is accomplished by extracting air from the cabin and mixing it with conditioned outside air. Recirculation provides two benefits: it allows the total air flow rate to be higher than the flow rate of outside air, so good circulation in the cabin can be maintained independently of the flow of outside air; and the conditioned air is mixed with the comparatively warm recirculated air before being introduced into the cabin. Consequently, the conditioned air can be supplied at a much lower temperature (T_s in Equation 2-4) without causing discomfort from cold drafts, and the maximal flow rates of conditioned air required for temperature control can be reduced to match those required for contaminant control more closely.

Because recirculation removes air from the cabin and returns it to the cabin, it has no effect on average contaminant concentrations throughout the cabin, if it is unfiltered. Equation 2-2 still applies, and the average concentration of contaminants in the cabin air is determined by the flow rate of outside air, independently of the amount of recirculation. If the recirculation air is cleaned of contaminants, it has the same effect on contaminant concentrations in the cabin air as does clean outside air.

Current practice is to use only particle filters on recirculation air. Filter efficiencies are at least 40% (Mil Std 282) on the MD-80 series, but exceeds 93% on all other aircraft[1] (Mil Std 282) (Boeing responses to NAS, April 10, 2001). Most aircraft manufactured more recently use HEPA filters. HEPA filters are highly effective (99.97% efficiency, Mil Std 282). Air filters are generally changed at the aircraft scheduled "c"-check, generally between 4,000 and 12,000 flight-hours (J. Lundquist, Pall Aerospace, personal communication, July 5, 2001). HEPA filters remove essentially all airborne pathogens and other particulate matter from an airstream that passes through them with a minimal efficiency of 99.97% for 0.3-μm particles. Although they are effective in removing particles, including bacteria and viruses, from the recirculated air and preventing their spread through the cabin by this route, they do not remove gaseous contaminants. Chemical adsorption of gaseous contaminants with activated charcoal or other types of filters may be used to clean the

---

[1]Rating based on ASHRAE Standard 52.1-1992, *Gravimetric and Dust-Spot Procedures for Testing Air-Cleaning Devices Used in General Ventilation for Removing Particulate Matter*, American Society of Heating, Refrigerating and Air Conditioning Engineers.

recirculation air. These filters are available as an option on some aircraft to adsorb organic gases that are not trapped by the HEPA filters, but they are not widely used.

It should be emphasized that recirculation air is not a substitute for outside air and that the flow rate of outside air required to maintain acceptable concentrations of gaseous contaminants is not reduced by using recirculation. Furthermore, the use of recirculation has no detrimental effect on cabin air quality for a given rate of flow of outside air and, when combined with effective HEPA filtration, does not contribute to the spread of infectious agents in the cabin.

Recirculation air is obtained from the area above the cabin, under the floor, or both. Only air from the passenger cabin is recirculated. Air from the cargo bay, lavatories, and galleys is not recirculated (Boeing Co. 1988) but is separately vented overboard so that odors and cargo-bay fire-fighting chemicals that could be used in the event of a fire are not introduced into the cabin (Boeing Co. 1995).

The flow rate of outside air per seat ranges from 5.9 to 9.6 L/s (12.4 to 20.4 cfm) on older aircraft without recirculation and from 3.6 to 7.4 L/s (7.6 to 15.6 cfm) on aircraft with recirculation (M. Dechow, Airbus, personal communication, Feb. 16 and March 8, 2001; Boeing responses to NAS, April 10, 2001). The percentage of recirculated air distributed to the passenger cabin typically is 30-55% of the total air supply. Some models of aircraft can use different amounts of recirculation or turn recirculation off. Other models are programmed to use different amounts of recirculated air during climb or descent compared with cruise.

Table 2-1 summarizes the characteristics of the ventilation systems of various aircraft models (adapted from Lorengo and Porter 1986). It shows that most aircraft use bleed air rather than ram air as the source of cabin air, and it demonstrates that as aircraft have moved from 100% bleed air to recirculated air, the amount of outdoor air has decreased.

The use of recirculation has been common in the design of building environmental control systems for many years. In contrast with aircraft, in which the total air flow to the cabin varies from 0% to 55% of recirculated air, building environmental control systems are commonly designed and operated with up to 90% recirculated air. However, the buildings still must maintain flow rates of outside air of about 0.5 kg/min (1.1 lb/min) per person or more to meet ventilation requirements regardless of the amount of recirculation (ASHRAE 1999b).

Relative Humidity

Humidity in the aircraft is controlled both for human comfort and for air-craft safety. The two needs are sometimes compatible, sometimes in conflict. High humidity in the cabin air (e.g., greater than 70% relative humidity), espe-cially when accompanied by high temperature leads to passenger discomfort. High humidity can also lead to condensation, dripping, and freezing of moisture on the inside of the aircraft shell, which can lead to a variety of safety prob-lems, including corrosion on the shell. Condensation can lead to biological growth and potential adverse effects on cabin air quality. The ECS must be able to prevent excessive humidity in the cabin air by removing moisture from the outside air before it is supplied to the cabin.

At cruise altitudes, the outside air contains very little moisture, and the main sources of humidity in the cabin air are respiration and evaporation from the skin of occupants. The steady supply of dry outside air is more than suffi-cient to flush the human-generated moisture from the cabin and maintain a low moisture content in the air, typically 10-20% relative humidity at cruise alti-tudes. Such values of relative humidity are below comfort guidelines (ASHRAE 1992).

Theoretically, at least, cabin air could be humidified to comfortable values. But a number of problems are associated with such humidification, including the weight penalty associated with the water that would need to be carried, the biological growth that is often associated with humidifiers, and the maintenance requirements of humidification systems. In addition, the humidity required for passenger comfort might exceed that which generates some of the safety concerns for the aircraft operations described previously. Whole-cabin humidi-fication systems are therefore not normally included on aircraft. Air supplied to the cockpit is humidified on a small fraction of the current aircraft fleet, but on most aircraft the cockpit is normally drier than the passenger cabin air because of the higher ventilation rates in the cockpit.

Humidity that is too low can be avoided to some extent by using the lowest feasible flow rate of outside air at cruise altitudes. There is an inherent con-flict between humidity control and contaminant control: increasing the flow of outside air to reduce contaminant concentrations in the cabin reduces the humidity level, and decreasing the flow of outside air to raise the humidity increases contaminant concentrations.

Although water vapor is not considered a contaminant in this context, Equation 2-2 can be used to evaluate its concentration in cabin air. Relation-ships developed by Fanger (1982) can be used to estimate the typical moisture generation by sedentary people in comfort (not sweating) in a low-humidity environment as 0.013 g/s per person. With that amount of moisture generation and the FAA minimal design flow rate of outside air of 0.042 kg/s, the water vapor concentration in the cabin air will be 0.0050, or 0.5%; this corresponds to a relative humidity of about 18% at typical cabin air temperatures.

That value might be a slight underestimate of the humidity because some occupants, particularly the cabin crew, will be more active than others. In addition, because of individual variations, some of the occupants might be sweating, even though the temperature is comfortable for the average person. Nevertheless, the value is a good indication of the humidity that can be expected with the FAA minimal design flow rate of outside air and is consistent with the humidity measured in aircraft (see Table 1-2).

Whether the FAA minimal design flow rate of outside air is appropriate for ventilation might be controversial. However, any attempt to increase cabin humidity by decreasing the ventilation flow rate will increase contaminant concentrations in the cabin air, and any attempt to reduce contaminant concentrations in the cabin by increasing the ventilation flow rate above the FAA minimum reduces humidity in the cabin further. The dilemma could be resolved, in theory, by using humidifiers; but they have their own problems, as discussed above. As a result, low humidity in aircraft cabins is not readily corrected.

ENVIRONMENTAL CONTROL SYSTEMS

All large commercial passenger aircraft manufactured today and nearly all such aircraft in service use ECSs based on engine bleed air. Figure 2-5 presents an overview of bleed-air-based aircraft ECSs and equipment. Compressed air called bleed air is extracted from propulsion engine compressors and supplied to one or more air-conditioning "packs," where it is further compressed, cooled, and then expanded in a rotating air-cycle machine to produce low-temperature air that is supplied to the aircraft cabin. The conditioned air from the packs is supplied to a mixing manifold that distributes it to zones in the cabin. Recirculation fans extract air from the cabin, pass it through filters, and supply it to the mixing manifold, where it mixes with the conditioned air from the packs. Trim air is hot bleed air that bypasses the air-conditioning packs. Small amounts of trim air are mixed with the air supplied to the cabin from the mixing manifold to provide independent fine temperature control in each zone. The bleed air from the engines is at a pressure sufficient to operate the air-

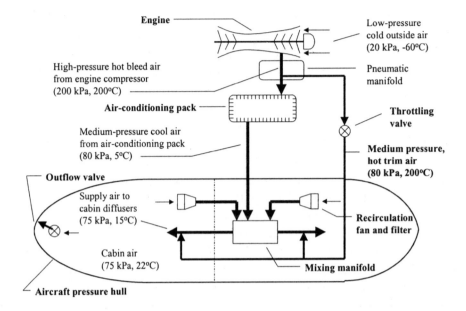

FIGURE 2-5 Simplified depiction of aircraft environmental control systems. (Temperatures and pressures listed are examples only and are typical of some aircraft in cruise conditions.)

conditioning packs and pressurize the cabin. Accurate cabin pressure is maintained by one or more outflow valves that automatically regulate the flow of air out of the aircraft pressure hull to the ambient environment to maintain the desired cabin pressure.

Each of the ECS component systems is discussed in more detail below. It should be understood that aircraft models vary in their systems and that, although the descriptions presented here are for typical aircraft, some of the details might not apply to a specific aircraft model.

Bleed-Air Supply

Figure 2-6 shows the key features of a modern fan-jet engine as is used on most commercial passenger aircraft. All the air entering the engine passes through the fan at the front and divides into two flow paths. The outer flow, which receives no compression other than that provided by the fan, is used directly for propulsion. The inner flow is used by the gas turbine that powers

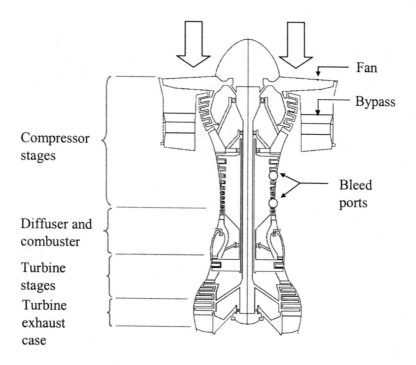

Fan

Bypass

Compressor
stages

Bleed
ports

Diffuser and
combuster

Turbine
stages

Turbine
exhaust
case

FIGURE 2-6 Basic components of a fan-jet engine.

the engine. A multiple-stage compressor compresses the inner air flow to high pressure. From the compressor, it flows into the combustion chamber, where fuel is burned. The combustion chamber does not increase the pressure, but it heats the air to a high temperature, which causes it to expand. The high-pressure, high-temperature air then passes through the turbine, which expands the air and extracts energy from it to drive the compressor. The air leaves the turbine at high velocity and provides additional thrust for propulsion.

The compressor consists of many sets of rotating blades; each set is referred to as a stage. The pressure of the air is increased as it passes through each stage. The pressure ratio relevant to ambient pressure at any stage of the compressor depends on the speed at which the compressor is rotating. At a given stage, the pressure increases as the speed increases.

Bleed air is compressed air that is extracted from the engine compressor. Figure 2-7 shows a typical bleed-air system. Most bleed-air systems have at least two extraction ports, one near the end of the compressor to get the highest possible pressure when the engine is operating at low speed and an inter-

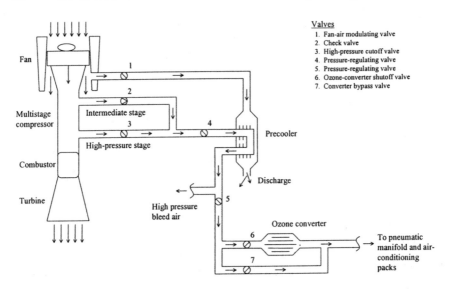

FIGURE 2-7 Common bleed air supply from engine.

mediate stage where the pressure is adequate during normal cruise and at high-power conditions. The bleed-air extraction from the high-pressure stage is automatically turned off when the pressure at the intermediate stage is adequate and is automatically turned on when the pressure from the intermediate stage is not adequate. The high-pressure port is used only during taxi and descent.

The bleed air coming off the engine is at high temperature and pressure; the exact conditions depend on the mode of operation of the engine. Table 2-2 shows typical temperatures and pressures of bleed air from the engine.

The high temperature of the bleed air is a result of compression. There is no combustion of fuel in the compressor. Under most conditions, the bleed air extracted from the engine is hotter than necessary and can be too hot to pipe safely to other parts of the aircraft; therefore, it passes through a precooler immediately after it is extracted from the engine. The precooler is a heat exchanger that uses comparatively cool air from the fan stage of the engine (fan air) to cool the bleed air. The temperature of the bleed air supplied to the aircraft is regulated by controlling the flow of fan air through the heat exchanger. The temperature is typically controlled to about 175°C (350°F). If there is a malfunction of the precooler and the temperature exceeds about 240°C (470°F), a fault will be indicated.

Not all the bleed air is used by the ECS. Some is used to heat leading-

TABLE 2-2 Typical Conditions of Bleed Air from Engine

Mode of Operation	Temperature, °C (°F)	Absolute Pressure, kPa (psi)	Extraction Stage
Takeoff—maximal power[a]	350 (660)	1170 (170)	Low pressure
Top of climb	310 (590)	690 (100)	Low pressure
Cruise	250 (480)	340 (50)	Low pressure
Initial descent	185 (365)	200 (29)	High pressure
End of descent (ground level)	230 (445)	460 (67)	High pressure
Switchover from high to low pressure[b]	280 (535)	480 (70)	High pressure
Ground operations	170 (340)	—	Auxiliary power unit

[a] Some engines (e.g., Avro/4GRJ) have only one bleed port; during takeoff, temperatures might exceed values given.
[b] Maximal temperature and pressure for high-pressure stage occur just before bleed air system automatically switches from high-pressure port to low-pressure port.
Source: Responses from Boeing to committee questions, July 13, 2001.

edge surfaces to prevent ice accumulation and for engine-starting. High-pressure bleed air is tapped off for those purposes before it is supplied to the ECS. The pressure of the bleed air can be as high as 1,170 kPa (170 psi) when it is extracted from the engine. Pressure regulators reduce the pressure to 410-480 kPa (60-70 psi) gauge pressure for the high-pressure bleed air and reduce it further to 310 kPa (45 psi) gauge pressure in the pneumatic distribution manifold, which supplies the air-conditioning packs.

At cruise altitudes, the air can contain excessive O_3. The bleed-air systems of some aircraft are equipped with O_3 converters to reduce the O_3 in the bleed air to acceptable concentrations before it is supplied to the ECS. The O_3 converter consists of a honeycomb lattice, which provides for a large surface area on which O_3 decomposition reactions can occur. The lattice is made of a chemically inert substrate coated with a proprietary catalyst, which speeds the kinetics of the decomposition reaction.

With the exception of the O_3 converter, all the components of the bleed-air system shown in Figure 2-7 are as close as practical to the engine to avoid ducting very hot air around the aircraft. The location of the O_3 converter depends on the aircraft; it can be in the wing or with the air-conditioning packs. It should be noted that some aircraft models have an O_3 converter bypass feature that permits bleed air to enter the cabin without passing through the converter, but most do not. This bypass feature can be used to prevent poisoning of the converter catalyst at low altitudes when the outside air might be contaminated, such as when the aircraft is sitting in a takeoff queue. Other aircraft models use in-line O_3 converters, which cannot be bypassed.

Auxiliary Power Unit

In addition to the engines for propulsion, most aircraft have an auxiliary power unit (APU). The APU is a small gas turbine mounted in the tail cone of the aircraft with an electric generator to supply electric power to the aircraft. Compressed, unfiltered air from the APU is used to supply the pneumatic system of the aircraft, which supplies air to the ECS packs. The APU is normally operated to supply air when the aircraft's main engines are not running (e.g., when the aircraft is sitting at the gate) or are not running at power adequate to generate the necessary bleed air or electric power. The location of the air inlet relative to the APU varies with the model of aircraft; it is generally near the APU.

There are several suppliers of APUs, but there are two basic designs. In one, compressed air is extracted from the APU turbine engine compressor before it reaches the combustion chamber in much the same way that bleed air is derived from the propulsion engines. In the second, the APU drives a second compressor that has no function other than to supply compressed air to the pneumatic system and can be optimized for that purpose.

Air-Conditioning Packs

Bleed air from the engines or the APU is used to pressurize, heat, cool, and ventilate the aircraft cabin. Air-conditioning packs are used to cool and, if necessary, dehumidify the bleed air from the engines or APU before it is supplied to the aircraft cabin. Commercial passenger aircraft typically have two to four air-conditioning packs. Figure 2-8 shows a typical air-conditioning system; this figure is for an aircraft with two engines and two air-conditioning packs, but the basic principles of operation are the same with more engines or packs.

The air-conditioning pack includes two heat exchangers cooled by ram (outside) air, a compressor, and a turbine. The inlet for the ram-air duct faces the airflow over the aircraft, thereby pressurizing the ram air to flow through the heat exchanger when the aircraft is in flight. When the aircraft is on the

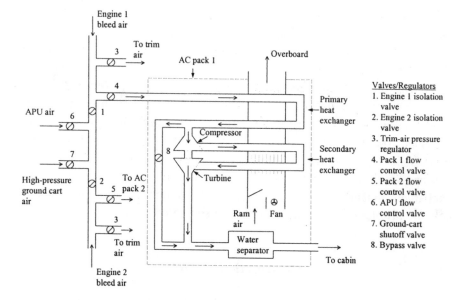

FIGURE 2-8 Common configuration of air-conditioning packs.

ground, however, a fan is required to drive the cooling air through the heat exchanger. The functions of the compressor and turbine in the pack are different from those in the engine. The air-conditioning pack has no fuel or combustion and the power to operate the pack is derived from the pressure of the conditioned air stream. Also, the pack compressor and turbine use air bearings to eliminate the need for lubricants.

Bleed air from the aircraft engines or APU enters a pneumatic manifold and is supplied to the air-conditioning packs. It passes through the primary heat exchanger where it is cooled, and then is compressed to a higher pressure in the compressor, increasing its temperature. That hot, high-pressure air is cooled again in the secondary heat exchanger. The air is then expanded, and its pressure is reduced in the turbine. Sufficient energy to operate the compressor is extracted from the pressurized air by the turbine. As a result, the temperature of the air drops markedly and is cold enough to provide cooling for the cabin even in hot environments.

Under typical cruise conditions, the ram air is cold enough to cool the bleed air, and the bleed air is sufficiently dry so that no moisture has to be removed. Under these conditions, a bypass valve is opened, and without further conditioning, the bleed air goes directly from the primary heat exchanger in the air-conditioning pack to the cabin.

When the aircraft is on or near the ground in a humid environment, moisture will condense from the air when it expands in the turbine. That moisture is removed from the air stream by a water separator downstream of the turbine. The water separator cannot be allowed to freeze, and this requirement often establishes the lower temperature limit for the air from the air-conditioning pack. The need to prevent freezing in the water separator is an important limitation on the system described in Figure 2-8. Many aircraft are now being equipped with newer, more sophisticated packs that include additional heat exchangers that allow air to be supplied at a lower temperature and that eliminate the problem of freezing in the water separator.

Air Distribution System

Figure 2-9 shows a typical air distribution system for an aircraft passenger cabin. Of all the systems described in this chapter, the cabin air distribution system probably varies most from aircraft to aircraft, so the specific details of the system described here do not necessarily apply to every aircraft.

Conditioned air from one or more air-conditioning packs is supplied to a central mixing manifold (there is more than one manifold on some aircraft), where it is mixed with recirculation air. The aircraft cabin can have up to seven separate ventilation zones, and the resulting mixture of conditioned and recirculation air is supplied to each zone of the cabin as needed. The amount of air distributed in the cabin varies for each zone and is controlled by orifices (vents) in the supply ducts. Air supplied to any zone depends on the number of seats and on other heat sources, such as the lighting and entertainment systems. The metabolic heat loads from passengers and cabin crew may at times account for only about half the heat load in the cabin, and the remainder is generated by solar radiation, lighting, entertainment systems, and heat exchange with the outside environment.

The temperature of the air distributed to the cabin and cockpit is controlled by mixing hot trim air with the conditioned air from the mixing manifold before it enters the cabin. There is a separate trim-air valve for each zone. The trim-air valves regulate the flow of trim air to control the temperature of the zones. Although the cockpit crew sets the cabin temperature, the temperature in each

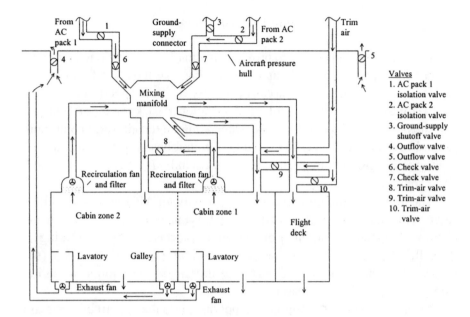

FIGURE 2-9 Common cabin air distribution system.

zone can be separately controlled by means of a thermostat that governs the amount of hot trim air mixed with the bleed air.

Additional thermal control is available to each passenger on some aircraft through gaspers. Gaspers are small outlets above the passenger seats that supply high-velocity air directly to individual passengers, who can control the quantity and direction of the air flow. The air supply for gaspers can be outside air from the air-conditioning system, recirculated air, or air from the mixing plenum, depending on the make and model of the aircraft. A separate gasper fan is used on some aircraft to circulate air through the gasper system. In some aircraft, air for the gaspers is extracted from the cabin and is independent of the recirculated air supplied to the mixing plenum. A separate gasper-air filter, similar to the filter for recirculated air, is used in these applications. The flow of air to the gasper system is controlled by the flight crew in some aircraft, and its operation is indicated by a light on the cabin-temperature control panel. Although the gaspers can influence individual thermal comfort, they play a relatively minor role in overall cabin ventilation.

Ground-Based Systems

It is sometimes desirable to provide pneumatic or conditioned air to the aircraft from ground-based sources rather than use the aircraft engines or APU. Ground-based equipment can be less expensive to operate, can prevent overheating of aircraft due to prolonged operation of packs on the ground, and can substitute for inoperable ECSs. Depending on the airport, one or more ground-based systems may be available. Ground-based systems include the high-pressure ground cart, the low-pressure ground cart, and the low-pressure fixed source.

The high-pressure ground cart is essentially an APU on wheels with a filter on the inlet to protect the cart from debris that might be entrained in the airflow. A pneumatic air line from the ground cart is attached to a fitting on the belly of the aircraft that connects to the aircraft pneumatic manifold (see Figure 2-8). This air is conditioned by the aircraft air-conditioning packs before being supplied to the cabin.

Airlines also have the option of supplying air to the aircraft downstream of the packs by using a low-pressure ground cart, which is essentially a conventional air conditioner on a mobile cart. With this method, air is supplied only to the cabin air supply system, not to the pneumatic system, and no filter is used (see Figure 2-9). The temperature of the air is adjustable and controlled by the cart. The flow rate is fixed but generally is nearly equal to the normal total output of the packs. A variation of the low-pressure ground cart operates by replacing the air conditioner with a heater. This approach is more economical and is typically used when only warm air is needed. Again, the air is normally not filtered.

Airlines have also begun to install fixed, low-pressure systems at each gate to replace the low-pressure ground carts. Rather than use an air conditioner mounted on a cart, the fixed system uses an air conditioner that is permanently mounted on the ground or in the terminal building and is connected to the aircraft by a flexible duct. The fixed system connects to the aircraft in the same manner as the low-pressure ground cart.

ALTERNATIVE ENVIRONMENTAL CONTROL SYSTEMS

Aircraft manufacturers and aircraft systems manufacturers might be investigating the use of alternatives to bleed-air-based ECSs, but any such work is highly confidential because of the competitive nature of this industry, and no information about alternative systems was provided to the committee. Nor did the committee encounter any information about new alternative systems in its literature review. Consequently, the committee had little information to review in preparing this report. However, a brief examination of the potential for alternative systems is given below.

As explained earlier in this chapter, the ECS functions to pressurize the cabin, provide an adequate supply of clean air throughout the cabin, control the air temperature in the cabin, and to some degree regulate the humidity in the cabin. Any alternative to a bleed-air-based ECS must meet the same requirements.

The key features that distinguish the bleed-air-based ECS from other possible systems are the use of engine bleed air as the source of outside air and the use of rotating air-cycle equipment to cool this air. Those two features are fully integrated; not only does the high-pressure bleed air provide enough pressure to pressurize the cabin, but the pressure is high enough to operate the rotating air-cycle cooling equipment without external power sources.

An ECS that uses an alternative source of outside air (e.g., ram air) must, at a minimum, include compression equipment that can compress the outside air enough to meet the pressurization needs of the cabin. The compression equipment must be lubricated, and it must have power sources to drive it, such as electric motors, gas-turbine engines, or compressed-air turbines that use bleed air as the power source. Just as the engine compressors have the potential for contaminating the supply air, any alternative means of compressing the air also has the potential for contamination, although the nature of the potential contamination can be different. Thus, using an alternative to the bleed air as a source of outside air does not automatically ensure uncontaminated air. Because of the need for a separate compressor and compressor drive mechanism, they can impose a substantial weight penalty as well.

If the air is compressed just enough to pressurize the cabin rather than being compressed to a point where it can operate rotating air-cycle equipment, the temperatures attained in the compression could possibly be limited enough to avoid pyrolysis of contaminants, such as hydraulic fluid or lubricating oils. However, an alternative means of cooling the air would be required. The most obvious method is vapor-compression air-conditioning equipment similar to that used in most buildings and in many land vehicles. That equipment works well in ground-based applications, but its size and weight might be unacceptable for aircraft.

The first passenger jet aircraft, introduced in the early 1960s, did not use bleed air as the source of outside air. Ram air was compressed and used to pressurize and ventilate the cabin. Bleed air was used to power a turbocompressor, which compressed the ram air to the proper pressure. That air was cooled by a Freon vapor compression-cycle air conditioner for temperature control before being distributed to the cabin. The arrangement was heavy, expensive, and inefficient because of the inefficiencies of the turbocompressor (E. Marzolf, retired, Douglas Aircraft Co., personal communication, April 29, 2001). Even more important was the high amount of maintenance that the systems required (R. Kinsel, retired, AlliedSignal, personal communication, May 7, 2001). For those reasons and because increasing the altitude of flights would have required an even larger turbocompressor and there was no discernable difference in quality between ram air and bleed air, it was decided in the late 1960s to use bleed air directly (E. Marzolf, op. cit.). New large passenger aircraft have since used bleed-air-based ECSs.

Essentially every large commercial passenger aircraft in use today is equipped with an alternative system, the APU. The APU provides a source of air that is independent of the propulsion engines. Although some APUs provide compressed air in much the same manner that the main propulsion engines provide bleed air, by extracting compressed air from the turbine-engine compressor, some APUs use a compressor that is independent of the engine compressor, as described previously. Definitive data are not available, but the committee saw no evidence that contamination from the APU or brought into the APU is any less likely to affect cabin air adversely than is bleed air.

The committee did not investigate alternative ECSs in depth. To do so would take person-years of effort because reliable information is not readily available in the public domain, if at all. Although it is possible that alternative systems are being developed or could be developed, the committee saw no evidence that bleed-air-based systems cannot be designed, maintained, and operated to provide adequate clean air to the aircraft cabin. That statement does not imply that they always provide uncontaminated air to the cabin. As described in Chapter 3, there is evidence that air is sometimes contaminated; but the measurements that have been taken during routine operation show no contamination of concern arising from bleed-air systems (Nagda et. al. 2001). Data are available on only a few aircraft and conditions, but they are adequate to make the point that bleed air can be clean.

The committee does not want to discourage the pursuit of alternative systems, but it finds that the best method to ensure good cabin air is to have the Federal Aviation Administration (FAA) focus on ensuring that bleed-air-based systems are designed, maintained, and operated properly and that problems with them are identified and resolved. The reasons for the committee's conclusion as follows:

1.   Many uncertainties and potential disadvantages are associated with alternative ECSs, the alternative systems might have problems as yet unknown.
2.   Alternative systems have been used and abandoned.
3.   There is a slow turnover of the aircraft fleet, and aircraft with existing systems almost certainly will be in service for a long time.
4.   There is good reason to believe that bleed-air systems can be designed and operated as to provide uncontaminated air to the cabin.

ENVIRONMENTAL CONTROL SYSTEM DESIGN AND OPERATIONAL STANDARDS

In this section of the report, the committee reviews guidelines, standards, and specifications that can have a bearing on aircraft ECS design and operation. Some of the guidelines and standards were developed primarily for building environments but might have relevance to aircraft as well. The FARs that are applicable to aircraft ECSs are discussed in Chapter 1. The FARs determine the design and operation requirements for aircraft; none of the guidelines and standards described below are legally enforceable for commercial aircraft.

For buildings, the primary indoor air-quality guideline in the United States is ASHRAE Standard 62-1999 (ASHRAE 1999b). As with all ASHRAE standards, it is a voluntary consensus guideline and is not legally binding unless adopted by the applicable regulatory body for a particular application. FAA has not adopted it for aircraft.

Standard 62-1999 was developed for and applies primarily to buildings. However, its scope claims applicability to all occupied indoor or enclosed spaces. There is no specific mention of aircraft or aircraft systems anywhere in the standard. Table 2 of the standard, which sets outdoor air requirements for various applications, has a single entry for vehicles but with no explanation as to type of vehicles (automobile, trains, buses, aircraft, and so on). In a requested interpretation, ASHRAE confirmed that aircraft are not specifically excluded from the scope of the standard but also stated: "Whether the requirements of Standard 62 should be applied to specific vehicles would be a decision for the authority having enforcement jurisdiction" (ASHRAE 2000a). In the case of aircraft, FAA is that authority.

If ASHRAE Standard 62-1999 were applied to aircraft, it would require outside air at 7 L/s (15 cfm) for each occupant. The standard does not specify the temperature and pressure that apply for this requirement. For most buildings, that omission is not critical, inasmuch as the overwhelming majority of buildings are at altitudes between sea level and 1,000 m (3,300 ft) and the density of outside air does not vary greatly over this range. However, the standard does not have any special requirement for buildings at high altitudes where the density of the air can be substantially less than that at sea level. For aircraft, that omission is more serious because they routinely operate at very high altitudes.

Using standard thermodynamic conditions of 101 kPa and a temperature of 25°C (14.7 psi and 77°F) (Howell and Buckius 1992), the Standard 62-1999 requirement translates to outside air at 0.50 kg/min (1.1 lb/min) per occupant. At the minimal allowed cabin pressure of 75 kPa (10.9 psi) and a temperature of 25°C (77°F), that would be 0.37 kg/min (0.82 lb/min). For outside conditions at a cruise altitude of 12,000 m (39,000 ft) and the standard atmospheric temperature of -63°C (-81°F) for that altitude, it would mean only 0.067 kg/min (0.15 lb/min). Basing the requirement on outside conditions at that altitude is unrealistic, but Standard 62-1999 does not specifically prohibit it. The numbers highlight the potential folly of applying to aircraft a standard developed for terrestrial applications.

Assuming that the cabin temperature and pressure apply, and not the outside conditions, it is seen that ASHRAE Standard 62-1999 would require 50-100% more outside air than the current requirement in FAR 25. However, a provision in Section 6.1.3.4 of Standard 62-1999 might allow the outdoor air flow rate to be as little as half that listed above for some flights less than 3 h long. The provision is meant to reflect the fact that human-generated contaminants in the indoor air take some time to rise in response to occupancy. That phenomenon probably is not effective in aircraft, because of their high ventilation rates, but there is nothing in current interpretations of the provision that would prevent it from being applied to aircraft (see, for example, ASHRAE 2000b or ASHRAE 2000c)

ASHRAE Standard 62-1999 is continuously maintained. A revision being considered would divide the ventilation requirements into occupant-generated and building-generated requirements. If the revision is adopted and if aircraft are treated similarly to how buildings are treated, the ventilation requirements for aircraft might drop considerably, with the minimal outside-air requirement for a fully loaded aircraft possibly falling to 2.4-3.3 L/s (5-7 cfm) per person.

Standard 62-1999 sets maximal concentrations of contaminants in the outside air used for ventilation. If the concentrations are exceeded, the outside air must be cleaned before it is used for ventilation. Most of the requirements are based on averages over periods of 24 h to 1 yr and so have little meaning for an aircraft that spends much of its time cruising in pristine air or on the ground not operating. However, ASHRAE's limits on carbon monoxide (CO) and O_3 could apply. The limits for CO are averages of 35 ppm for 1 h and 9 ppm for 8 h. The O_3 limit is an average of 0.12 ppm for 1 h.

It is not possible to compare the above requirements from ASHRAE Standard 62-1999 with those in FAR 25 directly; because the FAR 25 requirements are for cabin air and the ASHRAE requirements are for inlet air. O_3, in particular, will continue to react after the air is in the cabin. An optional provision of Standard 62-1999 does have a guideline for the maximal allowed concentration of O_3 in the indoor air: 0.05 ppm on a continuous basis. CO, in contrast, is mostly inert in the cabin; it is present continuously in the inlet air, the cabin concentration ultimately will match that of the inlet air. Thus, ASHRAE Standard 62-1999 is generally more restrictive than FAR 25 with respect to both O_3 and CO.

The optional provisions in Standard 62-1999 also include a guideline for the maximal concentration of CO_2: 700 ppm above the outside air concentration, or about 1,000-1,100 ppm in most situations. That limit is strictly for the purpose of limiting discomfort related to odors from human bioeffluents and is not meant to be a limit on exposure to CO_2 (ASHRAE 2000d). It uses respiration-generated CO_2 as a proxy for the human-generated odors and is approximately comparable with the 7 L/s (15 cfm) per person requirement at sea level. Obviously, it is not applicable if other sources of CO_2, such as dry ice, are present. The guideline of 1,000-1,200 ppm that comes from Standard 62-1999 should not be compared with the 5,000-ppm limit in FAR 25; the latter is a limit on exposure to CO_2 itself and is not intended to be a measure of ventilation.

ASHRAE Standard 62-1999 does not mandate temperature or humidity requirements. Such requirements are, however, covered in ASHRAE Standard 55-1992, *Thermal Environmental Conditions for Human Occupancy* (ASHRAE 1992), which sets ranges of temperature and humidity that are generally found comfortable as a function of activity level and clothing. The standard's scope claims applicability to environments "at atmospheric pressure equivalent to altitudes up to 3,000 m (10,000 ft) in indoor spaces designed for human occupancy for periods not less than 15 minutes" and thus appears to apply to aircraft. The lower relative-humidity limits range from about 20% to

30%, depending on the temperature, and might be difficult to attain in most aircraft on all but the shortest flights. A public review draft of a proposed revision of the standard removes the lower humidity limits (ASHRAE 2001).

ASHRAE is developing a standard specifically for aircraft cabin air quality and has established the special-project committee on Air Quality within Commercial Aircraft, SPC 161P, to complete the task. The scope under which the committee is working states that it "applies to commercial passenger air-carrier aircraft carrying 20 or more passengers and certified under Title 14 CFR Part 25." Although the committee has been working on the standard for several years, the first public review draft is not expected before to 2002.

The Society of Automotive Engineers (SAE) recommended practice *Procedure for Sampling and Measurement of Engine Generated Contaminants in Bleed Air Supplies from Aircraft Engines Under Normal Operating Conditions*, ARP4418 (SAE 1995), includes a table from AIR4766, *Air Quality for Aircraft Cabins* that specifies the maximal concentrations of contaminants in engine bleed air. Those limits are presented in Table 2-3. AIR4766 is under development by SAE and has not been released to the public (SAE, personal communication, May 7, 2001).

Military Specification MIL-E-5007D, *General Specifications for Aircraft Turbojet and Turbofan Engines* (1973), also includes limits on contaminants in bleed air, which are presented in Table 2-4. Neither the limits in ARP4418

TABLE 2-3 Limits on Engine-Generated, Bleed-Air Contaminants Under Normal Operating Conditions

Contaminant	Maximal Allowable Concentration Above Ambient
CO_2	400 ppm
CO	5 ppm
Hydrogen fluoride	1 ppm
Oxides of nitrogen (nitrogen dioxide equivalent)	1 ppm
Formaldehyde	0.3 ppm
Acrolein	0.05 ppm
Organic material (synthetic oil equivalent, MW 600)	0.4 ppm
Respirable particles	0.5 mg/m^3

Source: Adapted from SAE (1995).

TABLE 2-4 Limits on Engine-Generated, Bleed-Air Contaminants Under Normal Operating Conditions

Substance	Maximal Allowable Concentration Above Ambient, ppm
CO_2	5,000
CO	50
Ethanol	1,000
Fluorine (as hydrogen fluoride)	0.1
Hydrogen peroxide	1.0
Aviation fuels	250
Methyl alcohol	200
Methyl bromide	20
Nitrogen oxides	5
Acrolein	0.1
Oil breakdown products (e.g., aldehydes)	1.0
O_3	0.1

Source: Military Specification (MIL-E-5007D) (1973).

nor the limits in MIL-E-5007D appear to be a factor in the commercial aircraft industry, because modern engines are associated with much lower contaminant concentrations under normal operating conditions when there are no equipment failures or malfunctions. Note that both specifications are for engine-generated contaminants; any contaminants in the engine inlet air would not be covered.

CONCLUSIONS

- The adequacy of oxygen in the cabin is determined by the PO_2. The cabin air pressure is the dominant factor in determining the PO_2 in the cabin. The ventilation rate has little effect on PO_2, and ventilation rates well below those normally present in aircraft would not seriously affect PO_2.
- The flow of outside air must be adequate for contaminant control in the cabin, whether or not recirculation is used. As long as the outside-air flow rate is appropriate for contaminant control in the cabin and the recirculation system is properly designed, operated, and maintained, recirculation does not normally affect cabin air quality adversely.

- As aircraft have moved from using 100% bleed air for cabin ventilation to incorporating recirculated air into ECS design, the amount of outside air supplied to the cabin has decreased.
- The mixing of the air that occurs in the cabin, with or without recirculation, and the proximity of cabin occupants makes it impossible to eliminate exposure to infectious agents and other contaminants in the cabin air.
- The low humidity in aircraft cabins cannot be increased through ventilation controls without raising questions about the effect on air quality. A number of problems are associated with humidification of cabin air. Consequently, the low humidity common in aircraft cabins is not readily corrected.
- The environmental conditions in an aircraft cabin respond quickly to changes in ECS operation. Consequently, it is important that the ECS not be shut down for a long period when the aircraft is occupied except in the case of an emergency, because excessive contaminant concentrations and uncomfortably high temperatures can occur quickly.

RECOMMENDATIONS

- FAA should rigorously demonstrate in public reports the adequacy of current and proposed FARs related to cabin air quality and should provide quantitative evidence and rationales to support sections of the FARs that establish air-quality-related design and operational standards for aircraft (standards for CO, CO_2, O_3, ventilation, and cabin pressure). If a specific standard is found to be inadequate to protect the health and ensure the comfort of passengers and crew, FAA should revise it. For ventilation, the committee recommends that an operational standard consistent with the design standard be established.
- The committee reiterates the recommendation of the 1986 NRC report that a regulation be established that requires removal of passengers from an aircraft within 30 min after a ventilation failure or shutdown on the ground and that full ventilation be maintained whenever on-board or ground-based air-conditioning is available.

REFERENCES

ASHRAE (American Society of Heating, Refrigerating and Air-Conditioning Engineers). 1992. Thermal Environmental Conditions for Human Occupancy. ANSI/

ASHRAE 55-1992. American Society of Heating, Refrigerating and Air-Conditioning Engineers, Atlanta, GA.

ASHRAE (American Society of Heating, Refrigerating and Air-Conditioning Engineers). 1997a. Ventilation and Infiltration. Chapter 25 in 1997 ASHRAE Handbook: Fundamentals. American Society of Heating, Refrigerating and Air-Conditioning Engineers, Atlanta, GA.

ASHRAE (American Society of Heating, Refrigerating and Air-Conditioning Engineers). 1997b. Non-Residential Heating and Cooling Load Calculations. Chapter 28 in 1997 ASHRAE Handbook: Fundamentals. American Society of Heating, Refrigerating and Air-Conditioning Engineers, Atlanta, GA.

ASHRAE (American Society of Heating Refrigerating and Air-Conditioning Engineers). 1999a. Aircraft. Chapter 9 in 1999 ASHRAE Handbook: Heating, Ventilating, and Air-Conditioning Applications. American Society of Heating, Refrigerating, and Air-Conditioning Engineers, Atlanta, GA.

ASHRAE (American Society of Heating Refrigerating and Air-Conditioning Engineers). 1999b. ASHRAE Standard—Ventilation for Acceptable Indoor Air Quality. ANSI/ASHRAE 62-1999. American Society of Heating Refrigerating and Air-Conditioning Engineers, Atlanta, GA.

ASHRAE (American Society of Heating Refrigerating and Air-Conditioning Engineers). 2000a. Ventilation for Acceptable Indoor Air Quality. Interpretation IC 62-1999-37 of ANSI/ASHRAE 62-1999. American Society of Heating, Refrigerating, and Air-Conditioning Engineers, Atlanta, GA.

ASHRAE (American Society of Heating Refrigerating and Air-Conditioning Engineers). 2000b. Ventilation for Acceptable Indoor Air Quality. Interpretation IC 62-1999-18 of ANSI/ASHRAE 62-1999. American Society of Heating, Refrigerating, and Air-Conditioning Engineers, Atlanta, GA.

ASHRAE (American Society of Heating Refrigerating and Air-Conditioning Engineers). 2000c. Ventilation for Acceptable Indoor Air Quality. Interpretation IC 62-1999-20 of ANSI/ASHRAE 62-1999. American Society of Heating, Refrigerating, and Air-Conditioning Engineers, Atlanta, GA.

ASHRAE (American Society of Heating Refrigerating and Air-Conditioning Engineers). 2000d. Ventilation for Acceptable Indoor Air Quality. Interpretation IC 62-1999-05 of ANSI/ASHRAE 62-1999. American Society of Heating, Refrigerating, and Air-Conditioning Engineers, Atlanta, GA.

ASHRAE (American Society of Heating Refrigerating and Air-Conditioning Engineers). 2001. Thermal Environmental Conditions for Human Occupancy. BSR/ASHRAE 55-1992R. ASHRAE Standard, Proposed Revision to American National Standard. First Public Review Draft. American Society of Heating, Refrigerating, and Air-Conditioning Engineers, Atlanta, GA. February.

Boeing Company. 1988. Air Conditioning—Description and operation. P. 1 in Boeing 747-400 Maintenance Manual. 21-00-00. Boeing Company. October 10, 1988.

Boeing Company. 1995. Ventilation—Description and operation. P.1 in Boeing 747-400 Maintenance Manual. 21-26-00. Boeing Company. February 10, 1995.

Fanger, P.O. 1982. Thermal Comfort: Analysis and Applications in Environmental Engineering. Malabar, FL: R.E. Krieger Publishing.

Howell, J.R., and R.O. Buckius. 1992. Fundamentals of Engineering Thermodynamics. New York: McGraw-Hill.

Jones, B.W., M.H. Hosni, and H. Meng. 2001. The Interaction of Air Motion and the Human Body in Confined Spaces. ASHRAE Research Project 978. Final Report. Institute for Environmental Research, Kansas State University, Manhattan, KS. May 1, 2001.

Lorengo, D., and A. Porter. 1986. Aircraft Ventilation Systems Study. Final Report. DTFA-03-84-C-0084. DOT/FAA/CT-TN86/41-I. Federal Aviation Administration, U.S. Department of Transportation. September 1986.

Military Specification. 1973. General Specifications for Engines, Aircraft, Turbojet and Turbofan. MIL-E-5007D. October 15, 1973.

Nagda, N.L., H.E. Rector, Z. Li, and E.H. Hunt. 2001. Determine Aircraft Supply Air Contaminants in the Engine Bleed Air Supply System on Commercial Aircraft. ENERGEN Report AS20151. Prepared for American Society of Heating, Refrigerating, and Air-Conditioning Engineers, Atlanta, GA, by ENERGEN Consulting, Inc., Germantown, MD. March 2001.

NASA (National Aeronautics and Space Administration). 1976. U.S. Standard Atmosphere, 1976. National Oceanic and Atmospheric Administration, National Aeronautics and Space Administration, and the U.S. Air Force, Superintendent of Documents, U. S. Government Printing Office, Washington, D.C.

Nishi, Y. 1981. Measurement of thermal balance of man. Pp. 29-40 in Bioengineering, Thermal Physiology, and Comfort, K. Cena, and J. A. Clark, eds. New York: Elsevier.

SAE (Society of Automotive Engineers). 1995. Procedure for Sampling and Measurement of Engine Generated Contaminants in Bleed Air Supplies from Aircraft Engines Under Normal Operating Conditions. SAE ARP4418. Society of Automotive Engineers, Warrendale, PA.

Weschler, C.J., and H.C. Shields. 2000. The influence of ventilation and reactions among indoor pollutants: Modeling and experimental observations. Indoor Air 10(2):92-100.

3

Chemical Contaminants and Their Sources

Passengers and crew have expressed concerns regarding exposure to various chemical contaminants in the aircraft cabin and have linked adverse health effects to specific potential exposures. Whether the exposures actually occur in the cabin is a critical question. Accordingly, this chapter addresses chemical contaminants, their sources, and potential exposure to them in aircraft. Unless otherwise noted, the term contaminants in this chapter refers to chemical contaminants. Biological agents are discussed in detail in Chapter 4.

Contaminants that originate outside the aircraft are discussed first, and then contaminants that originate inside the aircraft are addressed. Finally, contaminants that can result from the environmental control system (ECS), including the main engines and the auxiliary power unit (APU), are discussed.

CONTAMINANTS WITH EXTERNAL SOURCES

Ventilation air provided to the cabin by the ECS is drawn from ambient air around the aircraft. Any pollutants in this air can be introduced into the passenger cabin. During the gate-to-gate course of a flight, an aircraft generally encounters the following types of ambient air:

- Ground-level air at the departure or arrival airport.
- Urban air aloft in the air basin of the departure or arrival city.

73

- Tropospheric air above the mixed surface layer.
- Air in the upper troposphere or lower stratosphere.

Therefore, the ventilation supply air can be contaminated by background urban pollution and by emissions from local airport sources when the aircraft is on the ground. Urban air pollution is also encountered shortly after takeoff and before landing; however, these periods are usually small fractions of an entire flight. Finally, when flying in the upper troposphere or lower stratosphere, an aircraft can encounter high ozone (O_3) concentrations. The issues noted are explored in the following sections.

Ground-Level Pollution

Most airports are near large metropolitan areas, where pollution can exceed health-based standards. For example, in 1999, 105 million U.S. residents lived in areas that were designated "nonattainment" with respect to at least one of the criteria pollutants (EPA 2000). On a population-weighted basis, the most serious problems were posed by O_3, 90 million residents; particulate matter (PM), 30 million residents; and carbon monoxide (CO), 30 million residents. In addition to urban air pollution, substantial amounts of combustion-generated pollutants are emitted on the ground at airports by, for example, aircraft jet engines and diesel-powered service vehicles. Because emissions at airports are important contributors to urban and regional air pollution, modeling and measurement programs have been established to measure the emissions and the resulting concentrations (Moss and Segal 1994). Some of the information is summarized in the following paragraphs, but it does not appear to have been used to investigate the effect of emissions at airports on air quality in passenger cabins of aircraft.

During the middle 1980s, the Emissions and Dispersion Modeling System (EDMS) was developed to assess air-quality effects of proposed airport development projects (Moss and Segal 1994). It was developed by the Federal Aviation Administration (FAA) in cooperation with the U.S. Air Force and is supported as a Windows-based simulation tool.[1] Specifically, the EDMS is

---

[1]Background information on the EDMS tool is available at http://www.aee.faa. gov/aee-100/aee-120/edms/banner.htm where it can be purchased.

designed to emphasize the effect of emissions from aviation sources—especially aircraft, APUs, and ground support equipment—on air quality in the surrounding areas. It incorporates aircraft engine emission factors from a data bank maintained by the International Civil Aviation Organization. The effects of emissions on ambient air concentrations are predicted by means of dispersion algorithms validated by the Environmental Protection Agency. In 1998, FAA designated the EDMS as the required model to perform air-quality analyses for aviation-related air pollution.

Examples of emission data available through the EDMS are presented in Table 3-1. Emission factors for three combustion conditions are provided for six species. The emission factors express the mass of a species emitted per mass of fuel burned. The last row in the table indicates typical fuel combustion rates of a single jet engine during idle and takeoff. The product of the combustion rate and the related emission factor is the mass emission rate for the species. For example, an idling jet engine would be estimated to emit CO as follows: (fuel at 0.205 kg/s)(CO at 25 g/kg of fuel) = CO at 5.1 g/s.

The data in Table 3-1 show that emission factors for products of incomplete combustion (e.g., CO and hydrocarbons [HC]) are greatest when the jet engine is idling. However, nitrogen oxides, which are produced by high-temperature oxidation of nitrogen in combustion air, are emitted at the greatest rate during the high-power takeoff period. Sulfur oxide emissions are a consequence of sulfur in jet fuel and are independent of combustion conditions. Similarly, because carbon dioxide (CO_2) and water vapor are the major products of combustion, they are emitted at rates that reflect the prevalence of carbon and hydrogen in the fuel rather than the combustion conditions.

The data in Table 3-1 do not reveal what other products of incomplete combustion and unburned fuel constituents are emitted from jet engines. Such components would generally be emitted at lower mass rates than the products listed in the table. However, some have important adverse health effects as toxic air pollutants and may be of concern even at lower rates of emission. Recent studies have begun to explore the problem of exposure of ground personnel and passengers to some pollutants (e.g., volatile organic compounds [VOCs], polycyclic aromatic hydrocarbons, and soot) at airports (Pleil et al. 2000; Childers et al. 2000). Because these studies are in their early stages, no conclusions can be drawn yet.

With respect to pollutant exposure in passenger aircraft cabins at the airport, a key factor is the duration of time on board while the aircraft is on the ground. Flight attendants have higher potential exposures because they board

TABLE 3-1 Typical Emission Factors for Selected Gaseous Species From Jet Engine Operating Regimes

Species	Emission Factor, g/kg		
	Idle	Takeoff	Cruise
Carbon dioxide	3,160	3,160	3,160
Water	1,230	1,230	1,230
Carbon monoxide	25 (10-65)	<1	1-3.5
Hydrocarbons (as methane)	4 (0-12)	<0.5	0.2-1.3
Nitrogen oxides (as nitrogen dioxide)— short haul	4.5 (3-6)	32 (20-65)	7.9-11.9
Nitrogen oxides (as nitrogen dioxide)— long haul	4.5 (3-6)	27 (10-53)	11.1-15.4
Sulfur oxides (as sulfur dioxide)	1.0	1.0	1.0
Fuel combustion rate[a]	0.205 kg/s	2.353 kg/s	—

[a] Fuel combustion rate for a high-bypass GE (CF6-80) turbofan engine.
Source: Data from Penner et al. (1999).

before the passengers and deplane after them. Delays on the ground after boarding because of traffic, inclement weather, or equipment malfunction would also be associated with potentially greater exposures.

Pollution During Ascent and Descent

Ambient air pollution varies markedly with altitude. Almost all air pollution of anthropogenic origin has ground-level or low-altitude sources. The pollution is mixed through the lower troposphere by meteorological processes. Specifically, wind blowing over rough ground surfaces causes turbulent mixing, and heating of the earth's surface by the sun induces vertical motion. The well-mixed layer of the lower troposphere typically extends from a few hundred to a few thousand meters above the earth's surface. For short-lived pollutants that are characteristic of photochemical-smog constituents, the pollutant concentrations are highest in the well-mixed layer and decline steeply above it.

Under ordinary flight circumstances, the duration of exposure of the aircraft to polluted air in the well-mixed layer is relatively brief, lasting for several minutes after takeoff and for several minutes before landing. Sometimes, because of airport traffic, planes are placed in a holding pattern near the airport; in such cases, the duration of exposure to polluted urban air would be increased.

Ozone During Cruise

At altitudes no greater than a few thousand meters above the ground, the contribution of anthropogenic emissions to air pollution is small, at least with respect to pollutants that could affect cabin air quality. Accordingly, the primary ambient air pollutant of concern at cruise altitude is O_3. In contrast with O_3 formed from pollutant emissions near the earth's surface, the O_3 at altitude has natural sources. Oxygen molecules (O_2) undergo photodissociation triggered by ultraviolet radiation from the sun. The oxygen atoms combine with other oxygen molecules to produce O_3. O_3 itself is reactive and decomposes fairly rapidly in the stratosphere either by photodissociation, by reaction with oxygen atoms, or by catalytic destruction (e.g., reactions with nitrogen oxides or chlorine oxides). The persistence and prevalence of stratospheric O_3 is a consequence of the dynamic balance between rates of production and destruction (Seinfeld and Pandis 1998a).

The following subsections review various aspects of O_3: atmospheric concentrations that might be encountered on flights, reactivity and possible reactions of O_3 with materials present in the passenger cabin, methods of controlling O_3 concentrations in passenger aircraft cabins, and studies that have measured O_3 concentrations in passenger aircraft cabins.

Atmospheric Ozone Concentrations

Figure 3-1 presents a summary of annual average O_3 concentrations as a function of latitude and altitude over North America. The figure shows that at cruise altitudes (9,000-12,000 m [29,500-39,400 ft]), the average O_3 concentration is much higher at high latitudes (greater than approximately 60°N) than at low latitudes (approximately 30°N). For example, the O_3 concentration of 35×10^{11} molecules per cubic centimeter at 12,000 m (39,400 ft) at a latitude of 70°N corresponds to a partial pressure of 1.0×10^{-7} atm (1.0×10^{-4} mbar) assuming a temperature of 216.7 K as in the U.S. standard atmosphere (Bolz

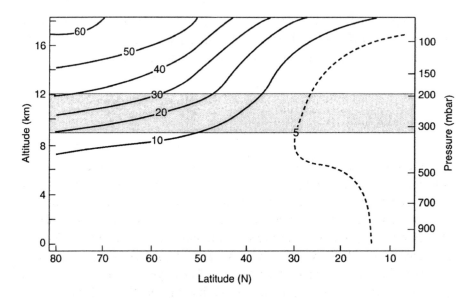

FIGURE 3-1 Annual mean vertical distribution of O_3 (10^{11} molecules per cubic centimeter) over North America. Shaded band represents range of common cruise altitudes. Range of latitudes for continental U.S. is approximately 24°- 49°N (e.g., Miami, Florida, is at a latitude of 25.77°N, and Seattle, Washington, is at 47.61°N). Source: Adapted from Wilcox et al. (1977).

and Tuve 1973). Given a total air pressure of 0.19 atm (190 mbar) at this altitude, the O_3 mole fraction would be 0.5 ppm.[2] In contrast, the value of 5 × 10^{11} molecules per cubic centimeter at 9,000 m (29,500 ft) at a latitude of 30°N would correspond to only 0.05 ppm.

In addition to the effects of latitude and altitude, O_3 varies with season and fluctuates over relatively short periods because of meteorological processes that cause air exchange between the lower stratosphere and the upper troposphere (Seinfeld and Pandis 1998b). Thus, although Figure 3-1 illustrates overall annual trends, O_3 concentrations at any altitude and latitude fluctuate substantially.

In the lower troposphere, outdoor air contains only trace amounts of O_3

---

[2] The mole fraction is defined as the ratio of the number of moles of a constituent to the total number of moles of air in a given parcel. Therefore, a mole fraction of 1 ppm implies that there is 1 mole of O_3 in each million moles of air.

(typically, 0.01-0.1 ppm). However, exposure to low concentrations of O_3 has been associated with adverse health effects, including decreases in pulmonary function (e.g., decreases in lung capacity and increased airway resistance), inflammation of lung tissue, and increased mortality (see Chapter 5). Accordingly, various national and international government organizations have established upper limits on the concentration of O_3 in the air that people breathe (see Chapter 1, Table 1-1). Specifically, the U.S. national ambient air-quality standards for O_3 are 0.12 ppm for a 1-h duration and 0.08 ppm for an 8-h duration.

Ozone Chemistry

In addition to affecting human health directly, O_3 can react with chemicals in aircraft to produce potentially irritating contaminants (Weschler and Shields 1997a). Products of indoor O_3-alkene reactions include short-lived, highly reactive radicals, quasistable compounds (e.g., secondary ozonides), and stable aldehyde, ketones, and organic acids (see Table 3-2). The substances produced can be more irritating than their precursors (Wolkoff et al. 2000); therefore, preventing their formation and accumulation is another reason to limit O_3 in the aircraft cabin.

The relative likelihood of those reactions in an aircraft depends on whether they occur in the gas phase or on surfaces. For a gas-phase (homogeneous) reaction between O_3 and an indoor pollutant to have important consequences, it must occur at a rate that is at least as great as the air-exchange rate (the rate at which the cabin air is replaced with outdoor air) (Weschler and Shields 2000). Outdoor-air exchange rates tend to be much greater in aircraft than in homes or commercial buildings. Specifically, exchange rates in aircraft range from 9.7 to 27.3 exchanges per hour (Hocking 1998), and in U.S. residences and office buildings from 0.2 to 2 exchanges per hour (Murray and Burmaster 1995; Persily 1989). The greater air-exchange rate in an aircraft limits the consequences of homogeneous O_3 chemistry. However, the potentially high O_3 concentrations in aircraft cabins compared with ordinary buildings somewhat offset the effect of faster air exchange.

The compounds that are known to react with O_3 at a rate competitive with aircraft air-exchange rates and that are most likely to be encountered in an aircraft environment include *d*-limonene, α-pinene, and isoprene. Sources of these compounds include solvents, cleaning fluids, and "synthetic" natural rubber materials (Budavari et al. 1989).

TABLE 3-2 Products of O_3/Alkene Reactions Identified or Suspected in Indoor Settings

Product	Reference
Hydroxyl radical	Nazaroff and Cass 1986; Weschler and Shields 1996; Weschler and Shields 1997b
Hydroperoxy and alkylperoxy radicals	Nazaroff and Cass 1986; Weschler and Shields 1997b
Stabilized Criegee biradicals	Finlayson-Pitts and Pitts 1999; Tobias and Ziemann 2000; Tobias et al. 2000
Unidentified radical	Clausen and Wolkoff 1997
Hydrogen peroxide	Li 2001
Organic hydroperoxides	Tobias and Ziemann 2000; Tobias et al. 2000
Peroxyhemiacetals	Tobias and Ziemann 2000; Tobias et al. 2000
Ozonides	Tobias and Ziemann 2000; Tobias et al. 2000; Morrison 1999
Formaldehyde	Finlayson-Pitts and Pitts 1999
Other volatile aldehydes and ketones	Finlayson-Pitts and Pitts 1999
Fine and ultrafine particles	Weschler and Shields 1999; Wainman et al. 2000; Long et al. 2000
Condensed-phase constituents containing carbonyl, carboxylate, and/or hydroxyl functional groups	Yu et al. 1998; Jang and Kamens 1999; Griffin et al. 1999; Virkkula et al. 1999; Glasius et al. 2000

Table 3-3 compares the half-lives of d-limonene, α-pinene, and isoprene at three O_3 concentrations with their half-lives in the cabin at three air-exchange rates representative of those encountered on aircraft. The table indicates that the reactions of O_3 with d-limonene and α-pinene are fast enough to compete with cabin air exchange under conditions of high cabin O_3.

Although the high ventilation rates on aircraft limit the time available for gas-phase (homogeneous) chemical reactions to occur, evaluating the importance of surface (heterogeneous) O_3 reactions is less straightforward. Surfaces in the cabin encounter more O_3 at high ventilation rates than at moderate ventilation rates. That situation promotes the O_3 oxidation of chemicals associ-

TABLE 3-3 Comparison of Half-Lives of Selected Pollutants at Different O_3 Concentrations and in Cabin Air at Different Air-Exchange Rates

Pollutant	2nd Order Rate Constant[a] (ppb^{-1}h^{-1})	Pollutant Half-Life (h) at O_3 Concentration (ppb)			Pollutant Half-Life (h) in Cabin Air at Air-Exchange Rate (air changes/h)		
		100	*200*	*300*	*7.5*	*10*	*12.5*
d-Limonene	1.8×10^{-2}	0.38	0.19	0.13	0.09	0.07	0.05
α-Pinene	7.6×10^{-3}	0.92	0.46	0.31	0.09	0.07	0.05
Isoprene	1.1×10^{-3}	6.4	3.2	2.1	0.09	0.07	0.05

[a] Data from Mallard et al. (1998)

ated with surfaces, especially chemicals with unsaturated carbon-carbon bonds. Such reactions generate products with a range of volatilities. Volatile products desorb from the surface and enter the gas phase, in which they are diluted with ventilation air. Therefore, although the increased ventilation rates favor the formation of reaction products, the greater production rate is offset by a greater dilution rate.

Higher ventilation rates can lead to the accumulation of semivolatile products of heterogeneous O_3 chemistry on surfaces depending on their vapor pressure and rate of volatilization. In other words, the surfaces can become a reservoir for such products, which later volatilize from them. Volatilization can occur for extended periods after the initial production of the semivolatile species (Morrison and Nazaroff 1999); this means that semivolatile oxidation products can continue to be emitted from surfaces in the aircraft even when the O_3 concentrations in the cabin are close to zero. Factors like a sudden change in relative humidity in the cabin could alter the rate of desorption of some compounds from surfaces. See Box 3-1 for further discussion and examples of heterogenous O_3 reactions.

Methods of Controlling Ozone in Aircraft

Concern about O_3 in aircraft was first expressed in the middle 1950s. In the early 1960s, Brabets and co-workers conducted a monitoring survey of O_3 aboard commercial aircraft (Brabets et al. 1967). They measured real-time cabin O_3 on 285 commercial jet flights. Little or no O_3 was detected on flights

BOX 3-1 Examples of Heterogeneous O_3 Reactions

Consider a February flight between New York City and Chicago (about 40°N latitude). Many of the aircraft that fly this route are not equipped with O_3 converters (M. Dechow, Airbus, personal communication, April 13, 2001; R. Johnson, Boeing, personal communication, April 25, 2001). Nonetheless, because of the high volume of traffic, a substantial fraction of flights on this route are assigned a cruise altitude of 10,700 m (35,100 ft). The average ambient O_3 concentration at this altitude and latitude in February is approximately 260 ppb ± 200 ppb (Law et al. 2000). Accordingly, if one assumes that 30% of the O_3 in the ventilation air is removed by indoor surfaces (see Box 3-2 for a discussion of retention ratio) and that 15% of the O_3 removed by surfaces produces formaldehyde (a middle-range estimate derived from O_3 interactions with recently painted latex surfaces, Reiss et al. 1995), then a central-tendency estimate of formaldehyde in cabin air as a consequence of O_3-driven heterogeneous reactions could be calculated according to the following equation:

$$(260 \text{ ppb})(0.30)(0.15) = 12 \text{ ppb}.$$

However, if one assumes that 50% of the O_3 in the ventilation air is removed by indoor surfaces, on the basis of data on the 747-100 aircraft (Nastrom et al. 1980), and that 30% of the O_3 removed by surfaces produces formaldehyde (an upper-bound estimate derived from O_3 interactions with recently painted latex surfaces, Reiss et al. 1995), then a plausible upper-bound estimate of formaldehyde in cabin air could be calculated according to the following equation:

$$(460 \text{ ppb})(0.50)(0.30) = 69 \text{ ppb},$$

where the ambient O_3 concentration of 460 ppb is considered to be 1 standard deviation above the mean. To put these estimates into perspective, the American Conference of Governmental Industrial Hygienists has established a threshold limit value ceiling of 300 ppb for formaldehyde. This value is recommended for occupational exposure of healthy workers. The California Air Resources Board has proposed indoor air-quality guidelines for formaldehyde with 100 ppb set as the "action" value and 50 ppb set as the "target" value. These guidelines apply to the general population.

Next, consider a heterogeneous process that produces a semi-volatile product, cis-2-nonenal, on carpeted and upholstered surfaces in an aircraft. The odor

(Continued)

> **BOX 3-1** *Continued*
>
> threshold for this compound is 2 ppt (Devos et al. 1990). At a ventilation rate of 10 air exchanges per hour, the emission rate of this compound from surfaces would have to be 20 ppt/h for its concentration to reach 2 ppt. Assume that the interior of an aircraft cabin has a volume of 200 m^3 (slightly larger than an MD-80, Hocking 1998) and a surface area of 100 m^2 covered with carpet and upholstery. The emission rate of cis-2-nonenal from these surfaces would have to be 0.23 $\mu g\ m^{-2}\ h^{-1}$ to achieve the odor threshold. As demonstrated by the studies of Morrison and Nazaroff (1999), such emission rates are easily achieved. They reported emissions of 2-nonenal of up to 200 $\mu g\ m^{-2}\ h^{-1}$ from carpets exposed to O_3 at 100 ppb.
>
> The examples provided show that the volatile products of heterogeneous reactions (e.g., formaldehyde) can reach meaningful, but not necessarily excessive, concentrations in cabin air. The semivolatile products of heterogeneous reactions can accumulate on surfaces, and the surface concentrations can eventually become large enough for the surface emission rates to exceed odor thresholds for some compounds.

below the tropopause. At altitudes above the tropopause, the O_3 was commonly above 0.1 ppm. The study found a maximal O_3 concentration of 0.35-0.40 ppm averaged over a 20-min period. Later, Bischof (1973) pointed out that the highest cabin O_3 would be experienced during high-altitude, long-distance flights at high latitude in the spring. He measured O_3 in the cabin air on 14 flights over polar areas. He reported concentrations greater than 0.1 ppm for 75% of the flight time and maximal concentrations of 0.4 ppm averaged over 4 h and 0.6 ppm averaged over 1 h.

By the middle 1970s, concern about O_3 in high-altitude flights had become widespread. Studies conducted in the late 1970s confirmed that high O_3 concentrations could be encountered in the passenger cabin during flight (Nastrom et al. 1980). Furthermore, symptoms that have been associated with O_3 exposure were more prevalent in flight attendants on long-range, high-altitude flights than on short-haul flights (Reed et al. 1980).

The results of those studies and others precipitated regulatory action by FAA. Two regulations introduced in 1980 established O_3 concentration limits for aircraft cabins (FAR 25.832 and FAR 121.578). The regulations are discussed in Chapter 1, and the regulatory language is in Appendix C. Airlines and aircraft manufacturers comply with the regulations through a combination of air treatment, which reduces O_3 in ventilation supply air, and flight planning, which reduces the likelihood of high ambient O_3. Another consideration in meeting the regulations is the fact that the mole fraction of O_3 in cabin air is lower than that in ambient air due to ozone's reactivity. The retention ratio quantifies this effect (see Box 3-2 for a discussion of retention ratio and related issues).

As discussed in Chapter 2, the use of O_3 converters is the only active treatment technology commonly used for reducing O_3 concentrations in aircraft passenger cabins. According to SAE International (2000), a 1997 survey indicates that about 50% of the world fleet of wide-body aircraft are equipped with O_3 converters. Such converters are essentially nonexistent on narrow-body commercial aircraft. More recent information from Airbus and Boeing provides greater detail on the number of aircraft equipped with O_3 converters. Airbus reports that two-thirds of narrow-body aircraft (A319, A320, and A321) produced in 2000 are equipped with converters. However, Airbus was not able to estimate the fraction of its wide-body aircraft (A300, A310, and A300-600) that have O_3 converters because operators may have retrofitted them with converters. All long-range aircraft (A330 and A340) are equipped with O_3 converters (M. Dechow, Airbus, personal communication, April 13, 2001). Boeing reports that all 777s and 767-400 come with O_3 converters as standard equipment. Converters are available as optional equipment on other Boeing models (Hunt et al. 1995), and various proportions of other aircraft models have such converters. Table 3-4 shows the approximate percentages of active Boeing aircraft with O_3 converters (personal communication, Richard Johnson, Boeing, April 25, 2001). When new, converters can decompose 90-98% of the O_3 present in the air flowing through them (SAE 2000). The useful life of O_3 converters is 10,000-20,000 flight hours (R. Lachelt, Engelhard, personal communication, June 12, 2001).

The introduction and use of O_3 converters in some aircraft have apparently reduced the frequency and severity of high-O_3 episodes in aircraft cabins, inasmuch as the widespread complaints about O_3 in the 1970s have not recurred. Some concerns remain, however. Some aircraft that are not equipped with the converter may have elevated O_3 concentrations. Furthermore, there is no process to ensure a high level of converter performance although FAA stipulates that the effectiveness of O_3 converters should be spot-checked (Fed. Regist. 45(14):3880-3883, January 21, 1980).

BOX 3-2 Retention Ratio and Related Issues

The *retention ratio* is defined as the O_3 mole fraction in cabin air normalized by the O_3 mole fraction in ambient air in the absence of deliberate control devices (e.g., O_3 converters). The retention ratio is intended to account for inadvertent O_3 decomposition in the ventilation system and in the aircraft due to reactions on surfaces. It also accounts for reactions that occur spontaneously at the high temperature of the compressor. The retention ratio is a function of the rate of reaction on these surfaces and the ventilation rate. It can be expressed as the ratio of the rate of ventilation (air-exchange rate in units of inverse time) to the sum of the rates of decomposition and ventilation.

The best direct data on the O_3 retention ratio are based on simultaneous measurements of cabin and ambient O_3 on many flights on two aircraft—a Boeing 747-100 and a Boeing 747SP (Nastrom et al. 1980). The average for the 747-100 was 0.465, and that for the 747SP was 0.825. For demonstrating compliance with FAA regulations, airplanes can be assigned a default retention ratio of 0.7 (cited in SAE AIR910 Rev. B page 5 as Final Rule Preamble, Amdt Nos 25-56 and 121-181). The use of a lower value would require the support of direct experimental testing. Such testing would be done on the ground by introducing O_3 into the supply air and measuring the resulting interior concentration.

The phenomenon of O_3 decomposition on surfaces has been widely studied to understand O_3 in buildings; see the reviews by Nazaroff et al. (1993) and Weschler (2000). Studies have shown three important points in the present context. First, the rate of reaction of O_3 on some surfaces depends on humidity, with higher rates of reaction occurring at higher humidity. That observation raises concern about the use of ground-based studies to measure O_3 retention ratios on aircraft because the higher humidity on the ground may lead to lower retention ratios than would prevail during flight. Second, materials exhibit an aging effect: the rate of reaction slows after a period of consistent exposure. That observation also raises concern about the ground-based testing procedure: a short-term test on the ground at low O_3 may not properly account for the aging that could occur during long-term exposure to higher O_3 on a flight. Third, O_3 decomposition on surfaces can produce secondary reaction products, such as aldehydes, that can have low odor and irritation thresholds (see Box 3-1).

TABLE 3-4 Percentage of Boeing Aircraft with O_3 Converters[a]

Aircraft	No. Delivered and Ordered Aircraft	Percentage with O_3 Catalytic Converters
B717	60	0
B727-100/200	1,380	0
B737-200/300/400/500	2,910	0
B737-600/700/800/900	1,100	22
B747-100/200/300/SP	700	20
B747-400	578	64
B757-200	1000	7
B757-300	35	34
B767-200	250	41
B767-300	600	42
B767-400	20	100
B777-200/300	335	100
DC-9	750	0
MD-80	1,175	0
MD-90	115	0
DC-10	400	88[b]
MD-11	195	78

[a] Data in table represent approximate number of active aircraft and approximate percentages with O_3 catalytic convertors. Table is snapshot of information as of May 2001.
[b] Percentage with ordered retrofit kits.

Maintenance procedures for O_3 converters specify cleaning or replacement of the catalyst according to a schedule of flight-hours; however, the committee found no evidence that converters are subject to performance testing after cleaning and maintenance. Concern exists because catalytic converters are generally subject to poisoning by deposits of PM or chemicals (Rodriguez and Hrbek 1999; SAE 2000). Of particular concern is the potential for fouling caused by an air-quality incident. For example, any sizable exposure of the bleed-air system to engine lubricating oil, hydraulic fluid, or even jet exhaust could contribute to fouling that would degrade the performance (destruction efficiency) of the O_3 converter. A lower destruction efficiency could lead to exceeding the regulatory limit. Specifically, a 95% O_3 destruction efficiency appears to be sufficient to maintain cabin O_3 below regulatory limits, but a 50% destruction efficiency would not suffice (see Box 3-3). Although the committee found no hard evidence that fouling of the O_3 converter occurs, it also did not find any evidence to the contrary.

As noted above, flight planning can be used to control cabin O_3 and is based on statistical summaries of atmospheric O_3 as a function of altitude, latitude, and season. Illustrative data are presented in Figure 3-2. The data reinforce the point from Figure 3-1 that O_3 increases with cruise altitude and tends to be higher at high latitudes than at middle latitudes. Figure 3-2 also shows that O_3 varies seasonally, with higher concentrations occurring during winter and spring than during summer and fall.

By regulation (14 CFR §121.578c1; see Chapter 1 or Appendix C), flight planning to limit cabin O_3 is based on the 84th percentile atmospheric O_3 for a given altitude, latitude, and month. O_3 is expected to exceed the planning value 16% of the time. Therefore, under the regulation, cabin O_3 is permitted to exceed the numerical limits of 0.25 ppm (peak) and 0.1 ppm (time-weighted average) in a substantial fraction of flights. Depending on the relationship between the absolute maximal atmospheric O_3 and the 84th percentile, permissible cabin O_3 concentrations under this regulation could be considerably higher than the nominal numerical limits.

BOX 3-3 Effect of Decreased Destruction Efficiency of O_3 Converters on Regulatory Compliance

Consider a sustained encounter (greater than 3 h) of an aircraft to O_3 at 0.5 ppm at an altitude of 12,000 m (39,400 ft). Assume that the cabin pressure is 0.78 atm (795 mbar), corresponding to an altitude of 2,000 m (6,560 ft). Assume a default value of 0.7 for the O_3 retention ratio. In the absence of an O_3 converter, the O_3 mole fraction in the cabin would be 0.35 ppm, and the sea-level equivalent value would be 0.27 ppm. Addition of an O_3 converter with a 50% destruction efficiency would reduce the values by 50%, which would exceed the regulatory limit of 0.1 ppm (14 CFR § 25.832 [b]; see Appendix C). However, an O_3 destruction efficiency of 95% would yield a predicted sea-level equivalent cabin O_3 concentration of 0.014 ppm, well below the regulatory limit.

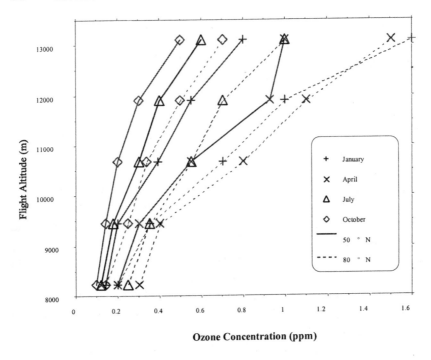

FIGURE 3-2 O_3 concentrations at 84% upper-bound confidence interval, as a function of altitude, for selected months and latitudes. The 0.1-ppm concentration corresponds to 200 μg/m^3, where the volume is determined at sea-level pressure, and temperature is 20°C. Source: Data from SAE (2000).

Measurements of Ozone in Aircraft

Since the introduction of FAA O_3 regulations, few systematic studies of cabin air quality have included O_3 measurements (Nagda et al. 1989; Nagda et al. 1992; Eatough et al. 1992; Sussell and Singal 1993; Spengler et al. 1997; Pierce et al. 1999; Waters et al. 2001). The lack of studies is due partly to limitations in the instruments available to measure O_3 (see Chapter 7 for a brief discussion of O_3 monitoring aboard aircraft). Only ultraviolet (UV) photometric and chemiluminescence instruments have sufficient sensitivity and accuracy to measure O_3 reliably and accurately in the concentration range of interest. Furthermore, only UV photometric instruments are acceptable aboard aircraft because chemiluminescence instruments use a combustible consumable (ethylene), which precludes their use. The disadvantages of the commercially available UV photometric instruments are that they weigh more than 10 kg and require a 110-V AC power source. Lighter, smaller instruments compatible with the power systems aboard aircraft are feasible but are not commercially available. Regardless, the post-1980 studies are summarized below, along with a brief critique of their methods and findings.

The most thorough post-1980 study of cabin air quality was conducted for the Department of Transportation (DOT) by Geomet (Nagda et al. 1989; Nagda et al. 1992). It was conducted during the transition period when smoking was common on flights and when it was almost completely banned, so the emphasis was on environmental tobacco smoke and its constituents. The researchers measured many air-quality characteristics in the passenger cabins of 92 randomly selected flights between April and June 1989. Only 12 of 92 flights were longer than 5 h, and only eight were international. The proportion of aircraft equipped with O_3 converters was not reported. O_3 was measured with an integrated sampling technique. A pump was used to draw air at a fixed rate over glass-fiber filters that had been treated with 3-methyl-2-benzothiazolinone acetone azine (MBTH) and 2-phenylphenol according to the method of Lambert et al. (1989). Time-integrated samples were collected beginning 15 min after takeoff and concluding 30 min before scheduled arrival. The results showed relatively low O_3 concentrations: averages of 0.010 ppm on smoking flights and 0.022 ppm on nonsmoking flights.[3] The highest reported time-weighted average was 0.078 ppm, which is less than the regulatory limit of 0.1 ppm.

Although the study suggests that average O_3 in aircraft cabins is low, the committee has several concerns regarding the study and believes that general

---

[3]The investigators did not report whether the difference between the smoking and nonsmoking flights was statistically significant. The emissions of nitric oxide (NO) from smoking could have reduced the O_3 in the cabin by means of the rapid reaction $NO + O_3 \rightarrow NO_2 + O_2$ (Seinfeld and Pandis 1998c). Some evidence indicates that NO emissions from cigarettes may be as high as 1-6 mg/cigarette (NRC 1986a; Jenkins et al. 2000a,b), with most emitted in the form of NO (Jenkins et al. 2000b). If ventilation data for a Boeing 727-100 (7 L/sec per person, Hocking 1998) are used, and a full flight with 33% of the passengers smoking an average of 1 cigarette per hour each is assumed, the average NO in the cabin is estimated to be up to 0.2-1 ppm, which is more than sufficient to account for the reported difference in average O_3 concentrations between smoking and nonsmoking flights.

conclusions based on the results should be avoided. The concerns include the accuracy and precision of the sampling method, the use of integrated samplers instead of real-time measurements, and the small number of the flights evaluated. The concerns noted are discussed further in the following paragraphs.

The MBTH sampling method chosen for the DOT study (Nagda et al.1989) was developed shortly before the study started. The primary reference (Lambert et al. 1989) is limited in scope, focusing on the colorometric change associated with the MBTH reagent and potential interferences, and does not thoroughly explore issues associated with sampling. Even with its limited scope, the study by Lambert and co-workers found that reagent response, measured by reflection absorbance, was significantly nonlinear. Furthermore, the response was not simply a function of exposure (the product of O_3 concentration and time). For example, exposure at 0.2 ppm for 500 s produced a reflection absorbance 33% higher than exposure at 1.0 ppm for 100 s. Some concern exists because the aircraft samples were collected over periods of approximately 1-10 h, much longer than the sampling periods of 1-10 min used by Lambert and coworkers.

Despite considerable continuing attention to the broad issue of environmental sampling of O_3, the MBTH method does not appear to have been used in any sampling studies after the DOT airliner investigation. Of the recent papers that report development of improved sampling methods, only one cites the work by Lambert and co-workers (Zhou and Smith 1997); others do not even acknowledge it (Monn and Hangartner 1990; Grosjean and Hisham 1992; Koutrakis et al. 1993; Black et al. 2000).

In addition to the concerns about the accuracy and precision of the sampling method, concerns regarding the use of integrated sampling have been raised. The evidence from real-time monitoring clearly demonstrates that O_3 in aircraft is highly variable with time. Thus, integrated sampling can fail to detect important occurrences of increased cabin O_3. For example, if in a 10-h flight O_3 was very low for 9 h but was high (above 0.5 ppm) for 1 h, an accurate time-integrated measurement would reveal a low concentration (approximately 0.05 ppm). However, the peak (0.5 ppm) would represent a real health concern and, if it occurred above a flight altitude of 32,000 ft (9,750 m), would be above the regulatory limit (14 CFR §121.578).

The authors of the DOT study concluded, on the basis of their O_3 measurements, that "all values were consistently below flight, occupational, and environmental standards by the Federal government. . . . This and current scientific knowledge leads to the conclusion that ozone does not pose a health hazard to cabin crew members or passengers."

The committee notes that in at least two respects the evidence does not support the authors' conclusion. First, high ambient O_3 encountered during cruise clearly can lead to cabin O_3 concentrations that exceed the DOT standard and ambient air-quality standards in the absence of effective controls. Therefore, at a minimum, the conclusion should be qualified to state "provided that control measures are functioning effectively." Second, high ambient O_3 occurs episodically, so one cannot conclude on the basis of flight-integrated measurements that the peak concentrations are below the appropriate standard.

O_3 measurements on aircraft have been reported in several other post-1980 studies. However, because of methodological concerns or small sampling scope, they add little to our knowledge of O_3 concentrations in the aircraft passenger cabins.

Eatough et al. (1992) measured O_3 in four 4- to 5-h flights on DC-10 aircraft in which smoking was permitted. The authors used sorbent tubes, which measure time-weighted average concentrations. The flight-average values ranged from less than 0.002 to 0.020 ppm, with an overall average of 0.009 ppm. The use of integrated samplers limits the utility of the data.

In 1990, in response to a request from the Association of Flight Attendants (AFA), the National Institute for Occupational Safety and Health (NIOSH) conducted a health-hazard evaluation of cabin air quality (Sussell and Singal 1993). The investigation included real-time measurements of O_3 on three Alaska Airlines short-haul flights on MD-80 aircraft during July 1990 along the West Coast of the United States. Because details of the measurement techniques were not included in the report, the accuracy of the data is difficult to judge. Nonetheless, the data indicate that episodic peaks were encountered in each flight, and peak concentrations were much larger than average O_3 concentrations. The limited data reinforce the finding that the use of time-averaged samplers for O_3 is inadequate to demonstrate compliance with DOT regulations.

In a study conducted for the Boeing Company, Spengler et al. (1997) made air-quality measurements in several modes of commercial transport. Measurements in aircraft passenger cabins were made on four flight segments in 1996. Integrated average O_3 concentrations were "consistently below the limit of detection for the method used (i.e., 1.8 to 9.8 ppb)." Again, the use of integrated samplers limits the utility of the data.

In a research project funded by the American Society of Heating, Refrigerating and Air-Conditioning Engineers (ASHRAE), O_3 measurements were made on eight commercial flights of Boeing 777 aircraft (ASHRAE/CSS 1999;

Pierce et al. 1999). As noted above, O_3 converters are standard on ventilation systems on the Boeing 777. Four segments were domestic (1,000-1,500 miles), and four covered international routes (over 3,000 miles). Monitoring was done during July 1998. O_3 was measured with a continuous, direct-reading electro-chemical sensor. The reported in-flight mean O_3 concentrations on domestic and international flight segments were 46 and 53 ppb, respectively.

Two technical aspects of the measurement system used by Pierce and co-workers are troubling. First, the authors reported that the instrument was calibrated with 2 ppm of NO_2 in air (ASHRAE/CSS 1999). It is highly unusual to calibrate a pollutant-monitoring device with a pollutant different from the one to be measured. Second, the reported accuracy of the monitoring instrument is ±0.075 ppm. Because that value is comparable with the lower concentration in the standard and is higher than the mean measured, the entire set of sampling results must be called into question. If the data were accurate, they would provide some evidence of degraded performance of the O_3 converter relative to the rated efficiency of 95%. On the basis of a default retention ratio of 0.7 and converter efficiency of 95%, the inside O_3 concentration should only be 3.5% of the outside concentration—0.7(1-0.95) = 0.035. Performance would not appear to be as good in these flight segments. The maximal ambient O_3 encountered by an aircraft is likely to be about 1.0 ppm (Nastrom et al. 1980). Therefore, with the O_3 converter operating at its rated efficiency, the maximal concentration in the cabin should be no higher than about 0.04 ppm. The reported maximum of 3 times that suggests a converter efficiency of approximately 85%.

Finally, a substantial study of air quality in aircraft cabins is in its late stages. The work is being conducted by NIOSH in partnership with the FAA Civil Aeromedical Institute (Waters et al. 2001). The project includes sampling on 37 flight segments in 11 aircraft types of four airline companies. O_3 is measured in real time with an electrochemical sensor. As noted earlier, such sensors have limited sensitivity and resolution. Preliminary results indicate substantially higher O_3 concentrations than reported for the other post-1980 studies. On the basis of gate-to-gate measurements, the reported average O_3 is 0.20 ppm, which is well in excess of the DOT standard of 0.1 ppm. The reported average peak O_3 on these flights is 0.28 ppm. The reported maximal time-weighted average and peak are 0.55 ppm and 1.0 ppm, respectively. These values are high enough to trigger serious concern; however, the investigators have indicated that the results are preliminary. Because the results have not been subjected to peer review, further evaluation of the data will not be warranted until a written report has been issued.

In summary, unacceptably high O_3 concentrations can occur in passenger cabins of commercial aircraft in the absence of effective controls. The 1986 National Research Council (NRC) report (NRC 1986b) on the aircraft cabin environment included this recommendation:

> The Committee could find no documentation of the effectiveness of the various methods being used by the airlines to control O_3. Therefore, the Committee suggests that FAA carry out a carefully designed program to ensure that cabin O_3 concentrations comply with Department of Transportation regulations.

In a 1987 DOT report to Congress, a commitment was made to establish such a program:

> The FAA will issue biennial action notices requiring FAA inspectors to report on the present status of all U.S. air carriers' compliance with the existing O_3 regulations. The response to the action notices will be summarized and published. Identified deficiencies will be corrected.

However, perhaps because of the low concentrations of O_3 reported in the 1989 DOT study (Nagda et al.1989; Nagda et al. 1992), no such program appears to have been implemented. For reasons discussed above, the present committee does not find that there is any basis for confidence that the DOT O_3 standard is regularly met. Nor does it appear possible to demonstrate that without a continuing program that incorporates regular real-time O_3 monitoring on aircraft. Therefore, the potential for high O_3 in aircraft cabins remains a concern. A program of regular monitoring of cabin O_3, particularly on high-altitude, high-latitude flights, is necessary to document the effectiveness of control measures. Such monitoring must use reliable and accurate measurement equipment that is capable of making real-time O_3 measurements.

CONTAMINANTS WITH INTERNAL SOURCES

Sources of contaminants inside the cabin are associated with the passengers and crew in the form of bioeffluents, viruses, bacteria, allergens, and spores; these contaminants are shed from clothing or skin or expelled from oral, nasal, or rectal orifices. Structural components of the aircraft, luggage, personal articles, food, and sanitation fluids can also be sources of vapors or particles. Furthermore, surface residues on aircraft components can be sources of cleaning compounds, pesticides, or simply accumulated debris. The sources and the mechanisms of their emission and dispersion are not specific to the aircraft cabin environment. Most public transportation conveyances will have similar sources. However, aircraft environments are somewhat different given the high surface-to-volume ratios and the relatively small volume-to-passenger ratios.

The following subsections discuss contaminants associated with passengers and their belongings, aircraft component materials, cleaning materials and dust, and pesticides. In several cases, few published data are available on the aircraft environment, and data on similar indoor environments are presented. For a discussion of similarities between building and aircraft environments, see Appendix B. As noted in the introduction to this chapter, biological agents (e.g., infectious agents and allergens) are discussed in detail in Chapter 4; they are not addressed further here.

Passengers and Belongings

Passengers and crew are sources of bioeffluents, allergens, and infectious agents. Through metabolic activity and personal sanitary habits, people can emit odors that others perceive as unpleasant. Bioeffluent emission rates are presented in Table 3-5 for several common compounds as prepared by the NRC Subcommittee on Guidelines for Developing Spacecraft Maximum Allowable Concentrations (SMACs) for Space Station Contaminants (NRC 1992).

As suggested by data in Table 3-5, the passengers and crew are the primary sources of CO_2 in the cabin. Studies conducted in the aircraft environment indicate that CO_2 can range from 293 to 4,238 ppm (see Table 1-2). CO_2 is often used as a marker of the adequacy of ventilation, which is critical in the aircraft environment. For the building environment, Wargocki et al. (2001) found on review of the current literature that "increasing the ventilation rate improves perceived air quality, decreases the prevalence of SBS [sick-building syndrome] symptoms, improves clinical symptoms, reduces absenteeism and improves the performance of office work."

In addition to producing bioeffluents, passengers may apply odor-producing products including nail polish, nail-polish remover, cologne, and perfume. Measurements of VOCs during cruise have identified many compounds, which are

TABLE 3-5 Human Sources of Bioeffluent Aircraft Cabin Contaminants[a]

Substance	Metabolic Generation Rate (mg/day per person)
Alcohols	
Methanol (methyl alcohol)	1.42
Ethanol (ethyl alcohol)	4.00
2-Methyl-1-propanol (isobutyl alcohol)	1.20
1-Butanol (*n*-butyl alcohol)	1.33
Aldehydes	
Ethanal (acetaldehyde)	0.08
Pentanal (valeraldehyde)	0.83
Hydrocarbon	
Methane	600.00
Ketone	
2-Propanone (acetone)	0.13
Mercaptans and sulfides	
Methanethiol (methyl mercaptan)	0.83
Ethanethiol (ethyl mercaptan)	0.83
1-Propanethiol (*n*-propyl mercaptan)	0.83
Organic acids	
2-Oxopropanoic acid (pyruvic acid)	208.30
n-Pentanoic acid (valeric acid)	0.83
Octanoic acid (caprylic acid)	9.17
Organic nitrogen compounds	
1-Benzopyrrole (indole)	25.00
3-Methylindole (skatole)	25.00
Miscellaneous	
Hydrogen	50.00
Ammonia	250.00
Carbon monoxide	33.30
Carbon dioxide	8.8×10^5 [b]

[a] IUPAC or accepted name is provided with common name in parentheses, where relevant.

[b] Estimate based on generation rate of 0.31 L/min, which was used by ASHRAE to determine ventilation standards (ASHRAE 1999).

Source: Data from NRC (1992).

listed with their concentration ranges in Table 3-6; data on other transportation modes are presented for comparison.

TABLE 3-6 Volatile Organic Compounds Frequently Detected in Passenger Compartments of Transportation Vehicles and Their Concentrations

Compound	Commercial Uses or Sources	Concentration, $\mu g/m^3$ [a]				
		Aircraft (1994)	Aircraft (1996), Boeing 777	Trains	Buses	Subways
Ethanol	Bioeffluent, distilled spirits	280-4,300	290-2,600	170-1,700	50-260	130-300
2-Propanol	Distilled spirits, solvent		12-43	0-33	7-63	9-23
Acetone	Bioeffluent, sealants, adhesives, solvent for cellulose acetate	74-150	52-140	49-92	30-73	30-92
2-Butanone	Solvent in resins, adhesives, nitro-cellulose coatings and vinyl films, cleaning fluids; printing catalyst; in lubricating oils	3-16	4-8	3-11	4-18	4-17
d-Limonene	Scent in cleaners	12-24	2-45	1-17	190-490	1-6
Pentadiene isomer	Combustion exhaust	10-30	—	0-30	0-20	10-20
Benzene	Aviation fuel, gasoline, perfumes	1-6	—	2-4	2-6	4-7
Toluene	Gasoline, solvent for paint, thinner, coatings and rubber, cosmetics	0-29	9-19	7-54	15-39	13-27
m- & p-Xylene	Gasoline, solvent in cosmetics	0-8	2-4	3-9	6-48	5-50
Trichlorofluoromethane	Aerosol sprays, blowing agent for polyurethane foam, refrigeration, fire extinguisher	0-2	3-6	3-6	1-4	3-150
Ethyl acetate	Whiskey fermentation	0-4	0-26	2-20	2-10	0-4
Decane	Gasoline, solvent	37016	36926	36971	6-34	36966

Compound	Use					
1,1,1-Trichloroethane	Solvent for chlorinated rubber, various organic materials (oil, grease, etc.)	0-3	0-5	2-10	1-2	2-5
Tetrachloroethene	Dry cleaning; solvent and degreaser for oils, wax, etc.	0-16	5-28	3-29	2-180	5-12
1,2,4-Trimethylbenzene	Gasoline	0-4	0-2	2-5	4-15	4-8
n-Undecane	Gasoline	0-20	4-20	6-61	6-27	5-31
o-Xylene	Aviation gasoline, solvent for alkyd resins, lacquers, insecticides, drugs	0-3	0-2	1-3	3-17	3-25
n-Hexane	Gasoline, solvent	—	0-20	0-3	2-6	0-6
MTBE (methyl tert-butyl ether)	Fuel additive	—	—	2-5	5-11	5-21
Naphthalene	Mothballs, lubricant, solvents	—	0-2	2-8	2-6	3-14
Vinyl acetate	Plastic resin	—	0-2	0-4	0-6	0-7
Chloromethane	Silicones, tetraethylead, synthetic rubber and methyl cellulose, refrigerant, fumigants, herbicide	—	3-4	1-4	3-4	3-11
Dichlorofluoromethane	Refrigerant	—	4-9	1-12	4-22	7-42

[a] Blank cell indicates below limit of detection.

Source: Adapted from Dumyahn et al. (2000).

Aircraft Component Materials

Many aircraft components are made of lightweight plastics. For example, the overhead luggage compartments, sidewalls, ceilings, lavatories, and bulkhead separators are made of formed plastics. Accordingly, passengers and crew can be exposed to various plasticizers, such as phthalates. Animal and human studies indicate that exposure to such plasticizers as mono(2-ethylhexyl) phthalate (MEHP), which is the primary hydrolysis product of di(2-ethylhexyl) phthalate (DEHP), may cause adverse health effects. Øie et al. (1997) suggested that phthalates play a role in asthma by mimicking prostaglandins and thromboxanes in the lungs. Doelman et al. (1990) demonstrated that MEHP induced bronchial hypersensitivity in rats, and Roth et al. (1988) concluded that DEHP from respiratory tubing used to ventilate preterm infants induced lung damage in them. More recently, Jaakkola et al. (2000) reported adjusted odds ratios for persistent wheeze of 3.42 (95% confidence interval, 1.13-10.36) and for cough of 2.41 (95% confidence interval, 1.04-5.63) in a cross-sectional study of 2,568 Finnish preschool children that evaluated the relationship between the presence of plastic wall materials in the home and respiratory health. These findings are consistent with earlier studies by Jaakkola et al. (1997). Although the health evidence linking phthalates to respiratory effects is still limited, exposures to these chemicals may be associated with several respiratory symptoms, including bronchial obstruction, asthma, and respiratory infections.

Passengers and crew may be exposed to coatings used on aircraft components. For example, because water condenses between the wrapped insulation barrier and the colder metal surface of the plane, the internal surface of the fuselage is coated with anticorrosive and antimicrobial materials. Fabric seat and floor coverings, like other commercial materials, may be treated with stain-resistant and antifungal-antibacterial chemicals. Although the potential for exposure to these coating materials exists, no published data were available to determine the extent or degree of exposure of passengers and crew to these materials.

Other materials associated with the aircraft can be sources of VOCs (see Table 3-6). Some measurements of VOCs in planes suggest that foaming agents, plastic resins, and cleaning materials contribute to the contaminant burden. However, measurements of VOCs and of other organic compounds (e.g., aldehydes) are limited. Concentrations reported to date are integrated samples that do not indicate short-term concentrations exceeding odor thresholds or occupational limits. However, further investigation of VOCs in aircraft may be warranted in that several chamber studies conducted by Mølhave (2001) and Otto et al. (1990) demonstrate that individual VOCs and mixtures of VOCs can lead to progressive eye, nose, and throat discomfort over several hours of continuous exposure.

Although the concentrations for individual compounds may be well below their individual odor thresholds, occurrence as mixtures may degrade the perceived air quality or lead to sensory irritation (Cometto-Muñiz 2001; Mølhave and Neilsen 1992; Ten Brinke et al. 1998). Cabin air quality can also be perceived as unacceptable because of interactions with temperature, humidity, and the combined effects of mixtures. Fang et al. (1998a, 1998b) conducted a series of sensory-perception tests in which subjects evaluated the quality of various air samples (clean air versus air with common indoor sources present) at different temperatures and humidities. The sources were scaled to approximate normal loadings found in office buildings. The study results indicated that as air becomes warm and moist, people cannot discriminate among emissions from different sources, and they uniformly rate the quality as low. Drier, cooler, clean air is perceived as better, and the introduction of odorants is recognized and discriminated more readily in cooler and drier air (see Figure 3-3). Although the experimental conditions used by Fang and co-workers do not extend to the low humidity range experienced on commercial aircraft, their results suggest why passengers and crew may be able to distinguish specific odors in the dry air of aircraft that cause them concern regarding air quality.

Cleaning Materials and Dust

The interiors of aircraft are cleaned between flights. One airline reported that aircraft on the ground overnight are thoroughly cleaned (e.g., floors, flight deck, and closets vacuum cleaned; tray tables, walls, and windows washed; debris collected; and galleys and lavatories wiped and sanitized), and aircraft between flights receive spot cleaning (e.g., debris collected and galleys and lavatories wiped and sanitized) (K. Vailu'u, United Airlines, personal communication, June 19, 2001). During more extensive scheduled maintenance, the interior passenger compartment components (e.g., seats, flooring, and overhead storage compartments) are removed; carpet and seat upholstery may be replaced, and portions of the air duct distribution system are removed. On the basis of committee tours of maintenance facilities, debris and soot appear to accumulate on the inside of planes.

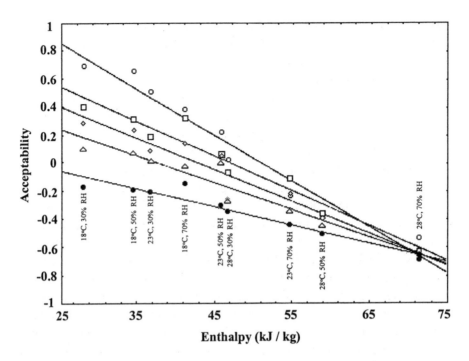

FIGURE 3-3 Acceptability of air samples with different pollutant loadings versus clean air as function of enthalpy. Enthalpy is defined as energy content of air where cool and/or dry air has low enthalpy and warm and/or moist air has high enthalpy. O, clean air; □, wall paint; ◇, carpet; △, floor varnish; ●, sealant. Source: Adapted from Fang et al. (1998a).

As indicated above, minimal information was provided by one airline regarding cleaning practices and materials; the committee could not reach any conclusions regarding the thoroughness or adequacy of cleaning on the basis of that information. Nor could the committee determine the potential for cleaning-product residues or the debris remaining in the carpet and upholstery to be resuspended or the potential for exposure of passengers and crew. Accordingly, the following is a general discussion of cleaning practices, cleaning products, and possible effects of dust exposure; it is not specific to aircraft.

Wolkoff et al. (1998) reviewed the chemical and physical properties of cleaning agents, which are designed to facilitate the removal of debris from surfaces without damaging materials. Active components include surfactants, corrosion inhibitors, solvents, disinfectants, complexing agents, and water softeners. Pigments, fragrances, and preservatives are also present in some products. Cleaning agents contain both volatile and nonvolatile components (see Table 3-7). When cleaning agents are applied, the emissions of volatile components peak within 2 h after application and then decay exponentially over many hours.

Denmark has required registration of washing and cleaning agents since 1986. Chemical composition data on over 2,500 agents are available. Products have been categorized into six distinct groups. Table 3-8 summarizes data on the hazards associated with those categories. The disinfectant category

TABLE 3-7 Components or Constituents of Cleaning Agents

Component or Constituent	Examples
Volatile Substances	
Fragrance	Terpenes (α-pinene, limonene)
Solvent	Glycols, glycol ethers (dipropylene glycol, 2-ethoxyethanol, 2-methoxyethanol)
Biocide (disinfectant)	Formaldehyde and releasers
Plasticizer (softener)	Phthalates
Residual monomer from polymer (film)	Styrene, methacrylate
Decomposed product	Impurities in raw products
Nonvolatile Substances	
Tensides	Soap, detergent
Film formers	Wax, polish, acrylate polymer
Complexing agents	Ethylenediamine tetraacetic acid (EDTA), citrate salts
Acids	Phosphoric acid
Bases	Sodium or potassium hydroxide
Fillers	Sodium chloride or sulfate
Biocides (disinfectants)	Benzalkonium chloride, formaldehyde releasers
Colors, pigments	Different substances

Source: Adapted from Wolkoff et al. (1998).

TABLE 3-8 Number of Products Classified As Irritant (Xi), Harmful to Health (Xn), or Corrosive (C) and Content of CRAN[a] Substances in Selected Danish Cleaning Agents

Product Category	Classified Products (%)	Number of Products Classified as			Number of CRAN Substances in Products
		Xi	Xn	C	
Disinfectants	75	39	22	38	22
Floor polishes	2	—	1	1	18
Furniture polishes	33	—	2	—	4
Universal cleaning agents	13	10	—	2	12
Wash and care cleaners with wax	0	—	—	—	27
Wash and care cleaners without wax	13	13	2	10	17

[a] CRAN: carcinogenic substance, reproductive toxicant, allergen, or neurotoxic substance.
Source: Borglum and Hansen 1994 (as cited in Wolkoff et al 1998). Reprinted with permission from *The Science of Total Environment*; copyright 1998, Elsevier Science.

had the highest percentage (75%) of products classified; many were identified as irritants, corrosives, or potential causes of serious adverse, chronic health effects.

Case studies over the last 25 years have associated respiratory and eye symptoms with carpet cleaning in buildings (Persoff and Koketsu 1978, as cited in Kreiss et al. 1982; Robinson et al 1983; Schmitt 1985; Berlin et al. 1995). Excessive use or improper dilution or application was the probable cause, but these cases provide evidence that inhaled carpet-cleaning residues can cause dry cough and irritation of the throat, skin, and eyes.

More recently, workplace exposures to cleaning compounds have been clinically associated with asthma and pulmonary-function changes. McCoach et al. (1999) examined 17 cases of occupational asthma associated with floor-cleaning materials. Provocation bronchial-challenge tests of six of the workers confirmed that floor-cleaning compounds caused the asthma observed. Rosins (e.g., pine oil and tall oil) and benzalkonium chloride were the constituents identified as bronchial irritants; they had been associated with occupational asthma in earlier studies (Innocenti 1978; Burge et al. 1986; Burge and Richardson 1994). In addition to respiratory effects, Taylor and Hindson (1982) have linked exposure to allyl phenoxyacetate in dry carpet shampoo to occupational dermatitis.

Another possible source of contaminants to which passengers and crew may be exposed is dust or PM. Vacuuming, mopping, and abrasion can resuspend particles in indoor dust. Resuspension can create personal PM exposures 50% higher than area measurements when averaged over 12-24 h (Clayton et al. 1993). However, active walking on surfaces can raise airborne dust to over 10 times preagitation conditions (Hambraeus et al. 1978). Active vacuum cleaning can also greatly increase concentrations of fine particles (Lioy et al. 1999) and cat allergens (Woodfolk et al. 1993). However, because the air-exchange rates on aircraft are high, airborne-particle exposure resulting from the activities noted would probably persist for only minutes.

Airplane dust should be similar in composition to dust in homes and offices. Dust contains minerals, metals, textile, paper and insulation fibers, combustion soot, nonvolatile organics, and various materials of biological origin (e.g., hair, skin flakes, and dander). Dust can be a vehicle for absorbed organic compounds, such as polycyclic aromatic hydrocarbons. In addition, airplane dust can include residues from cleaning agents and pesticides (see discussion below).

Little information on particle mass and number counts on airplanes exists (see Table 1-2). The previous NRC report (1986a) noted that mass concentrations of respirable particles (mass median aerodynamic diameter approximately 3.5 μm) routinely exceeded 100 μg/m^3 while passengers were actively smoking in the designated areas. In the absence of smoking, mass concentrations were usually less than 50 μg/m^3 and often below 25 μg/m^3. More recently, measurements made on a Boeing 777 (nonsmoking flight) yielded 10-min PM$_{2.5}$ concentrations of 3-10 μg/m^3 during cruise and 11-90 μg/m^3 during boarding (Spengler et al. 1997). Measurements reported by Nagda et al. (2001) also support findings of low concentrations of respirable particles. They reported a mean PM$_{10}$ and PM$_{2.5}$ concentration of less than 10 μg/m^3 in the cabin during cruise. In comparison with particle mass concentrations (PM$_{10}$ and PM$_{2.5}$) in U.S. office buildings—indoor concentrations of 3.0-35.4 μg/m^3 for PM$_{10}$ and 1.3-24.8 μg/m^3 for PM$_{2.5}$, according to Burton et al. (2000)—airplanes would rank at the lower end of the concentration ranges during the nonsmoking cruise portion of a flight.

Ultrafine particles (aerodynamic diameters less than 0.1 μm) are formed in copious amounts in pyrolysis of oils, combustion of fossil fuels, and chemical reactions between O_3 and unsaturated hydrocarbons. Episodic events of high ultrafine-particle counts would be expected when engine or ground-equipment exhaust enters the passenger cabin. In contrast with catastrophic engine oil-seal failures, which are accompanied by visible smoke in the cabin, slowly leaking oil that comes into contact with hot ECS or engine components would result in ultrafine-particle formation that might not be perceived in the cabin air. Similarly, O_3 reactions known to produce ultrafine particles (see Table 3-2) could create imperceptible but large amounts of ultrafine particles. Although ultrafine particles may be introduced into the aircraft cabin, no data are available on counts.

Although no published data on health effects of particle exposure on aircraft are available, dustiness and surface soiling have been associated with complaints of comfort and irritation in schools and offices (Wallace et al. 1991; Skov et al. 1989; Norbäck and Torgen 1989; Roy et al. 1993). Skyberg et al. (1999) conducted an intervention study in which the effect of office cleaning on mucosal symptoms was investigated in 104 nonsmoking office workers. The office spaces of 49 workers (intervention group) were thoroughly cleaned. The office spaces of the other 55 workers (control group) received the usual superficial cleaning. The intervention group experienced a 27% decrease in mucosal symptoms, whereas the control group experienced an increase of 2% (p = 0.02). Nasal volume increased in the intervention group by 15% but decreased in the control group by 6% (p = 0.02). The indoor airborne-particle concentrations were lower by more than 20 μg/m^3 in the intervention group.

Wyon et al. (2000) conducted a double-blind reverse-intervention study of enhanced filtration and air cleaning in a London office building. Occupants reported (p < 0.05) less fatigue, less eye ache, clearer thinking, and improved productivity with better-filtered air. Electronically cleaned air did not have as impressive an effect on symptoms.

Mendell et al. (1999) explored enhanced submicrometer particle filtration on an office buildings without known complaints. In a double-blind study, conventional filters were periodically replaced with HEPA filters. Submicrometer-particle mass was significantly lowered, but 2-μm-particle concentrations did not change. Of 16 reported symptoms, 13 improved slightly. However, improvements reached statistical significance in only three of the symptoms (confusion, "too stuffy," and "too humid"). This study also showed that an increase in temperature by 1 °C, although well within the comfort range, significantly increased complaints. A 1 °C increase in temperature had offset the improvements in filtration by a factor of 2-5, indicating that many factors affect perceptions of occupants.

The emerging literature from well-designed double-blind studies and clinical investigations strengthens the observations reported for almost 15 years. Dust—perhaps because of the biological agents, fibers, VOCs, cleaning residues, or other components—can adversely affect the airways, skin, and eyes of occupants. Accordingly, because of the absence of published data specific to aircrafts, systematic sampling of dust on aircraft floors and seats should be conducted. Information should include dust loadings (gram per unit area) and concentration of the components (gram per gram). If problems are identified, the airlines should review their cleaning practices (e.g., cleaning frequency and materials and equipment used) and incorporate newly available techniques to assess the quality of cleaning (Kildesø and Schneider 2001).

Pesticides

The use of pesticides in aircraft for the purpose of insect control is commonly referred to as disinsection (Naumann and McLachlan 1999). The countries that require disinsection on aircraft are listed in Table 3-9 according to disinsection procedure and include Australia, New Zealand, India, and many island nations (DOT 2001). Some of the countries require disinsection only on selected flights on the basis of the origin of in-bound flights; others require disinsection on all in-bound flights. Disinsection is conducted because it is believed to protect public health, agriculture, and the native ecosystems from unwanted insect pests. However, the United States eliminated the practice of disinsection on in-bound flights in 1979 because the Centers for Disease Control and Prevention concluded that disinsection of aircraft was ineffective in preventing insect pests from entering a country and that it would pose a potential health risk to passengers and crew (Anonymous 1999).

The countries that continue to disinsect aircraft base their decisions on reported incidents in which an unwanted vector of human disease has entered the country, established itself, and caused an outbreak of disease (Naumann and McLachlan 1999; WHO 1995). Airport malaria and runway malaria have been reported. In airport malaria, a person contracts malaria from an infected mosquito transported on an aircraft from a malarious region; in runway malaria, a person contracts malaria during a stopover in a region where malaria

TABLE 3-9 Countries That Require Disinsection

Countries That Treat All In-Bound Aircraft		Countries that Treat Selected In-Bound Aircraft (Origin of Aircraft Targeted for Disinsection)[c]
Passengers on Board[a]	Passengers Not on Board (Method)[b]	
Grenada	Australia (residual treatment)	Czech Republic (areas of contagious diseases)
India	Barbados (residual treatment)	Indonesia (infected areas)
Kiribati	Fiji (residual treatment)	South Africa (areas of malaria or yellow fever)
Madagascar	Jamaica (residual treatment)	Switzerland (intertropical Africa)
Trinidad and Tobago	New Zealand (residual treatment)	United Kingdom (malarial countries)
Uruguay	Panama (spray treatment)	

[a] Aerosolized spray used as treatment.
[b] DOT (2001) also mentioned that American Samoa sprayed aircraft while passengers were not on board but did not include this country in list.
[c] No information provided on method of treatment or passenger presence during treatment. DOT (2001) noted that Guam requires disinsection on all flights from Commonwealth of the Northern Mariana Islands, Thailand, Philippines, Korea, Indonesia, Malaysia, the Federated States of Micronesia, Papua New Guinea, Solomon Islands, and the Republic of the Marshal Islands. Furthermore, flights from Taiwan, Korea, and Japan are disinsected during some months. Residual treatment was listed as disinsection method for Guam.
Source: Information from DOT (2001).

is endemic (WHO 1995). Karch and co-workers (2001) summarized cases associated with malaria vectors in European aircraft and estimated that 78 cases of airport malaria have occurred in western Europe since 1977. Of the 78, 28 cases occurred in France, primarily at the Paris Roissy Airport, which is the port of entry for aircraft from tropical Africa. The fatality rate was estimated to be about 5%. Gratz et al. (2000) reviewed cases of airport malaria and other related cases of malaria (see Table 3-10) and noted that airport malaria can be problematic if an infected person (e.g., a person who lives near an airport) has no history of travel to a malarious areas; proper diagnosis can be delayed, and the risk of death increased.

As indicated in Table 3-10, countries that continue to disinsect aircraft

TABLE 3-10 Countries in Which Confirmed or Probable Cases of Airport Malaria Have Been Reported, 1969-August 1999

Country	Period					Total
	1969-77	1978-86	1987-95	1996-98	1999	
France	9	3	11	3	—	26
Belgium	0	9	7	1	—	17
Switzerland	3	0	5	1	—	9
United Kingdom	4	3	0	7	—	14
Italy	0	1	3	0	—	4
USA	0	0	3	1	—	4
Luxembourg	—	—	—	2	3	5
Germany	0	0	2	1	1	4
Netherlands	0	2	0	0	—	2
Spain	0	1	1	0	—	2[a]
Israel	0	0	0	1	—	1
Australia	0	0	0	1	—	1
Total						89

[a] Original table mistakenly reported zero.
Source: Gratz et al. (2000). Reprinted with permission from *Bulletin of the World Health Organization*, copyright 2000, World Health Organization.

justify the practice by pointing to data that show that insects are transported by aircraft (Naumann and McLachlan 1999). Data on the numbers of insects transported by aircraft appear to differ considerably and to depend on the place of origin and the time when the assessments were conducted. For example, an average of 11.4 insect per aircraft in Australia was estimated for the 1970s, while an average of 0.05 insect, including 0.04 mosquito, per aircraft in the Philippines was reported for the same period. Furthermore, an average of 0.1 mosquito per inbound flight in the United States in 1963 was estimated, while an average of 0.18 mosquito per aircraft arriving in the United Kingdom from tropical countries in 1985 was reported. The high number in the Australian studies may have been the result of accumulation from a large number of flights or the result of stopping and refueling at overseas airports at night when bright lights, which attract insects, are used to service aircraft.

Disinsection Procedures

Regardless of the controversy surrounding disinsection, some countries, as stated above, continue to disinsect aircraft. Two publications (WHO 1985; WHO 1995) form the basis of current disinsection practices that are used, or required, by various countries. For passenger cabins, five disinsection procedures appear to be used and include "blocks-away" spraying, "top-of-descent" spraying, "on-arrival" spraying, residual treatment, and pre-embarkation spraying (Naumann and McLachlan 1999).[4] All procedures with the exception of pre-embarkation spraying have been approved by the World Health Organization (WHO) (WHO 1995; Naumann and McLachlan 1999). The methods are discussed below.

Blocks-away, top-of-descent, and on-arrival spraying are similar in that the passengers are on board the aircraft while the spraying is conducted (Naumann and McLachlan 1999). The procedures differ in the time of spraying. Blocks-away spraying is conducted when the aircraft is loaded, doors are closed, the aircraft is prepared for departure, and the wooden blocks are removed from the front of the aircraft tires, allowing the aircraft to taxi to the runway for takeoff. Top-of-descent spraying is conducted before descent to the port of destination. On-arrival spraying is conducted by a quarantine officer who boards the aircraft at the port of destination and sprays the cabin while the passengers are on board. The passengers must remain on board for 5 min after treatment. Top-of-descent and on-arrival spraying may result in lower total exposure of passengers and crew than blocks-away spraying because the time spent on the aircraft after the pesticide has been sprayed is shorter. WHO (1995) commented that "Member States should limit any routine requirement for disinsection of aircraft cabins and flight decks with an aerosol, while passengers are onboard, to aircraft operations originating in, or operating via, territories that they consider to pose a threat to their public health, agriculture or environment."

Residual treatment and pre-embarkation spraying are similar in that no passengers are on board while the spraying is conducted. Because passengers are not present, these procedures tend to be preferred by the airlines (WHO 1995). Residual treatment is conducted by trained operators who certify the aircraft for a period of 8 weeks (WHO 1995); high-use surfaces require touch-

---

[4] For cargo compartments, three options appear to be used and include preflight spraying, on-arrival spraying, and residual treatment (Naumann and McLachlan 1999).

up spraying during this period. Residual treatment of an aircraft has the potential to expose passengers who fly only domestically because aircraft that are treated can also be used on domestic routes. Pre-embarkation spraying, which is being investigated as a new procedure (WHO 1995), is conducted after catering of the aircraft, before passengers board the aircraft, and within 1 h of departure.

Composition of Pesticide Sprays

The pesticide sprays are composed of a pesticide (the active ingredient), a propellant, and possibly solvents. The most commonly used pesticides are *d*-phenothrin and permethrin (cis:trans ratio of 25:75), which belong to a class of pesticides known as pyrethroids, and are recommended by WHO for disinsection (WHO 1995). Permethrin is considered a residual pesticide and *d*-phenothrin a nonresidual pesticide. No pesticides are registered in the United States for use as disinsectants in occupied aircraft (Anonymous 1999).

The disinsection procedures use sprays that vary in composition (Naumann and McLachlan 1999; WHO 1995). For blocks-away, top-of-descent, and on-arrival spraying, the aerosol sprays typically contain 2% *d*-phenothrin and are applied at 1 g/s and aimed to disperse 10 g of formulation per 1,000 ft^3. The preflight application often used in conjunction with the procedures noted is typically a 2% permethrin formulation. For residual treatment, a 2% aqueous emulsion of permethrin is typically applied at 10 mL/m^2 and aimed to deliver permethrin at a rate of 0.2 g/m^2; a higher application rate may be required for carpets. For pre-embarkation spraying, an aerosol spray containing both 2% *d*-phenothrin and 2% permethrin is typically used with application designed to disperse 10 g of formulation per 1,000 ft^3. Both pesticides are used because the *d*-phenothrin kills insects present in the cabin at the time of spraying, and the permethrin kills insects that enter the cabin between spraying and departure. The permethrin also has a repellant effect that discourages insects from entering the cabin.

In addition to the active ingredients, a propellant is present in the aerosol sprays (WHO 1995). Two chlorofluorocarbons (CFC 11 and CFC 12) were recommended by WHO but are no longer recommended because of their O_3-depleting potential. Although partially halogenated chlorofluorocarbons have a lower O_3-depleting potential, they are considered to be only transitional substitutes. Hydrofluorocarbons are considered the ideal substitutes because they have minimal O_3-depleting potential. However, several of them (e.g., HFC 32,

HFC 143a, and HFC 152a) are flammable and therefore are not considered suitable for use in aircraft. HFC 134a and HFC 227ea are nonflammable and might be recommended by WHO for use in disinsection of aircraft.

In addition to the active ingredients and propellant, solvents (e.g., petroleum distillates) might be added to enhance the solubility of the active ingredients (WHO 1995). However, the amount of solvent added is typically small (0.067 %). A synergist (e.g., piperonyl butoxide) might also be added to the pyrethroid formulation to increase activity of the pesticide (WHO 1995).

Pesticide Exposure

No information on quantitative measures of exposures of passengers or crew to pesticides is available. However, passengers and crew clearly receive inhalation and dermal exposures if pesticides are sprayed while they are on board the aircraft. If pesticides are sprayed before a flight (blocks-away spraying) or during a flight (i.e., top-of-descent spraying), passengers may continue to receive dermal exposures after spraying because the aerosol spray settles onto interior surfaces, which serve as a reservoir for later dermal exposures. Oral exposures can occur as a result of hand-to-mouth activity.

Passengers and crew are also exposed to pesticides through residual or pre-embarkation spraying, but these exposures are primarily oral and dermal because the cabin is treated while the passengers and crew are not on board. Inhalation exposure would not be a primary route of exposure because *d*-phenothrin and permethrin have low volatility.

Wipe samples, taken by a flight attendant who appeared to use the proper equipment and procedure, showed concentrations of permethrin of 0.17-0.69 $\mu g/cm^2$ (J. Murawski, AFA, personal communication, March 5, 2001). The wipe samples were obtained from four surfaces after a residual spray treatment several hours before boarding of a 747-400 aircraft. The values cited may be underestimates; no data were provided on the efficiency of an alcohol-swipe method to capture such a highly lipophilic substance as permethrin.

CONTAMINANTS FROM AIRCRAFT SYSTEMS

Contaminants from the aircraft systems (e.g., engines) may enter the cabin in bleed air. The stages at which air is bled from the system depend on the aircraft and the flight segment. Bleed air can also be supplied by the APU, which is a small turbine engine usually in the tail cone of the aircraft. The APU provides bleed air to the aircraft when the main engines are not operating, or when main-engine power demands are high during specific flight segments and main-engine bleed air cannot be spared. See Chapter 2 for further details on the bleed-air system.

Problems can arise when fluids used in the operation and maintenance of the aircraft enter the bleed-air system. However, few data have been collected on contaminants that might be present in engine bleed air under normal operating or upset conditions. This section addresses the potential for lubricating oils, hydraulic fluids, deicing fluids, and their pyrolyzed products to enter the aircraft cabin.

Over the years, newspaper reports have documented incidents in which smoke, fumes, or mists have entered a passenger cabin or cockpit (Norton 1998, Acohido 2000, Arlidge and Clark 2001). Many of the incidents have been attributed to the entry of engine lubricating oils or hydraulic fluids into the cabin through the bleed-air system. The oils or fluids enter the ECS as a result of equipment failures (e.g., leaking engine-oil seals). Many estimates have been provided of the frequency of such events ranging from one in 22,000 flights (Winder 2000) to one in 1,000 flights (Hood 2001). In testimony before an Australian Senate inquiry, Jean Christophe Balouet stated that 70 "major smoke/haze events" and 500 "severe fume events" have been estimated to occur each year worldwide (Parliament of the Commonwealth of Australia 2000). The committee was not able to verify any of the estimates.

Balouet further stated that "some aircraft types, especially BAe 146, MD 80, B 737, A 300, and a limited number of companies . . . have been the cause of over 90% of the world wide problems identified today, whereas they represent less than 3% of world flights" (Parliament of the Commonwealth of Australia 2000). A recent study using information from several years of operation for three airlines attempted to quantify the relative frequency of air-quality incidents for aircraft (C. van Netten, University of British Columbia, personal communication, November 13, 2001). Although the results cannot necessarily be extended to other airlines, they demonstrate that the frequency of incidents can vary considerably from one aircraft to another—3.88 per 1,000 flight cycles for the BAe 146, 1.29 per 1,000 flight cycles for the A320, 1.25 per 1,000 flight cycles for the B747, 1.04 per 1,000 flight cycles for the DC-10, 1.02 per 1,000 flight cycles for the MD-80, 0.63 per 1,000 flight cycles for the B767, and 0.09 per 1,000 flight cycles for the B737. See Box 3-4 for further discussion of problems with air quality on BAe 146 and MD-80 aircraft.

The airlines do not publicize the frequency of those events. However, that

BOX 3-4 BAe 146 and MD-80 Aircraft

Some aircraft types have been associated with air-quality problems in which smoke or fumes enter the passenger cabin or cockpit. One such aircraft is the BAe 146, which was the focus of a recent investigation by the Australian Senate (Parliament of the Commonwealth of Australia 2000). Persistent problems with this aircraft have been noted since the early 1990s. Poor cabin air quality has been attributed to excessive oil-seal leaks in the main engines or the APU. These leaks allow lubricating oil to enter the bleed-air system, where high temperatures can volatilize or pyrolyze the oil and result in smoke in the aircraft cabin or cockpit. The frequency of the leaks in Australian BAe 146 aircraft has been reported as one in 66 flights in 1992 and one in 128 flights in 1997 (Winder 2000). The outcome of the Australian Senate investigation was the issuance of an airworthiness directive (AD) by the Australian Civil Aviation Safety Authority (CASA). The AD requires investigation and reporting of any suspected cabin air contamination with engine oil (Airworthiness Directive, March 31, 2001). Furthermore, defective components must be fixed before further flight or within 10 flight hours with the provision that the source of contamination is isolated from the passenger cabin and cockpit.

Another aircraft that has been associated with poor cabin air quality is the MD-80. An investigation by the AFA found that 73% of air-quality incidents in one airline were attributed to the MD-80 (Witkowski 1999). FAA, in a notice of proposed rule-making for an AD, described the problems with the MD-80 aircraft (Fed. Regist. 65(11), January 18, 2000). Smoke and odor have been reported in the passenger cabins of these aircraft. An investigation indicated that the problems result from the "failure of a hydraulic pipe in the aft fuselage accessory compartment." Failure of the component leads to leaking of hydraulic fluids in the bilge area of the tailcone and the drainage and ingestion of the fluids into the air intake of the APU, which results in the smoke and odors that have been observed in the passenger cabin and cockpit of these aircraft. The AD became effective on September 12, 2000, and requires replacement of faulty hydraulic components and installation of drain tubes and diverter assemblies in the area of the APU inlet (14 CFR Part 39).

leaks do occur is supported by information provided in the Boeing 737 maintenance manual, which describes how air can become contaminated and how to purge contaminants from the system (Boeing 1998). Specifically, the manual indicates that "smoke or fumes" may enter "the passenger cabin or the flight compartment through the air distribution system during flight." The problems are attributed to leaks of oil, fuel, hydraulic fluids, or deicing fluids from the engines or APU or ingestion of the fluids into the inlet of the engines or APU.

Because no published data are available on contaminants present in bleed air (or cabin air) during an air-quality incident (e.g., leakage of engine lubricating oil or hydraulic fluid into the cabin air-supply system), data on the composition of engine lubricating oils and hydraulic fluids are useful for determining potential exposures. Components of some engine lubricating oils and hydraulic fluids are summarized in Table 3-11.

As indicated in Table 3-11, engine lubricating oils contain cresyl phosphate isomers. Specifically, these oils typically contain 3% tricresyl phosphate (TCP) with the neurotoxic isomer tri-*o*-cresyl phosphate (TOCP) present at 0.1% (van Netten 2000). Although the Australian material safety data sheet (MSDS) for Mobil jet oil 291 indicates "no reportable ingredients," the U.S. MSDS for this oil indicates the presence of TCP isomers. Common ingredients in hydraulic fluids are tributyl phosphate, butyl diphenyl phosphate, and dibutyl phenyl phosphate. Although it is not indicated in the MSDSs, hydraulic fluids may contain as much as 1% TOCP (van Netten 2000).

Independent laboratory analyses confirmed the presence of the TCP isomers in Mobil jet oil 254, Mobil jet oil II, and Mobil jet oil 291 and of tributyl phosphates in Skydrol 500B-4, Skydrol LD-4, and HyJet IV-A⁺ (van Netten 2000; van Netten and Leung 2001). The only difference that could be identified between Mobil jet oil II and Mobil jet oil 291 was the absence of *N*-phenyl-1-naphthalenamine in Mobil jet oil 291. These analyses also indicated that Hyjet IV-A⁺ contained TCP isomers, which were not reported in the MSDS for this product.

Although information on the composition of engine lubricating oils and hydraulic fluids is useful, it does not positively identify components that would be present in bleed air as a result of equipment failures. Laboratory experiments have been conducted to determine constituents that might be generated when engine lubricating oils and hydraulic fluids are exposed to high temperatures (i.e., 525°C) like those in some aircraft engines (van Netten 2000, van Netten and Leung 2000, van Netten and Leung 2001). The results showed that Castrol 5000, Exxon 3280, Mobil jet oil II, Mobil jet oil 254, and Mobil jet oil 291 release TCP isomers that can be trapped in air at room temperature. These compounds appear to remain airborne and are probably associated with smoke particles. Many other compounds are also generated, including substantial amounts of CO. Hydrogen cyanide (HCN) was generated in very negligible amounts under the conditions of these experiments.

TABLE 3-11 Components of Some Engine Lubricating Oils and Hydraulic Fluids, as Reported in Material Safety Data Sheets

Type of Oil or Fluid	Reported Components (wt%)	CAS No.
Engine lubricating oils		
Mobil jet oil 254	Tricresyl phosphate (1-5%)	1330-78-5
Mobil jet oil II	Tricresyl phosphate (1-5%)	1330-78-5
	N-Phenyl-1-naphthalenamine (1-5%)[a]	90-30-2
Hydraulic fluids		
Skydrol 5	Triisobutyl phosphate	126-71-6
(Solutia Inc.)	Triphenyl phosphate	115-86-6
	Epoxy-modified alkyl ester	
Skydrol 500B-4	Tributyl phosphate	126-73-8
(Solutia Inc.)	Dibutyl phenyl phosphate	2528-36-1
	Butyl diphenyl phosphate	2752-95-6
	Epoxy-modified alkyl ester	
Skydrol LD-4	Tributyl phosphate	126-73-8
(Solutia Inc.)	Tributyl phosphate	126-73-8
	Dibutyl phenyl phosphate	2528-36-1
	Butyl diphenyl phosphate	2752-95-6
	2,6-Di-*tert*-butyl-*p*-cresol	128-37-0
	Epoxy modified alkyl esters	
HyJet IV-A+	Tributyl phosphate (79%)	126-73-8
(Chevron)	Cyclic aliphatic epoxide (<2.9%)	3388-03-2
	Additives (<21%)	

[a] Contains 1 wt% phenyl-α-naphthylamine (PAN).

The hydraulic fluids Skydrol 500B-4, Skydrol LD-4, and HyJet IV-A+ appeared to generate much less CO than the engine lubricating oils under identical conditions. Specifically, the oils produced CO at 56-141 ppm, and the hydraulic fluids at 2-18 ppm. Few compounds were detected in the air when the hydraulic fluids were heated. The constituents appeared to evaporate into the air before any pyrolysis took place.

The identification of CO as a degradation product when engine lubricating oils and hydraulic fluids are subjected to high temperatures indicates a potential acute hazard to passengers and crew.

Trimethylolpropane phosphate (TMPP) was not detected in any of the experiments. This compound has been shown to form from trimethylolpropane esters and TCP isomers at high temperatures (Wyman et al. 1993; Wright 1996).

Although the experiments indicate components that might be present during an air-quality incident, it is important to know whether a constituent could be present at a hazardous concentration. Simple calculations illustrate that only very small quantities of oils need to be pyrolyzed under conditions that occur in the bleed air system to exceed commonly accepted health standards. See Box 3-5 for a sample calculation. Therefore, possible contaminant concentrations (e.g., of TCP isomers) that might result from upset conditions should be evaluated.

The foregoing discussion focused on potential contaminants in bleed air during an air-quality incident. Another issue to address is the quality of bleed air during normal operating conditions. Nagda et al. (2001) conducted a study to investigate contaminants in bleed air under normal operating conditions. They obtained measurements on three wide-body Boeing 767s (two flights each), two standard-body Boeing 737s (one flight each), and two wide-body Boeing 747s (one flight each). Contaminants in bleed air appear to have been measured only on the Boeing 767. They measured CO_2, CO, PM ($PM_{2.5}$ and PM_{10}), VOCs, aldehydes, ketones, and semivolatile organic compounds (SVOCs) in bleed air.

Average CO_2 concentrations in the bleed air on the Boeing 767 during boarding, ascent, cruise, and descent were 680 ± 173 ppm, 402 ± 5 ppm, 340 ± 13 ppm, and 359 ± 4 ppm, respectively. Average CO_2 concentrations in the cabin air on the Boeing 737 and Boeing 747 aircraft during the different flight segments were 1,091-1,547 ppm. On the Boeing 767, CO_2 was measured in the cabin only for a 15-min period during cruise. PM and CO concentrations were extremely low. $PM_{2.5}$ and PM_{10} were consistently below 10 $\mu g/m^3$ in bleed air and cabin air after instrument calibration. CO measurements were typically below 1 ppm (see Table 3-12).

The authors noted that VOC concentrations were generally below those measured in buildings. Maximal concentrations of VOCs detected in bleed air are summarized in Table 3-13. For comparison, maximal concentrations in cabin air are also presented, where measured. The only VOCs detected in

BOX 3-5 How Much Does it Take?

Consider an incident in which a leaky engine seal in an aircraft results in oil pyrolysis in bleed air. Assume that one of the major pyrolysis products is formaldehyde. How much oil must be pyrolyzed during a 15-min period for the concentration of formaldehyde in the aircraft cabin to reach its threshold limit value-ceiling (TLV-C)?

We begin by calculating how much formaldehyde must be emitted during a 15-min period for its concentration in the cabin to reach its TLV-C. Under quasi-steady-state conditions, the concentration of a contaminant, C, emitted into the cabin of an aircraft can be calculated with the following equation:

$$C = (E/V)/\lambda_v,$$

where E is emission rate, V is volume of the cabin, and λ_v is ventilation rate. Rearranging yields the following equation:

$$E = CV\lambda_v.$$

The TLV-C for formaldehyde is 300 ppb or 370 μg/m³ (assuming 760 mm Hg and 25°C). Assume that the volume of the cabin is 150 m³ and the ventilation rate is 23 h⁻¹. Substituting yields

$$E = (370 \text{ μg/m}^3)(150 \text{ m}^3)(23 \text{ h}^{-1}) = 1.3 \times 10^6 \text{ μg h}^{-1} = 1.3 \text{ g h}^{-1}.$$

In a 15-min period (0.25 hr), this emission rate equates to 0.3 g of formaldehyde. Pyrolysis data on one of the oils typically used in aircraft indicate that at 370°C, 0.3 g of formaldehyde is produced for each gram of oil pyrolyzed (R. Fox, Honeywell, personal communication, April 4, 2001). Therefore, 1 g of this oil, pyrolyzed over a 15-min interval, would be sufficient to produce formaldehyde in the cabin that matched the TLV-C.

It should be emphasized that this is a simplified calculation. It assumes that the temperature of the bleed air is 370°C. Such a temperature would typically occur only during takeoff and climbing. It neglects the loss of formaldehyde in the airstream, especially to moist surfaces; moist surfaces may be present during takeoff and climbing. It assumes that pyrolysis occurs at a constant rate, that all the oil (1 g) is pyrolyzed and enters the ECS, and that the air in the cabin is perfectly mixed. Nonetheless, the calculation illustrates the feasibility of achieving contaminant concentrations in cabin air, during upset conditions, that exceed occupational-exposure limits. This calculation cannot be used to estimate oil seal leakage rates required to generate this concentration of formaldehyde because only a small, unknown fraction of oil leaking from a seal is likely to get into the bleed air system.

TABLE 3-12 Carbon Monoxide in Aircraft Air

		Carbon Monoxide, ppm			
		B767 Bleed	B767 Cabin	B737 Cabin	B747 Cabin
Boarding	Average	0.5		1.0	0.6
	SD	0.1		0.3	0.1
	Maximum	0.9		1.7	0.8
Ascent	Average	0.1		0.6	0.3
	SD	0.1		0.1	0.1
	Maximum	0.3		0.8	0.5
Cruise	Average	0.1	0.1	0.5	0.1
	SD	0.1	0.01	0.1	0.1
	Maximum	0.3	0.1	0.8	0.3
Descent	Average	0.3		0.5	0.2
	SD	0.1		0.01	0.01
	Maximum	0.7		0.6	0.3

Source: Adapted from Nagda et al. (2001).

cabin air that were not detected in bleed air were vinyl acetate, 4-methyl-2-pentanone, styrene, and 1,4-dichlorobenzene with maximal concentrations of 0.94-4.9 $\mu g/m^3$. The only VOC detected at a high concentration was methylene chloride in cabin air. Because bleed-air concentrations were below 10 $\mu g/m^3$, the methylene chloride in cabin air was attributed to a source in the cabin.

Aldehyde and ketone concentrations were low. The authors noted that aldehyde concentrations were "lower than those encountered in ground level buildings." For example, maximal concentrations of formaldehyde were 2.1-3.1 $\mu g/m^3$ in bleed air and 6.4-13.0 $\mu g/m^3$ in cabin air. Maximal concentrations of acetaldehyde were 26.4-30.7 $\mu g/m^3$ in bleed air and 20.8-70.2 $\mu g/m^3$ in cabin air. Acrolein, a strong respiratory irritant, was not detected in bleed or cabin air.

SVOC concentrations were extremely low. In fact, only naphthalene was detected in bleed or cabin air. Maximal concentrations were 2.0-3.4 $\mu g/m^3$ in bleed air and 0-1.6 $\mu g/m^3$ in cabin air. No TCP isomers, including TOCP, were detected, and no TMPP was detected.

On the basis of data obtained in this investigation, the authors rated the quality of the bleed air as high under normal operating conditions. However,

TABLE 3-13 Maximal Volatile Organic Compound Concentrations

Compound	Maximal Concentration, $\mu g/m^3$			
	Ascent-Bleed Air	Cruise-Bleed Air	Cruise-Cabin Air	Descent-Bleed Air
Chloromethane	2.4	0	0	2.1
Acetone	15	12	130	31
Trichlorofluoromethane	3.3	2.2	16	2.3
Methylene chloride	5.4	9.5	2,900	10
Trichlorotrifluoroethane	2	0	0	0.8
Carbon disulfide	0	2.2	9.2	0
Methyl *tert*-butyl ether	2.5	0	0	52
2-Butanone	1.7	1.7	9.4	6.1
Benzene	0	0	0	7.3
Trichloroethene	0	0	0	3.4
Toluene	9.3	10	21	45
Tetrachloroethene	0	1.1	12	2
Ethylbenzene	1.1	0	0	6.8
m- and *p*-Xylenes	10	1.5	24	28
o-Xylene	1.6	0	0	7.8

Source: Data from Nagda et al. (2001).

they noted that only 10 flights were monitored, and these flights "do not provide a statistically robust sample for representing the universe of all flights."

Other fluids that can potentially enter an aircraft ventilation system are deicing fluids. Deicing fluids are applied to aircraft on the ground before take-off under some weather conditions (e.g., freezing rain). Deicing fluids are usually aqueous solutions of ethylene glycol and/or propylene glycol containing proprietary additives (Ritter 2001). Because ethylene glycol is considerably more toxic than propylene glycol, the latter is being used more and more. The additives may include "a surfactant, polymer thickening agent, pH buffer, corrosion inhibitor, flame retardant, or dye." Some concern has been raised about the toxicity of several additives (e.g., urea and tolyltriazoles).

Aircraft deicing fluids are typically sprayed onto the aircraft at 150-180°F under high pressure to optimize the removal of snow or ice (Ritter 2001).

When the aircraft is scheduled to take off immediately after application, a type I deicing fluid is used; these typically contain 90% glycol and 8% water. When the aircraft is scheduled to remain on the runway for more than 15 min, it is sprayed with the deicing fluid and then with an anti-icing agent that contains 65% glycol with a polymer thickener. No published information is available on the fate of these fluids if they are taken into the engine or APU, heated to a high temperature, and released into the aircraft ventilation system. The committee notes that application does not necessarily lead to exposure.

CONCLUSIONS

- When an aircraft is on the ground, exposure to various pollutants (e.g., O_3, CO, and PM) under normal operating conditions is determined primarily by the concentrations in the ground-level air. Contaminants result from ambient air pollution and airport sources (e.g., jet engine exhaust).
- Exposure to ambient pollutants in the departure or arrival city's air basin is limited because of the small fraction of time spent in takeoff and landing under normal flight conditions.
- At cruise altitude, the primary ambient air pollutant of concern is O_3.
- O_3 concentrations in the aircraft cabin are controlled by active O_3 destruction by O_3 converters, by passive O_3 destruction that occurs on the interior surfaces of the aircraft, and by flight planning. Some aircraft use O_3 converters to control O_3, but no formal process exists for ensuring a high level of converter performance. There is concern because O_3 converters are susceptible to poisoning by deposits of PM and some chemicals.
- The committee found no data that provided confidence that the DOT O_3 standard is regularly met. Nor does it appear to be possible to demonstrate compliance without a program that incorporates regular real-time O_3 monitoring on aircraft.
- Although data on exposure to other contaminants in aircraft cabins are extremely sparse, there are some data on CO_2, CO, PM, VOCs, and SVOCs under normal operating conditions. CO_2 concentrations appear to be below the FAA regulatory limit, although often higher than the ASHRAE ventilation standard; CO and PM concentrations appear to be lower than health-based standards for ambient air; and VOC and SVOC concentrations appear to be similar to those in other public-transport vehicles.
- Pesticides are used on selected international flights, but the committee

could not locate any quantitative data on pesticide exposure of passengers or crew members.

• Incidents have occurred in which engine lubricating oils, hydraulic fluids, or their pyrolyzed products have entered the ECS and contaminated the cabin air. However, no available exposure data identify the contaminants present in cabin air during an air-quality incident.

• Controlled-pyrolysis experiments in the laboratory indicate that a large number of volatile and nonvolatile agents (e.g., TCP isomers) are released from engine lubricating oils and hydraulic fluids into the ambient air, where they can be measured at room temperature. However, the components released into the passenger cabin during air-quality incidents and their possible concentrations cannot be determined from the experiments.

RECOMMENDATIONS

• A surveillance program should be developed and conducted to monitor cabin O_3 on a representative number of flights and aircraft to determine compliance with existing federal aviation regulations for O_3. The program should accurately establish temporal trends in O_3 concentrations and determine the effectiveness of O_3 control measures. Continuing monitoring should be conducted to ensure accurate characterization of O_3 concentrations as new aircraft come on line, and aircraft equipment ages and is upgraded. Monitoring should be done with reliable and accurate instrumentation that is capable of making real-time O_3 measurements.

• FAA should develop procedures for ensuring O_3 converter performance. At a minimum, FAA should conduct spot checks to verify that O_3 converters are operating properly, according to the *Federal Register* (Vol. 45, No. 14, January 21, 1980).

• Pesticide-exposure data should be collected to determine exposures of passengers and crew.

• Wipe samples of aircraft cabin, cockpit, and ventilation ducts should be taken and analyzed after air-quality incidents to identify the contaminants to which passengers and crew were exposed. Filters from the aircraft ventilation system should also be analyzed to identify contaminants that have collected on them.

• Because CO is most likely produced during air-quality incidents involving leaks of engine lubricating oils or hydraulic fluids in the ECS, it should be monitored in the ducts that introduce air into the cabin or cockpit.

• More research should be conducted to determine products that might be generated when engine lubricating oils, hydraulic fluids, and deicing fluids are exposed to high temperatures that might be encountered in the ECS.

REFERENCES

Acohido, B. 2000. Alaska airlines jet flew repeatedly with fouled air. The Seattle Times. January 21, 2000.

Anonymous. 1999. The plane truth about disinsection. Environ. Health Perspect. 107(8):A397-398.

Arlidge, J., and T. Clark. 2001. British pilots overcome by fumes. The Observer. April 22, 2001.

ASHRAE (American Society of Heating Refrigerating and Air-Conditioning Engineers). 1999. ASHRAE Standard - Ventilation for Acceptable Indoor Air Quality. ANSI/ASHRAE 62-1999. American Society of Heating Refrigerating and Air-Conditioning Engineers, Atlanta, GA.

ASHRAE/CSS (American Society of Heating Refrigerating and Air-conditioning Engineers/ Consolidated Safety Services). 1999. Relate Air Quality and Other Factors to Symptoms Reported by Passengers and Crew on Commercial Transport Category Aircraft. Final Report. ASHRAE Research Project 957-RP. Results of Cooperative Research Between the American Society of Heating, Refrigerating and Air-Conditioning Engineers, Inc., and Consolidated Services, Inc. February 1999.

Berlin, K., G. Johanson, and M. Leindberg. 1995. Hypersensitivity to 2-(2-butoxy-ethoxy) ethanol. Contact Dermatitis 32(1):54.

Bischof, W. 1973. O_3 measurements in jet airliner cabin air. Water Air Soil Pollut. 2(1):3-14.

Black, D.R., R.A. Harley, S.V. Hering, and M.R. Stolzenburg. 2000. A new portable real-time O_3 monitor. Environ. Sci. Technol. 34(14):3031-3040.

Boeing. 1998. Removing smoke or fumes from the air conditioning system-trouble shooting. Pp. 101 in Boeing 737-300/400/500 Maintenance Manual. 21-00-01. Boeing Company. Nov. 15, 1998.

Bolz, R.E., and G.L. Tuve, eds. 1973. Environmental and Bioengineering. Pp. 649-705 in CRC Handbook of Tables for Applied Engineering Science, 2nd Ed. Boca Raton: CRC.

Borglum, B., and A.M. Hansen. 1994. A Survey of Washing and Cleaning Agents. AMI Repport #44. Copenhagen. (as cited in Wolkoff et al. 1998).

Brabets, R.I., C.K. Hersh, and M.J. Klein. 1967. O_3 measurement survey in commercial jet aircraft. J. Aircraft 4(1):59-64.

Budavari, S., M.J. O'Neil, and A. Smith, eds. 1989. The Merck Index: An Encyclopedia of Chemicals, Drugs, and Biologicals, 11th Ed. Rahway, NJ: Merck.

Burge, P.S., and M.N. Richardson. 1994. Occupational asthma due to indirect exposure to lauryl dimethyl benzyl ammonium chloride used in a floor cleaner. Thorax 49(8):842-843.

Burge, P.S., A. Weiland, A.S. Robertson, and D. Weir. 1986. Occupational asthma due to unheated colophony. Br. J. Ind. Hyg. 43(8):559-580.

Burton, L.E., J.G. Girman, and S.E. Womble. 2000. Airborne particulate matter within 100 randomly selected office buildings in the United States (BASE). Pp. 157-162 in Healthy Buildings 2000: Exposure, Human Responses, and Building Investigations, Proceedings, Vol. 1, O. Seppänen, and J. Säteri, eds. Helsinki, Finland: SIY Indoor Air Information.

Childers, J.W., C.L. Witherspoon, L.B. Smith, and J.D. Pleil. 2000. Real-time and integrated measurement of potential human exposure to particle-bound polycyclic aromatic hydrocarbons (PAHs) from aircraft exhaust. Environ. Health Perspect. 108(9):853-862.

Clausen, P.A., and P. Wolkoff. 1997. Degradation products of tenax TA formed during sampling and thermal desorption analysis: indicators of reactive species indoors. Atmos. Environ. 31(5):715-725.

Clayton, C.A., R.L. Perritt, E.D. Pellizzari, K.W. Thomas, R.W. Whitmore, L.A. Wallace, H. Ozkaynak, and J.D. Spengler. 1993. Particle Total Exposure Assessment Methodology (PTEAM) study: distributions of aerosol and elemental concentrations in personal, indoor, and outdoor air samples in a southern California community. J. Expo. Anal. Environ. Epidemiol. 3(2):227-250.

Cometto-Muñiz, J.E. 2001. Physicochemical basis for odor and irritation potency of VOCs. Pp. 20.1-20.21 in Indoor Air Quality Handbook, J.D. Spengler, J.M. Samet, and J.F. McCarthy, eds. New York: McGraw-Hill.

DOT (Department of Transportation). 2001. Aircraft Disinsection Requirements, Office of the Assistant Secretary for Transportation Policy, Department of Transportation. [Online]. Available: http://ostpxweb.ost.dot.gov/policy/safety/disin.htm [August 22, 2001].

Devos, M., F. Patte, J. Rouault, P. Laffort, and J.J. Gmert. 1990. Standardized Human Olfactory Thresholds. New York: Oxford University Press.

Doelman, C., P.J. Borm, and A. Bast. 1990. Plasticisers and bronchial hyperreactivity. [letter]. Lancet 335(8691):725.

Dumyahn, T.S., J.D. Spengler, H.A. Burge, and M. Muilenburg. 2000. Comparison of the Environments of Transportation Vehicles: Results of Two Surveys. Pp. 13-25 in Air Quality and Comfort in Airliner Cabins, N.L. Nagda, ed. West Conshohocken, PA: American Society for Testing and Materials.

Eatough, D.J., F.M. Caka, J. Crawford, S. Braithwaite, L.D. Hansen, and E.A. Lewis. 1992. Environmental tobacco smoke in commercial aircraft. Atmos. Environ. Part A Gen. Top. 26(12):2211-2218.

EPA(U.S. Environmental Protection Agency). 2000. National Air Quality and Emissions Trends Report, 1998. EPA 454/R-00-003. Office of Air Quality Planning and Standards, Emissions Monitoring and Analysis Division, Air Quality Trends Analysis Group, U.S. Environmental Protection Agency, Research Triangle Park, NC. March 2000.

Fang, L., G. Clausen, and P.O. Fanger. 1998a. Impact of temperature and humidity on the perception of indoor air quality. Indoor Air 8(2):80-90.

Fang, L., G. Clausen, and P.O. Fanger. 1998b. Impact of temperature and humidity on perception of indoor air quality during immediate and longer whole-body exposures. Indoor Air 8(4):276-284.

Finlayson-Pitts, B.J., and J.N. Pitts, Jr. 1999. Chemistry of the Upper and Lower Atmosphere. Orlando: Academic Press.

Glasius, M., M. Lahaniati, A. Calogirou, D. Di Bella, N.R. Jensen, J. Hjorth, D. Kotzias, and B.R. Larsen. 2000. Carboxylic acids in secondary aerosols from oxidation of cyclic monoterpenes by O_3. Environ. Sci. Technol 34(6):1001-1010.

Gratz, N.G., R. Steffen, and W. Cocksedge. 2000. Why aircraft disinsection? Bull. World Health Org. 78(8):995-1004.

Griffin, R.J., D.R. Cocker III, R.C. Flagan, and J.H. Seinfeld. 1999. Organic aerosol formation from the oxidation of biogenic hydrocarbons. J. Geophys. Res. D Atmos. 104(3):3555-3567.

Grosjean, D., and M.W.M. Hisham. 1992. A passive sampler for atmospheric O_3. J. Air Waste Manage. Assoc. 42(2):169-173.

Hambraeus, A., S. Bengtsson, and G. Laurell. 1978. Bacterial contamination in a modern operating suite. 3. Importance of floor contamination as a source of airborne bacteria. J. Hyg. 80(2):169-174.

Hocking, M.B. 1998. Indoor air quality: recommendations relevant to aircraft passenger cabins. Am. Ind. Hyg. Assoc. J. 59(7):446-454.

Hood, E. 2001. OPs cause of bad trips? Environ. Health Perspect. 109(4):A156.

Hunt, E.H., D.H. Reid, D.R. Space, and F.E. Tilton. 1995. Commercial Airliner Environmental Control System, Engineering Aspects of Cabin Air Quality. Presented at the Aerospace Medical Association Annual Meeting, Anaheim, CA.

Innocenti, A. 1978. Occupational asthma due to benzalkonium chloride. [in Italian]. Med. Lav. 69(6):713-715.

Jaakkola, J.J.K., L. Øie, P. Nafstad, G. Botten, K.C. Lødrup-Carlsen, S.O. Samuelsen, and P. Magnus. 1997. Interior surface materials and development of bronchial obstruction in young children. Epidemiology 8(suppl. 4):S54.

Jaakkola, J.J.K., P.K. Verkasalo, and N. Jaakkola. 2000. Plastic interior materials and respiratory health in young children. Pp. 139-141 in Healthy Buildings 2000: Exposure, Human Responses, and Building Investigations, Proceedings, Vol. 1, O. Seppänen, and J. Säteri, eds. Helsinki, Finland: SIY Indoor Air Information.

Jang, M., and R.M. Kamens. 1999. Newly characterized products and composition of secondary aerosols from the reaction of a-pinene and O_3. Atmos. Environ. 33(3):459-474.

Jenkins, R.A., M.R. Guerin, and B.A. Tomkins. 2000a. Mainstream and sidestream cigarette smoke. Pp. 49-75 in The Chemistry of Environmental Tobacco Smoke: Composition and Measurement, 2nd Ed. Boca Raton, FL: Lewis.

Jenkins, R.A., M.R. Guerin, and B.A. Tomkins. 2000b. Field studies-oxides of nitrogen. Pp. 239-252 in The Chemistry of Environmental Tobacco Smoke: Composition and Measurement, 2nd Ed. Boca Raton, FL: Lewis.

Karch, S., M.F. Dellile, P. Guillet, and J. Mouchet. 2001. African malaria vectors in European aircraft. Lancet 357(9251):235.

Kildesø, J., and T. Schneider. 2001. Prevention with cleaning. Pp. 64.1-64.18 in Indoor Air Quality Handbook, J.D. Spengler, J.M. Samet, and J.F. McCarthy, eds. New York: McGraw-Hill.

Koutrakis, P., J.M. Wolfson, A. Bunyaviroch, S.E. Froehlich, K. Hirano, and J.D. Mulik. 1993. Measurement of ambient O_3 using a nitrite coated filter. Anal. Chem. 65(3):209-214.

Kreiss, K., M.G. Gonzales, K.L. Comight, and A.R. Scheere. 1982. Respiratory irritation due to carpet shampoo: two outbreaks. Environ. Int. 8:337-342.

Lambert, J.L., J.V. Paukstells, and Y.C. Chiang. 1989. 3-methyl-2-benzothiazolinone acetone azine with 2-phenylphenol as a solid passive monitoring reagent for O_3. Environ. Sci. Technol. 23(2):241-243.

Law, K.S., P.H. Plantevin, V. Thouret, A. Marenco, W.A.H. Asman, M. Lawrence, P.J. Crutzen, J.F. Muller, D.A. Hauglustaine, and M. Kanakidou. 2000. Comparison between global chemistry transport model results and Measurement of O_3 and Water Vapor by Airbus In-Service Aircraft (MOZAIC) data. J. Geophys. Res. 105(1):1503-1525.

Li, T-H. 2001. Generation, Characterization, Aerosol Partitioning, and Indoor Measurements of Hydrogen Peroxide for Exposure and Toxicological Assessment. Ph.D. Dissertation. Graduate School-New Brunswick of Rutgers, The State University of New Jersey and The University of Medicine and Dentistry of New Jersey, New Brunswick, NJ.

Lioy, P.J., T. Wainman, J. Zhang, and S. Goldsmith. 1999. Typical household vacuum cleaners: the collection efficiency and emissions characteristics for fine particles. J. Air Waste Manage. Assoc. 49(2):200-206.

Long, C.M., H.H. Suh, and P. Koutrakis. 2000. Characterization of indoor particle sources using continuous mass and size monitors. J. Air Waste Manage. Assoc. 50(7):1236-1250.

Mallard, W.G., F. Westley, J.T. Herron, R.F. Hampson, and D.H. Frizzell. 1998. NIST Chemical Kinetics Database: Windows Version 2Q98, NIST Standard Reference Database 17. Gaithersburg, MD: U.S. Department of Commerce.

McCoach, J.S., A.S. Robertson, and P.S. Burge. 1999. Floor cleaning materials as a cause of occupational asthma. Pp. 459-464 in Indoor Air 99, Proceedings of the 8th International Conference on Indoor Air Quality and Climate, Edinburgh, Scotland, 8-13 Aug. 1999, Vol. 5, G. Raw, C. Aizlewood, and P. Warren, eds. London: Construction Research Communications.

Mendell, M.J., W.J. Fisk, M. Petersen, M.X. Dong, C.J. Hines, D. Faulkner, J.A. Deddens, A.M. Ruder, D. Sullivan, and M.F. Boeniger. 1999. Enhanced particle filtration in a non-problem office environment: summary findings from a double-blind crossover intervention study. Pp. 974-975 in Indoor Air 99, Proceedings of the 8th International Conference on Indoor Air Quality and Climate, Edinburgh, Scotland, 8-13 Aug. 1999, Vol. 4, G. Raw, C. Aizlewood, and P. Warren, eds. London: Construction Research Communications.

Mølhave, L. 2001. Sensory irritation in humans caused by volatile organic compounds

(VOCs) as indoor air pollutants: a summary of 12 exposure experiments. Pp. 25.1-25.28 in Indoor Air Quality Handbook, J.D. Spengler, J.M. Samet, and J.F. McCarthy, eds. New York: McGraw-Hill.

Mølhave, L., and D.N. Neilsen. 1992. Interpretation and limitations of the concept "total volatile organic compounds" (TVOC) as an indicator of human response to exposures of volatile organic compounds (VOC) in indoor air. Indoor Air 2:65-77.

Monn, C., and M. Hangartner. 1990. Passive sampling for O_3. J. Air Waste Manage. Assoc. 40(3):357-358.

Morrison, G.C. 1999. O_3-Surface Interactions: Investigation of Mechanisms, Kinetics, Mass Transport, and Implications for Indoor Air Quality. Ph.D. Dissertation. Department of Civil and Environmental Engineering, University of California, Berkeley.

Morrison, G.C., and W.W. Nazaroff. 1999. Emissions of odorous oxidised compounds from carpet after O_3 exposure. Pp. 664-669 in Indoor Air 99, Proceedings of the 8th International Conference on Indoor Air Quality and Climate, Edinburgh, Scotland, 8-13 Aug. 1999, Vol. 4, G. Raw, C. Aizlewood, and P. Warren, eds. London: Construction Research Communications.

Moss, M.T., and H.M. Segal. 1994. The emissions and dispersion modeling system (EDMS): its development and application at airports and airbases. J. Air Waste Manage Assoc. 44(6):787-790.

Murray, D.M., and D.E. Burmaster. 1995. Residential air exchange rates in the United States: empirical and estimated parametric distributions by season and climatic region. Risk Anal. 15(4):459-465.

Nagda, N.L., M.D. Fortmann, M.D. Koontz, S.R. Baker, and M.E. Ginevan. 1989. Airliner Cabin Environment: Contaminant Measurements, Health Risks, and Mitigation Options. DOT-P-15-89-5. NTIS/PB91-159384. Prepared by GEOMET Technologies, Germantown, MD, for the U.S. Department of Transportation, Washington DC.

Nagda, N.L., M.D. Koontz, A.G. Konheim, and S.K. Hammond. 1992. Measurement of cabin air quality aboard commercial airliners. Atmos. Environ. Part A Gen. Top. 26(12):2203-2210.

Nagda, N.L., H.E. Rector, Z. Li, and E.H. Hunt. 2001. Determine Aircraft Supply Air Contaminants in the Engine Bleed Air Supply System on Commercial Aircraft. ENERGEN Report AS20151. Prepared for American Society of Heating, Refrigerating, and Air-Conditioning Engineers, Atlanta, GA, by ENERGEN Consulting, Inc., Germantown, MD. March 2001.

Nastrom, G.D., J.D. Holdeman, and P.J. Perkins. 1980. Measurements of cabin and ambient O_3 on B747 airplanes. J. Aircraft 17:246-249.

Naumann, I.D., and K. McLachlan. 1999. Aircraft Disinfection. Report commissioned by the Australian Quarantine and Inspection Service. June 1999.

Nazaroff, W.W., and G.R. Cass. 1986. Mathematical modeling of chemically reactive pollutants in indoor air. Environ. Sci. Technol. 20(9):924-934.

Nazaroff, W.W., A.J. Gadgil, and C.J. Weschler. 1993. Critique of the use of deposition velocity in modeling indoor air quality. Pp. 81-104 in Modeling of Indoor Air Quality and Exposure, N.L. Nagda, ed. ASTM STP 1205. Philadelphia, PA: American Society for Testing and Materials.

Norbäck, D., and M. Torgen. 1989. A longitudinal study relating carpeting with sick building syndrome. Environ. Int. 15(1-6):129-135.

Norton, D. 1998. 7 treated after hydraulic-fluid leak on plane. The Seattle Times. February 21, 1998.

NRC (National Research Council). 1986a. The physicochemical nature of sidestream smoke and environmental tobacco smoke. Pp. 25-54 in Environmental Tobacco Smoke: Measuring Exposures and Assessing Health Effects. Washington, DC: National Academy Press.

NRC (National Research Council). 1986b. The Airliner Cabin Environment: Air Quality and Safety. Washington, DC: National Academy Press.

NRC (National Research Council). 1992. Guidelines for Developing Spacecraft Maximum Allowable Concentrations for Space Station Contaminants. Washington, DC: National Academy Press.

Øie, L., L.G. Hersoug, and J.Ø. Madsen. 1997. Residential exposure to plasticizers and its possible role in the pathogenesis of asthma. Environ. Health Perspect. 105(9):972-978.

Otto, D., L. Molhave, G. Rose, H.K. Hudnell, and D. House. 1990. Neurobehavioral and sensory irritant effects of controlled exposure to a complex mixture of volatile organic compounds. Neurotoxicol. Teratol. 12(6):649-652.

Parliament of the Commonwealth of Australia. 2000. Air Safety and Cabin Air Quality in the BAe 146 Aircraft. Report by the Senate Rural and Regional Affairs and Transport References Committee, Parliament House, Canberra. October 2000.

Penner, J.E., D.H. Lister, D.J. Griggs, D.J. Dokken, and M. McFarland, eds. 1999. Aviation and the Global Atmosphere. Cambridge: Cambridge University Press.

Persily, A. 1989. Ventilation rates in office buildings. Pp. 128-136 in The Human Equation: Health and Comfort, Proceedings of the ASHRAE/SOEH Conference IAQ89 San Diego, CA, April 17-20, 1989. Atlanta, GA: American Society of Heating Refrigerating, and Air Conditioning Engineers.

Persoff, R., and M. Koketsu. 1978. Respiratory irritation and carpet shampoo. Pp. 347-349 in Proceedings of the 5th International Conference of Medichem, San Francisco, CA, Sept. 5-10, 1977, C. Hine, and D.J. Kilian, eds.

Pierce, W.M., J.N. Janczewski, B. Roethlisberger, and M.G. Janczewski. 1999. Air quality on commercial aircraft. ASHRAE J. 41(9):26-34.

Pleil, J.D., L.B. Smith, and S.D. Zelnick. 2000. Personal exposure to JP-8 jet fuel vapors and exhaust at air force bases. Environ. Health Perspect. 108(3):183-192.

Reed, D., S. Glaser, and J. Kaldor. 1980. O_3 toxicity symptoms among flight attendants. Am. J. Ind. Med. 1:43-54.

Reiss, R., P.B. Ryan, P. Koutrakis, and S.J. Tibbetts. 1995. O_3 reactive chemistry on interior latex paint. Environ. Sci. Technol. 29(8):1906-1912.

Ritter, S. 2001. Aircraft deicers. Chem. Eng. News 79(1):30.

Robinson, P.A., R.V. Tauxe, W.G. Winkler, and M.E. Levy. 1983. Respiratory illness in conference participants following the use of a carpet shampoo. Infect. Control 4(3):158-160.

Rodriguez, J.A., and J. Hrbek. 1999. Interaction of sulfur with well-defined metal and oxide surfaces: unraveling the mysteries behind catalyst poisoning and desulfurization. Accounts of Chemical Research 32(9):719-728.

Roth, B., P. Herkenrath, H.J. Lehmann, H.D. Ohles, J.J. Hömig, G. Benz-Bohm, J. Kreuder, and A. Younossi-Hartenstein. 1988. Di-(2-ethylhexyl)-phthalate as plasticizer in PVC respiratory tubing systems: indications of hazardous effects on pulmonary function in mechanically ventilated, preterm infants. Eur. J. Pediatr. 147(1):41-46.

Roy, S.P., G.J. Raw, and C. Whitehead. 1993. Sick building syndrome: cleanliness is next to healthiness. Pp. 261-266 in Indoor Air 93, Proceedings of the 6th International Conference on Indoor Air Quality and Climate, Vol. 6, Seppänen, R. Ilmarinen, J.J.L. Jaakkola, E. Kukkonen, J. Sateri, and H. Vuorelma, eds. Helsinki: International Society of Indoor Air Quality and Climate (ISIAQ).

SAE International (Society of Automotive Engineers International). 2000. (R) O_3 in High Altitude Aircraft. Aerospace Information Report. SAE AIR910 Rev. B. Society of Automotive Engineers.

Schmitt, H.-J. 1985. Irritation of the airways following the use of a carpet shampoo [in German]. Offentl. Gesundheitswes. 47(9):458.

Seinfeld, J.H., and S.N. Pandis. 1998a. Chemistry of the stratosphere. Pp. 163-233 in Atmospheric Chemistry and Physics: From Air Pollution to Climate Change. New York: Wiley.

Seinfeld, J.H., and S.N. Pandis. 1998b. The atmosphere. Pp. 1-48 in Atmospheric Chemistry and Physics: From Air Pollution to Climate Change. New York: Wiley.

Seinfeld, J.H., and S.N. Pandis. 1998c. Chemistry of the troposphere. Pp. 234-336 in Atmospheric Chemistry and Physics: From Air Pollution to Climate Change. New York: Wiley.

Skov, P., O. Valbjorn, and B.V. Pedersen. 1989. Influence of personal characteristics, job-related factors, and psychosocial factors on the sick-building syndrome. Danish Indoor Climate Study Group. Scand. J. Work Environ. Health 15(4):286-295.

Skyberg, K., K.R. Skulberg, K. Kruse, P.O. Huser, F. Levy, and P. Djupesland. 1999. Dust reduction relieves nasal congestion: a controlled intervention study on the effect of office cleaning, using acoustic rhinometry. Pp. 153-154 in Proceedings of the 8th International Conference on Indoor Air Quality and Climate, Edinburgh, Scotland, 8-13 Aug. 1999, Vol. 4, G. Raw, C. Aizlewood, and P. Warren, eds. London: Construction Research Communications.

Spengler, J., H. Burge, T. Dumyahn, M. Muilenberg, and D. Forester. 1997. Environmental Survey on Aircraft and Ground-Based Commercial Transportation Vehicles. Prepared by Department of Environmental Health, Harvard University School of Public Health, Boston, MA, for Commercial Airplane Group, The Boeing Company, Seattle, WA. May 31, 1997.

Sussell, A., and M. Singal. 1993. Alaska Airlines, Seattle, Washington. NIOSH Health Hazard Evaluation Report. HETA 90-226-2281. U.S. Department of Health and Human Services, Public Health Service, Centers for Disease Control and Prevention, National Institute for Occupational Safety and Health, Cincinnati, OH.

Taylor, A.E., and C. Hindson. 1982. Facial dermatitis from allyl phenoxy acetate in a dry carpet shampoo. Contact Dermatitis 8(1):70.

Ten Brinke, J., S. Selvin, A.T. Hodgson, W.J. Fisk, M.J. Mendell, C.P. Koshland, and J.M. Daisey. 1998. Development of new volative organic compound (VOC) exposure metrics and their relationship to "Sick Building Syndrome" symptoms. Indoor Air 8(3):140-152.

Tobias, H.J., and P.J. Ziemann. 2000. Thermal desorption mass spectrometric analysis of organic aerosol formed from reactions of 1-tetradecene and O_3 in the presence of alcohols and carboxylic acids. Environ. Sci. Technol. 34(11):2105-2115.

Tobias, H.J., K.S. Docherty, D.E. Beving, and P.J. Ziemann. 2000. Effect of relative humidity on the chemical composition of secondary organic aerosol formed from reactions of 1-tetradecene and O_3. Environ. Sci. Technol. 34(11):2116-2125.

van Netten, C. 2000. Analysis of two jet engine lubricating oils and a hydraulic fluid: their pyrolytic breakdown products and their implication on aircraft air quality. Pp. 61-75 in Air Quality and Comfort in Airliner Cabins, N.L. Nagda, ed. West Conshohocken, PA: American Society for Testing and Materials.

van Netten, C., and V. Leung. 2000. Comparison of the constituents of two jet engine lubricating oils and their volatile pyrolytic degradation products. Appl. Occup. Environ. Hyg. 15(3):277-283.

van Netten, C., and V. Leung. 2001. Hydraulic fluids and jet engine oil: pyrolysis and aircraft air quality. Arch. Environ. Health 56(2):181-186.

Virkkula, A., R. van Dingenen, F. Raes, and J. Hjorth. 1999. Hygroscopic properties of aerosol formed by oxidation of limonene, alpha-pinene and beta-pinene. J. Geophys. Res. 104(3):3569-3579.

Wainman, T., J. Zhang, C.J. Weschler, and P.J. Lioy. 2000. O_3 and limonene in indoor air: a source of submicron particle exposure. Environ. Health Perspect. 108(12):1139-1145.

Wallace, L.A., C.J. Nelson, and G. Dunteman. 1991. Workplace Characteristics Associated with Health and Comfort Concerns in Three Office Buildings in Washington, DC. Atmospheric Research and Exposure Assessment Lab, U.S. Environmental Protection Agency, Research Triangle Park, NC. NTIS PB91-211342. 11pp.

Wargocki, P., J. Sundell, W. Bischof, G. Brundrett, P.O. Fanger, F. Gyntelberg, S.O. Hanssen, P. Harrison, A. Pickering, O. Seppänen, and P. Wouters. 2001. Ventilation and Health in Nonindustrial Indoor environments. Report from a European multidisciplinary scientific consensus meeting. Clima 2000/Napoli 2001 World Congress, Napoli, Italy, September 15-18, 2001.

Waters, M .,T. Bloom, and B. Grajewski. 2001. Cabin Air Quality Exposure Assessment. National Institute for Occupational Safety and Health, Cincinnati, OH. Federal Aviation Administration Civil Aeromedical Institute. Presented to the

NRC Committee on Air Quality in Passenger Cabins of Commercial Aircraft, January 3, 2001. National Academy of Science, Washington, DC.

Weschler, C.J. 2000. O_3 in indoor environments: concentration and chemistry. Indoor Air 10(2):269-288.

Weschler, C.J., and H.C. Shields. 1996. Production of the hydroxyl radical in indoor air. Environ. Sci. Technol. 30(11):3250-3258.

Weschler, C.J., and H.C. Shields. 1997a. Potential reactions among indoor pollutants. Atmos. Environ. 31(21):3487-3495..

Weschler, C.J., and H.C. Shields. 1997b. Measurements of the hydroxyl radical in a manipulated but realistic indoor environment. Environ. Sci. Technol. 31(12):3719-3722.

Weschler, C.J., and H.C. Shields. 1999. Indoor O_3/terpene reactions as a source of indoor particles. Atmos. Environ. 33(15):2301-2312.

Weschler, C.J., and H.C. Shields. 2000. The influence of ventilation and reactions among indoor pollutants: modeling and experimental observations. Indoor Air 10(2):92-100.

WHO (World Health Organization). 1985. Recommendations on the disinsecting of aircraft. Weekly Epidemiological Record 60(7):45-47.

WHO (World Health Organization). 1995. Report of the Informal Consultation on Aircraft Disinsection, WHO/HQ, Geneva, 6-10 November 1995, International Programme on Chemical Safety. Geneva, Switzerland: World Health Organization.

Wilcox, R.W., G.D. Nastrom, and A.D. Belmont. 1977. Period variations of total ozone and its vertical distribution. J. Appl. Meteorol. 16(3):290-298.

Winder, C. 2000. Cabin air quality in the Bae 146: the Senate Inquiry report. Hazard Alert 6:1-4. (December 1, 2000).

Witkowski, C.J. 1999. Remarks on Airliner Air Quality. Presentation at ASHRAE Conference, Chicago, IL, January 24, 1999.

Wolkoff, P., P.A. Clausen, C.K. Wilkins, and G.D. Nielsen. 2000. Formation of strong airway irritants in terpene/O_3 mixtures. Indoor Air 10(2):82-91.

Wolkoff, P., T. Schneider, J. Kildesø, R. Degerth, M. Jaroszewski, and H. Schunk. 1998. Risk in cleaning: chemical and physical exposure. Sci. Total Environ. 215(1-2):135-156.

Woodfolk, J.A., C.M. Luczynska, F. de Blay, M.D. Chapman, and T.A. Platts-Mills. 1993. The effect of vacuum cleaners on the concentration and particle size distribution of airborne cat allergen. J. Allergy Clin. Immunol. 91(4):829-837.

Wright, R.L. 1996. Formation of the neurotoxin TMPP from TMPE-Phosphate Formulations. Tribology Transactions 39(4):827-834.

Wyman, J., E. Pitzer, F. Williams, J. Rivera, A. Durkin, J. Gehringer, P. Servé, D. von Minden, and D. Macys. 1993. Evaluation of shipboard formation of a neurotoxicant (trimethylolpropane phosphate) from thermal decomposition of synthetic aircraft engine lubricant. Am. Ind. Hyg. Assoc. J. 54(10):584-592.

Wyon, D.P., K.W. Tham, B. Croxford, A. Young, and T. Oreszczyn. 2000. The effects on health and self-estimated productivity of two experimental interventions which reduced airborne dust levels in office premises. Pp. 641-646 in Healthy Buildings 2000: Exposure, Human Responses and Building Investigations. Proceedings, Vol.1., O. Seppänen, and J. Säteri, eds. Helsinki, Finland: SIY Indoor Air Information.

Yu, J., R.C. Flagan, and J.H. Seinfeld. 1998. Identification of products containing - COOH, -OH, and -C=O in atmospheric oxidation of hydrocarbons. Environ. Sci. Technol. 32(16):2357-2370.

Zhou, J., and S. Smith. 1997. Measurement of O_3 concentrations in ambient air: using a badge-type passive monitor. J. Air Waste Manage. Assoc. 47(6):697-703.

4

Biological Agents

This chapter provides general information on biological agents, identifies sources of bioaerosols that might be found in aircraft cabins, and summarizes the environmental sampling of biological agents that has been conducted on commercial aircraft. The chapter also describes the two health effects considered most likely to be associated with bioaerosol exposures in aircraft cabins—acute hypersensitivity disease and infectious respiratory disease.

GENERAL INFORMATION ON BIOAEROSOLS

The term bioaerosol includes a variety of airborne particles of biological origin (from plants, microorganisms, or animals). For this review, the committee will consider bioaerosols according to their associated health effects (e.g., aeroallergens, biological toxins, biological irritants, and infectious agents) (see Table 4-1).

Biological Agents in Outdoor Air Near the Ground

Outdoor air up to an altitude of about 1,500 m (4,900 ft) contains abundant biological material (Lighthart and Stetzenbach 1994; Lacey and Venette 1995), especially fungal spores and pollen grains. Airborne bacteria and fragments of insects, plants, and soil also are present at various concentrations. Depending on where the plane is located and the ambient conditions (e.g., the time of

TABLE 4-1 Examples of Common Biological Agents, Reservoirs, and Health Effects Related to Inhalation Exposure

Agents	Bioaerosol	Reservoirs	Possible Health Effects
Aeroallergens			
Plant allergens	Pollen grains, plant fragments	Vegetation, settled dust	Allergic conjunctivitis, rhinitis, sinusitis, asthma
Bacterial antigens	Cells, cell fragments	Growth in water or on organic matter, settled dust	Hypersensitivity pneumonitis
Fungal allergens	Spores, cell fragments	Growth in water or on organic matter, settled dust	Allergic conjunctivitis, rhinitis, sinusitis, asthma
Arthropod allergens	Dust mite or cockroach excreta and body fragments	Settled dust	Allergic conjunctivitis, rhinitis, sinusitis, asthma
Mammalian allergens	Particles of cat, dog, or rodent skin, saliva, or urine	Pets, pests, settled dust	Allergic conjunctivitis, rhinitis, sinusitis, asthma
Toxins and inflammatory agents			
Bacterial endotoxin	Gram-negative bacterial cells and cell fragments	Growth in water or on organic matter, settled dust	Humidifier fever, respiratory inflammation
Fungal toxins	Spores, cell fragments	Growth on organic matter	Toxic and irritant effects
Biological irritants and nuisance biological agents			
Microbial volatile compounds	Gases and vapors	Bacterial or fungal growth on organic matter	Irritant and nuisance effects (e.g., unpleasant or annoying odors)
Carbon dioxide, body effluents	Gases and vapors	Human breath, body odor	Perception of insufficient fresh air, irritant and nuisance effects
Infectious agents			
Infectious viruses	Droplets, droplet nuclei of sputum, saliva, nasal or throat secretions	Infected or colonized persons	Acute viral respiratory disease, influenza, varicella (chickenpox), measles

(Continued)

Infectious agents *(Continued)*			
Infectious bacteria	Droplets, droplet nuclei of sputum, saliva, nasal or throat secretions	Infected or colonized persons	Diphtheria, pertussis (whooping cough), bacterial pneumonia, meningococcal disease, tuberculosis
		Contaminated water	Legionnaires' disease
Infectious fungi	Spores	Growth in soil or on organic matter	Histoplasmosis, coccidioidomycosis (San Joaquin Valley fever)
		Growth in compost or on organic matter	Aspergillus infection

day, ground cover, wind speed, and precipitation), the outdoor air that enters a grounded aircraft through the ventilation system and open doors may contain a mixture of diverse bioaerosols. Exposure of passengers to outdoor bioaerosols in grounded aircraft are likely to be similar to what would be experienced in an airport terminal or in traveling to and from airports (see Chapter 3 for sources of contaminants outside the aircraft).

Biological Agents in Outside Air in Flight

At flight altitude, the air outside an aircraft contains many of the same bioaerosols found at ground level, but at much lower concentrations (Lighthart and Stetzenbach 1994). Therefore, the likelihood of allergic reactions from components of outside air during flight is less. Exposure to solar radiation, the presence of ambient air pollutants, and low air temperature and relative humidity greatly reduce the virulence of infectious agents that may be in outside air at flight altitude (Kim 1994; Muilenberg, 1995; Mohr 1997). Furthermore, the high compression temperatures (about 250°C) and pressure (about 3,000 kPa) of the bleed air used for ventilation are likely to alter allergenic proteins and inactivate microorganisms that still are viable when they enter an aircraft through an outside air intake (Withers and Christopher 2000). Therefore, in flight, biological agents in aircraft cabins arise virtually exclusively from inside sources rather than from the intake of outside air.

People

People are a primary source of airborne bacteria and are the most important reservoirs of infectious agents on aircraft (Masterton and Green 1991). Studies that have measured the concentration of airborne microorganisms have concluded that culturable bacteria and fungi in aircraft arise principally from the cabin occupants and furnishings. Most microorganisms that have been isolated from occupied spaces, including aircraft cabins, are human-source bacteria shed from exposed skin and scalp and from the nose and mouth. These microorganisms are found normally on the human body (they are normal flora) and only rarely cause infections. People also are indirect sources of some allergens, as discussed in the following subsections.

Arthropods, Pets, and Service Animals

Arthropods

Arthropods (e.g., flies, mosquitoes, spiders, dust mites, and cockroaches) can be found in aircraft, especially aircraft that spend time on the ground in tropical environments. Some of them are of concern because they are considered hazards to public health, agriculture, or native ecosystems, and this has lead to the practice in some countries of disinsection of aircraft (see Chapters 3 and 5). Dust mites and cockroaches are the only arthropods that might habitually infest aircraft and possibly become sources of allergens in the cabin environment.

Dust mites thrive on protein-containing material in dust (e.g., skin scales and fungal spores), especially in warm, humid indoor areas; they can be found in any enclosed space where such conditions exist (Lundblad 1991; Menzies et al. 1993; Hung et al. 1993; Janko et al. 1995; Squillance 1995; Arlian 1999). Dust mites require high humidity because they absorb water vapor through their exoskeleton rather than by drinking water.

Cockroaches repeatedly have been recognized as a common source of indoor allergens (Powell, 1994). They flourish where sufficient food, moisture, and warmth are available (e.g., kitchens, bathrooms, and similar areas), and they can survive low ambient humidity better than dust mites and, unlike mites, search actively for the water they need to survive (Squillace 1995). Allergenic proteins from cockroaches and mites are associated with particles that are 5 microns (μm) or greater and become airborne only when settled dust is disturbed (Platts-Mills and Carter 1997).

Pets and Service Animals

Depending on the air carrier, pets that are likely to be carried in the passenger cabins of commercial aircraft include cats and occasionally dogs, birds, rabbits, hamsters, guinea pigs, pot-bellied pigs, ferrets, and tropical fish. Most airlines limit the number of pets allowed in the cabin and the size of the animal carrier. The Air Carrier Access Act (ACAA, 14 CFR 382, Nondiscrimination on the Basis of Disability in Air Travel) protects the rights of air travelers with disabilities and requires that air carriers permit passengers to fly with their service animals in the cabin.

Although pets other than service animals must be confined while onboard and the container stowed under a seat during takeoff and landing, particulate matter, including bioaerosols (particularly allergenic particles), readily can escape animal cages. Of the animals that are permitted to travel in aircraft cabins, cats are of greatest concern (BRE 2001) as they are prolific allergen generators (Chew et al. 1998). Cat allergens exist on skin flakes and saliva particles, a significant proportion of which is small enough to remain suspended in air for long periods (Wood et al. 1993; Ormstad et al. 1995; Custovic et al. 1999b).

Dog allergens (from dander, saliva, and urine) might be carried on the clothing of dog owners or be shed by onboard pets or service dogs. The concentration of airborne canine allergens generally correlates well with the amount in dust, suggesting that dog allergens are carried on fairly large particles that settle quickly (Custovic et al. 1999a,b).

Settled Dust

Settled dust (house dust) is one of the most important reservoirs of indoor allergens and other biological agents, and the allergen content of dust is a primary indicator of exposure to many allergens (Platts-Mills et al. 1992; IOM 1993a; Trudeau and Fernández-Caldas 1994). In aircraft cabins, dust might contain soil particles, fabric fibers, human hairs and skin fragments, residues of cleaning products and pesticides, and particles from arthropods, mammals, and microorganisms. When analyzed by culture, the microbial population in dust is dominated by common spore-forming fungi and bacteria from soil or water (Logan and Turnbull 1999). Although some of the fungi in settled dust carry allergens and some of the bacteria have inflammatory properties, very few of these microorganisms are infectious, or they cause infection only in severely immunosuppressed persons.

Microbiological Growth

Most of the fungi and environmental bacteria in aircraft cabins enter with outside air while aircraft are on the ground or are carried in by the occupants (e.g., on their shoes, clothing, or hand luggage). Infectious agents that are transmitted from person to person generally grow poorly outside the human body, so contamination in an aircraft cabin is unlikely to be a source of them. Although some microorganisms may grow in cabin areas where moisture is routinely present (e.g., in water that condenses on the internal skin of an aircraft and collects in the bilge), exposure of cabin occupants to microorganisms resulting from environmental contamination has not been demonstrated. Furthermore, microbiological growth sufficient to result in bioaerosol release into cabin air would have to be fairly extensive, which is unlikely to go unnoticed on well-maintained aircraft.

Air filters used in ventilation systems collect large numbers of microorganisms and other organic material, but bacteria and fungi that are captured on filters remain dormant and eventually die if water is not supplied (Moritz et al. 1998). If sufficient water is available, fungi can grow and eventually penetrate the filter of a personal respirator (releasing spores on the downstream side of the filter) (Pasanen et al. 1993). Therefore, it is possible that filters in aircraft ventilation systems could be a source of microbes; however, microbial growth on the filters used to clean recirculated air in aircraft has not been reported.

Contaminated Fluids

Several studies have documented that all humidifiers support some microbial growth (Burge et al. 1980; Suda et al. 1995). The predominant contaminants in such reservoirs are waterborne bacteria, often Gram-negative species (which contain endotoxin) and thermophilic actinomycetes (filamentous bacteria that prefer very warm conditions and produce airborne spores).

Glines (1991) suggested that exposure to *Legionella* species could be associated with the operation of humidification systems on aircraft. *Legionella* species were identified in one study of the water in an onboard humidifier and the air in the cockpit served by the humidifier (Danielson and Cooper 1992; Cooper and Danielson 1992). It should be emphasized, however, that the cockpit is humidified on very few aircraft and the cabin air is not humidified on any aircraft (see the section on humidity control in Chapter 2). Therefore, passengers and cabin crews should have no opportunity for exposure to legionellae during air travel. The cockpit crew may be exposed to aerosolized bacteria if the water in a humidifier is contaminated and the unit is one that generates a water mist, but contamination of water used in humidifiers or for drinking is unlikely if the water storage tanks on aircraft are well maintained and the water supply is clean. Furthermore, there is no evidence that *Legionella* species are transmitted from person to person. Therefore, the risk of disease transmission would be negligible even if a person infected with legionellosis were on an aircraft.

Other Reservoirs of Potential Infectious Agents

Sewage is a potential source of enteric pathogens and other biological agents on aircraft. Shieh et al. (1997) detected enteroviruses in aircraft sewage. Airport waste handlers were identified as potentially at risk for exposure to enteric pathogens in aircraft sewage as a result of contact with waste material. Burton and McCleery (2000) investigated similar concerns for workers who clean, overhaul, and repair aircraft lavatory tanks. Live bacteria were isolated from waste tanks, but none of them were types that are associated with intestinal disease; the authors suggested changes in work practices to reduce inhalation exposures and wound contamination with potentially infectious agents.

Although no studies were identified that directly assess the aerosol-generating potential of sewage-waste disposal systems on aircraft during normal operation or the potential for inhalation exposure of cabin crew or passengers to infectious agents in sewage from aircraft toilets, no significant potential for aerosol release into the cabin and no infections due to the operation of onboard toilets have been reported. Air from the restrooms on aircraft is not recirculated, reducing the opportunity for volatile and particulate contaminants to enter the passenger cabin. Air carrier adherence to regulations related to aircraft sanitation and the practice of good hygiene by passengers should be sufficient to protect cabin crew and passengers from this potential source of exposure to enteric pathogens.

ENVIRONMENTAL SAMPLING ON AIRCRAFT

The 1986 NRC report recommended that the Federal Aviation Administration (FAA) establish a program for the systematic measurement of microbial aerosols on a representative sample of routine commercial flights. In the last 15 years, eight investigations of biological agents in aircraft cabins have sampled a total of more than 200 domestic and international flights on multiple airlines and 17 types of aircraft of various ages and designs (Table 4-2).

Sampling for Culturable Bacteria and Fungi

Bioaerosol samples on commercial aircraft typically have been collected in the breathing zone of a seated person (e.g., with a sampler placed on a tray table); a few air samples have been collected in galleys and lavatories and in supply and exhaust air streams (Wick and Irvine 1995; Dechow 1996; Dechow et al. 1997). Investigators have compared concentrations of biological agents found on aircraft with (1) ground-based public conveyances (Spengler et al. 1997; Dumyahn et al. 2000); (2) common locations in a southwestern city (municipal buses, a busy shopping mall, a sidewalk adjacent to a downtown street, and an airport departure lounge) (Wick and Irvine 1995); (3) hospital guidelines (Dechow 1996; Dechow et al. 1997); (4) a local indoor air-quality standard (Lee et al. 1999, 2000); and (5) an undocumented occupational exposure limit (attributed to National Institute for Occupational Safety and Health (NIOSH) without reference) (ATA 1994; Janczewski 2001). The major observations from the studies, some of which were noted in more than one investigation, are as follows:

- Concentrations and types of bacteria and fungi on aircraft:
  — Concentrations of culturable bacteria were higher than those of culturable fungi.
  — Bacteria were typical of those shed by humans or were common soil organisms.
  — Fungi were typical of those found in outdoor air and occupied indoor environments.

TABLE 4-2 Studies of Biological Agents on Commercial Aircraft[a]

Reference	Flights, Aircraft	Biological Agents	Sampler, Sampling Location, Flight Segment	Findings	Authors' Conclusions
Nagda et al. (1989, 1992)	92 flights (23 smoking, 8 international) 12 airlines Airbus; B727, B737, B747, B757, B767; DC8, DC9, DC10; L1011 (Study conducted in 1989)	Culturable airborne bacteria, fungi Rank-order comparisons	Multiple-hole impactor (SAS Compact) (90 L/min for 0.67, 1.0, 1.3, 2.0, 3.0 min) Smoking flights: middle of nonsmoking section, rear of aircraft (smoking section) Nonsmoking flights: middle of aircraft Near end of flight, before descent	Concentrations of both agents generally were higher with higher passenger loads. Concentrations of bacteria were higher than those of fungi and were somewhat higher in smoking sections, on wide-body aircraft, on aircraft with recirculation, and on flights with lower nominal air-change rates.	Bacteria and fungi typically encountered in indoor environments that are characterized as "normal" were also found in cabin air with similar prevalences and at similar concentrations. Neither bacteria nor fungi were present at concentrations generally thought to pose a risk of illness. No actions need be taken to reduce prevailing concentrations of bioaerosols.
ATA (1994)	35 domestic flights 8 airlines No recirculation, B727, DC9	Culturable airborne bacteria, fungi	Multiple-hole impactor (SAS Compact) (90 L/min for 1.3 min) 2 locations in first class; 2 locations in front, center, and rear of coach class	All concentrations were below 1000 CFU/m³ and were proportional to number of cabin occupants. Concentrations of both agents were somewhat higher earlier in flights and in coach.	People and their activities were source of most contaminants. Concentrations of contaminants in aircraft cabins are not likely to cause adverse health effects.

(Continued)

TABLE 4-1 *Continued*

Reference	Flights, Aircraft	Biological Agents	Sampler, Sampling Location, Flight Segment	Findings	Authors' Conclusions
ATA (1994)	Recirculated air, B757, MD80 (Year not specified)	Purported NIOSH recommended exposure limit of 1,000 CFU/m³	Early and late phases of flight segments	Concentrations were relatively low for all aircraft types, seating configurations, flight durations, and airlines. No agents of respiratory infections were isolated.	Removal of contaminants appears to be sufficient for both ventilation designs because contaminant concentrations were below those common in other indoor spaces.
Spengler et al. (1994, 1997)	22 domestic flights; 8 airlines; A300, A320; B727, B737, B747, B757, B767; DC9, DC10; MD80 (Study conducted in 1994)	Culturable airborne bacteria; culturable fungi in dust; cat and mite allergens in dust; endotoxin in dust; Concentrations in other public transportation-vehicles	Multiple-hole impactor (Burkard) (44-52 L/min for 2 min); Coach section, front and rear; Terminal before boarding, onboard before takeoff, during cruise, during taxiing to gate; occasionally during deboarding (ground); Dust samples (12 flights): 4-5 min samples from carpet and seat	(Findings applicable to both Spengler et al. 1994, 1997 and Dumyahn et al. 2000) Concentrations of bacteria and fungi in air and fungi in dust were not significantly different across vehicle type with exception of bacteria during aircraft deboarding. Concentrations of bacteria tended to be higher on planes with air recirculation. For ground-transportation vehicles, highest concentration of bacteria was observed in subways, but airborne fungi were found most often in trains.	(Authors' conclusions applicable to both Spengler et al. 1994, 1997 and Dumyahn et al. 2000) Concentrations of biological agents in vehicles were generally lower than those common in homes and outdoor air, and none was present at concentrations that can be considered to present an unusual exposure risk. Bacterial and viral respiratory agents were not measured.

Reference	Study description	Contaminants	Method	Results	Conclusions
Dumyahn et al. (2000)	27 segments: 6 domestic flights, 6 interstate train trips, 7 interstate bus trips, 2 commuter train trips, 6 subway trips; B777; (Study conducted in 1996)	Culturable airborne bacteria; culturable fungi in air and dust; Concentrations in other public-transportation vehicles	Multiple-hole impactor (44 L/min for 1 min); Rear of aircraft; center of buses, subway, train cars; Immediately after departure (ground), mid-trip (about 60 min prior to landing), immediately on touchdown; Dust samples: 1-3 m^2; 4-6 seats per travel segment per vehicle	Bacteria were typical of human sources. Similar species of fungi were isolated on all vehicle types. Cat allergen was detected in most samples and exceeded concentration considered high in two subway samples and one train sample.	Identifying effects of relative humidity, temperature, air movement patterns, and occupancy rates could be critical to understanding potential spread of infectious agents.
Wick and Irvine (1995)	44 flights (38 domestic, 4 intercontinental, 2 international); 1 airline; B727, B737, B767; BAe 146; DC10; MD80; SWM	Culturable airborne bacteria, fungi; Concentrations in common urban locations	Centrifugal impactor (RCS) (40 L/min for 4 min); 9 domestic, 2 international flights: first class, coach, galleys; 4 intercontinental flights: multiple time intervals in flight	Higher bacterial and fungal concentrations were observed during periods of higher passenger activity. Little difference was seen in concentrations of airborne bacteria or fungi between locations. Concentrations on aircraft were highest 30 cm above floor near exhaust vents.	Concentration of microorganisms in U.S. aircraft cabin air is much lower than in ordinary city locations and does not contribute to risk of disease transmission among passengers.

(Continued)

TABLE 4-1 Continued

Reference	Flights, Aircraft	Biological Agents	Sampler, Sampling Location, Flight Segment	Findings	Authors' Conclusions
Wick and Irvine (1995) (Continued)	(Study begun in 1987)		Various times: municipal buses, shopping mall, downtown sidewalk, airport departure lounge		
Dechow (1996), Dechow et al. (1997)	14 scheduled intercontinental flights; 2 airlines; A310 (trans-Mediterranean), A340 (trans-Atlantic); (Study conducted in 1995)	Culturable airborne bacteria, fungi; German guidelines for hospital operating rooms (<50 CFU/m³), dispensaries, nurseries, and intensive care units (<150 CFU/m³)	Slit impactor FH2 (50 L/min for 2 min); Cockpits, galleys, toilets, all classes of passenger cabins, supply air; Various times on ground and during flight	Fungal concentrations were extremely low, but bacterial concentrations frequently exceeded the limits for hospital air. Highest bacterial concentrations were observed early in flight and during deboarding. Bacterial concentrations were lowest in cockpits; concentrations in economy class and galleys were higher than in business or first class. Bacteria were mainly nonpathogenic types.	Occupants are main source of airborne culturable bacteria. Exposures to fungi on aircraft have no health significance. Only health risk posed by biological agents on aircraft is person-to-person contact as in sneezing or coughing, with transmission over short distances. Apprehension about increased infection risk compared with other crowded spaces is not reasonable.
Lee et al. (1999, 2000)	3 flights; 1 airline	Culturable airborne bacteria, fungi	Multiple-hole impactor (Burkard) (10 L/min; sampling time not reported)	All concentrations of bacteria and fungi were below 1000 CFU/m³.	In general, aircraft air quality was satisfactory.

B747 (Study conducted 1996-1997)	Hong Kong IAQ guideline (1000 CFU/m³)	One sampling location each flight (usually business class, otherwise center of economy class) Immediately after takeoff, middistance, before landing	Highest concentrations were measured at beginning and end of flights (i.e., during boarding and deplaning).	Biological concentrations were generally low, and highest ones were associated with passenger activity.
ASHRAE/ CSS (1999), Pierce et al. (1999), Janczewski (2001), B777 (Study conducted in 1998)	8 flights (4 domestic, 4 international) Culturable airborne bacteria, fungi Concentrations in buildings	Multiple-hole impactor (SAS Compact) (90 L/min; total air volume, 123 L) One coach location each flight (5 forward, 3 rear) During boarding, in flight, during deplaning	Concentrations of bacteria and fungi were relatively low compared with typical indoor concentrations. Concentrations were highest during boarding and deplaning. No infectious bacterial agents were isolated; two potentially infectious fungal agents (*Aspergillus niger*, *Paecilomyces variotii*) were isolated.	Data were not sufficient to draw definitive conclusions about air quality on commercial aircraft; however, available information indicated that no significant air quality health hazards were present for either passengers or crew. Risk of disease transmission via aircraft ventilation system is low, but there is potential for disease transmission because of proximity of passengers; increasing amount of outside air will not minimize this type of disease transmission.

[a]Culturable bacteria and fungi from air samples reported as colony-forming units per unit volume of air (CFU/m³).

— None of the bacteria and few of the fungi were considered potential infectious agents.
- Higher concentrations of culturable bacteria and fungi on aircraft were observed:
  — During periods of passenger activity (e.g., boarding and deplaning).
  — With higher passenger loads and in coach or economy class or the rear of aircraft, compared with first or business class.
  — When cabin air was recirculated or air change rates were lower.
  — Near the cabin floor (in the exhaust air stream).
- No statistically significant differences in concentrations of culturable bacteria and fungi were observed:
  — Among different aircraft, airlines, or flight durations.
  — Between aircraft cabins and other types of public-transportation vehicles.[1]
  — Between aircraft cabins and typical indoor and outdoor urban environments.[1]

Sampling for Allergens

Few studies have measured the concentrations of allergens in aircraft cabins. (The allergen content of settled dust is reported as the amount of an agent per unit amount of settled dust—typically micrograms per gram ($\mu g/g$)). Wickens et al. (1997) found that the concentration of dust mite allergens was much lower in public places (including aircraft) than in homes. Dumyahn et al. (2000) found similar concentrations of dust mite allergens in samples from aircraft and other transportation vehicles; the concentration of cockroach allergens on the four flight segments that were tested was close to the detection limit of the assay method—essentially nonquantifiable.

Cat allergen was found in low to moderate concentrations in all of the vehicles that Dumyahn et al. (2000) tested. The concentrations on trains and subways and in living rooms occasionally exceeded 8 $\mu g/g$, the concentration of cat allergen associated with sensitization (Chapman 1995; Gelber et al. 1993). In the four aircraft that Dumyahn et al. (2000) sampled, the average concentration of cat allergen in cabin air was higher than the average for 24

---

[1]Concentrations on aircraft often were lower than those in other locations.

homes even though no cats were on any of the flights. If a cat had been present, the concentration of cat allergen can be expected to be even higher (Chew et al. 1998) and conceivably could put sensitive individuals at risk of responding.

Evaluation of Available Information on Bioaerosols

Data on some biological agents in commercial aircraft cabins now are available as a result of the studies (see Tables 1-2 and 4-2) that were conducted after the publication of the previous NRC report (NRC 1986). Findings are consistent with patterns that have been observed in other occupied indoor environments. The bioaerosol data that have been collected on aircraft, however, are of little value for the estimation of acute hypersensitivity disease and respiratory infections, which might be the most important health risks associated with bioaerosol exposures in this environment. A discussion of acute hypersensitivity disease and respiratory infections related to bioaerosol exposures in commercial aircraft is presented later in this chapter.

Microorganisms Sampled in Cabin Air

All the bioaerosol samples described in Table 4-2 were grab samples (discrete, short-term samples of relatively small volume). The brief sampling times (up to 3 min) allow identification of rapid changes in bioaerosol concentrations but do not provide information on exposures throughout a flight unless sequential samples are collected. The number of samples collected per flight generally has been small, few replicates have been collected, and estimates of the variability within and between aircraft are poor.

As can be seen in Table 4-2, bioaerosols have only been measured with instruments that impact airborne particles directly onto agar and using culture media and incubation conditions that support the growth of broad groups of bacteria and fungi. Viruses have not been sampled because of practical difficulties.

Culture-based analysis was used in all studies of cabin air as an indication of exposure to microbial aeroallergens (especially fungal allergens). The viability of airborne bacteria and fungi was used as a surrogate measure of their infective potential, even though culture sampling is not a good method of measuring airborne exposure to microbial allergens or infectious agents. It seriously underestimates the concentrations of microorganisms in environmental samples because many microorganisms grow poorly under laboratory conditions and some important infectious agents require special growth media or incubation conditions. Most bacteria that have been isolated in aircraft cabins represent normal human flora, reflecting human occupancy and activity, but they do not necessarily predict the presence of infectious agents that people release. Furthermore, the allergenic, inflammatory, toxic, or irritant properties of environmental microorganisms are not related to cell viability. Therefore, bacteria or fungi that may have been present in cabin air and could have had serious health effects in earlier studies would have been missed because of the inappropriate detection methods that were used.

Other Biological Agents

Studies of bioaerosols on aircraft have been interpreted in terms of potential infection but have not considered the risks of inflammation, irritation, or toxicity. All but one of the studies in Table 4-2 also ignored the potential role of biological agents other than microbial agents, such as bacterial endotoxin, fungal toxins, and microbial volatile organic compounds.

Bacterial Endotoxin

Inhalation of endotoxin has been linked causally with acute airflow obstruction and airway inflammation (Rylander 1994; Milton 1996). Although dust samples have been collected for comparison of endotoxin content in aircraft and other vehicles (Spengler et al. 1997; Dumyahn et al. 2000), no information has been published on the endotoxin content of those samples. Because of a lack of data, it is not yet possible to determine whether endotoxin concentrations in aircraft cabins are high enough to be associated with adverse health effects. An ongoing NIOSH study is expected to provide relevant data on endotoxin concentrations in aircraft cabins (Waters et al. 2001).

Fungal Toxins

Low-molecular-weight fungal products that have toxic effects are called mycotoxins. There are no aircraft cabin air-quality studies in which airborne or dust-associated fungal toxins were measured, although some fungal species that are known to produce toxins (e.g., species of *Aspergillus* and *Penicillium*) have been isolated from this environment. Extensive fungal growth would be necessary for cabin occupants to be exposed to high concentrations of fungal toxins, and it is not known whether inhalation exposures to mycotoxins on aircraft ever reach the levels that have been associated with adverse health effects.

Microbial Volatile Organic Compounds (MVOCs)

MVOCs, which often have distinctive odors, are under study as possible markers of microbial growth (Batterman 1995; Ammann 1999; Fischer et al. 1999). When more is known about the production of these compounds and their health effects, measurement of them might allow identification of aircraft in which there are substantial reservoirs of microbial growth.

Recommended Sampling for Biological Agents

Bioaerosol sampling on commercial aircraft has been broadly based. If the prevalence of biological agents in the cabins of commercial aircraft is to be understood, rigorous studies focused on specific biological agents are needed. In this section, the committee offers suggestions for the types of studies that should be conducted.

NIOSH-Sponsored Review

NIOSH contracted with researchers from the University of Colorado to review the available studies on bioaerosol exposures in aircraft and to provide recommendations on the need for and design of studies (Hernandez and Swartz 2000). These recommendations covered what microorganisms should be sampled, and why? When, where, and how in an aircraft should the sampling be conducted? How many samples should be taken? What types of aircraft and duration of flights should be sampled.

Hernandez and Swartz (2000) recommended that microorganisms be collected in ways that allow analysis with a variety of methods (e.g., culture, staining and direct microscopic examination, immunoassay, chemical assay, nucleic acid amplification, and other molecular detection methods). High-volume air sampling (≥ 100 L/min) may be needed in aircraft cabins for detection limits to be low enough to yield useful exposure information. It was also recommended that studies of aircraft air quality not include endotoxin, fungal toxins, or other microbiological agents (e.g., fungal glucans) until assays for these materials are better developed and more evidence is available to suggest that these compounds are present in aircraft cabins at concentrations sufficient to cause adverse health effects.

Hernandez and Swartz (2000) further recommended that future bioaerosol monitoring focus on some of the newest and oldest aircraft in service (the Boeing 747-400 and 727 series aircraft, respectively) and account for the following factors: extremes in air recirculation rates and replenishment of fresh air, operation of air filtration systems, medium and long flight duration (1-4 h and over 5 h of stable cruise time, respectively), and passenger load. They suggested that monitoring be conducted in the coach cabin with composite samples collected throughout a flight.

Allergens

Studies of common allergens in aircraft cabins are insufficient to assess the risk posed by allergen exposures on aircraft. Allergens usually are measured through analysis of dust rather than air samples because few of the particles carrying allergens are fine enough to remain airborne for more than a few min. Dust samples, however, can be used only to estimate the potential for allergen exposure rather than actual exposure, because the concentration of an allergen in dust does not take into account other factors that may influence exposure, such as the total density of dust on surfaces or its resuspension rate. More-sensitive analytical methods soon may make air sampling feasible, yielding better measurements of airborne allergens.

Infectious Agents

More information is needed on the frequency with which people are exposed to human infectious agents during air travel. Environmental sampling for infectious agents is problematic, however, because of the unpredictable presence of infectious persons, the many agents that travelers may carry, the different volumes of air that would need to be collected to detect different microorganisms given their wide range of release rates and infectious doses (the number of cells required to cause infection), and difficulties related to capturing airborne bacteria and viruses and preserving their biological activity for accurate identification and enumeration. Nevertheless, it may be possible to assay concentrated, large-volume air samples with methods that do not rely on culture, such as nucleic acid amplification assays, which are available for many of the infectious agents to which people may be exposed on aircraft (Mastorides et al. 1999; Schafer et al. 1998; Aintablian et al. 1998; Echavarria et al. 2000). MacNeil et al. (1995) have reviewed these methods and their application to the evaluation of indoor air quality.

The presence of infectious agents in cabin air also can be recognized by indirect methods, such as surveillance for disease transmission or isolation of microorganisms from passengers and crew members. Clinical specimens (e.g., throat cultures or sputum samples) have been collected from recent air travelers to isolate specific infectious agents (Clayton et al. 1976; Moser et al. 1979; Brook, 1985; Klontz et al. 1989; Brook and Jackson 1992; Sato et al. 2000; CDC, 2001a). Immunological tests on blood samples can also be used to identify infection, and tuberculin skin testing is a valuable tool to identify recent infection in studies of possible tuberculosis transmission.

Release of tracer particles and modeling of particle dispersion in aircraft cabins also may play a role in the assessment of the spread of infectious agents and other biological materials generated by cabin occupants and their activities. NIOSH is modeling airflow and migration of airborne biological agents throughout typical aircraft cabins. With this model, exposures for crew members and passengers could be estimated and measures to minimize such exposures could be evaluated. Similarly, the effect of passenger-controlled air supply (gaspers) on the distribution of aeroallergens and infectious agents could be assessed (BRE 2001).

HEALTH EFFECTS OF EXPOSURE TO BIOAEROSOLS

The principal biological contaminants of potential concern in cabin air are allergens and infectious agents. Substantial numbers of individuals are sensitive to one or more airborne allergens. Several common allergens have been detected in aircraft cabins (e.g., cat and dust mite allergens). Peanut-allergic passengers have raised concerns about potential contact, accidental ingestion, and inhalation exposure to allergens released from peanuts that are served on aircraft.

Respiratory infections are fairly common in humans. For example, adults annually experience one to six colds and infants and children may have two to six episodes of acute, febrile, respiratory disease a year (Chin 2000). Persons with influenza, measles, tuberculosis, and meningococcal disease are known to have traveled on commercial aircraft while infectious and evidence of transmission of the causative agents during air travel is convincing for the first three diseases.

This section addresses the allergens of greatest potential concern in aircraft cabins, describes investigations that have been conducted of possible exposures to infectious agents on aircraft, and describes how exposures to biological agents in aircraft cabins can be controlled.

Hypersensitivity Disease

Immediate hypersensitivity involves stimulation of immunoglobulin E (IgE) antibodies. The condition occurs in persons who are genetically predisposed to mount an IgE response to specific allergens, have been exposed to a sensitizing dose of an allergen to which they are predisposed to respond, and are exposed appropriately to an allergen to which they previously were sensitized. An estimated 20% of Americans (over 50 million people) suffer from allergic rhinitis (hay fever) or other allergic diseases, and 8-17% of the population have asthma (IOM 1993b; Montealegre and Bayona 1996; Bellanti and Wallerstedt 2000). Sensitivity to food allergens may induce asthma or anaphylactic reactions in some people, with the reaction triggered by ingestion or airborne exposure.

For passengers on aircrafts, the initial sensitizing steps in the development of immediate hypersensitivity are unlikely to occur during travel because passengers spend relatively little time in this environment. The situation could differ for a genetically susceptible crew member who had never previously encountered a common allergen but who was exposed repeatedly on aircraft. It is unlikely, however, that cabin personnel would have avoided exposure during childhood or early adult life to the allergens most likely to be present in aircraft.

The major allergens that appear to be involved in the development and exacerbation of asthma are those associated with dust mites and cockroaches, cats, dogs, other small animals, *Alternaria* species and other fungi, and pollen (IOM, 2000). Patterns of allergen sensitivity often change with age so that young children are more likely to become sensitized and to respond to either dust mite, cockroach, or fungal allergens, whereas adults respond primarily to cat or pollen allergens (Silvestri et al. 1999; Dharmage et al. 2001).

A review of studies of health effects and pets concluded that cat and dog allergens are present everywhere as a result of people's carrying allergen on their clothing and that all exposure to pets involves some risk of sensitization (Ahlbom et al. 1998). Cat allergens have been detected on people's clothing (Tovey et al. 1995; D'Amato et al. 1997; DeLucca et al. 2000) and in places seldom visited by cats (e.g., office buildings, schools, new homes, allergists' offices, hospitals, and shopping malls) (Lundblad 1991; Enberg et al. 1993; Hung et al. 1993; Janko et al. 1995). Hundreds of thousands of animals travel by air each year, but it is not known how often animals (particularly cats) travel in the cabins of commercial aircraft rather than the cargo bays. Thus, it is not surprising that cat and dust mite allergens have been detected in aircraft cabins (Wickens et al. 1997; Dumyahn et al. 2000). It also has not been determined if the presence of a cat measurably increases the concentration of cat allergen either in cabin air or settled dust. No conclusions can be reached regarding routine inhalation exposures to common animal, arthropod, or fungal allergens in aircraft because so few data are available (see the earlier section on sampling for allergens in aircraft cabins).

The few case reports that link hypersensitivity responses with allergen exposure on aircraft focus on peanut allergens. Hypersensitivity to peanuts is a foodborne allergy, and even minute ingested amounts can be extremely dangerous (Hourihane et al. 1997). An estimated 1% of Americans are hypersensitive to ground nut (peanut) or tree nut (e.g., walnut, almond, and cashew) allergens, so there is a sizeable health concern (Sicherer et al. 1999a). Aircraft passengers and crew members with life-threatening peanut allergies understandably are concerned about potential exposures on aircraft, where access to medical care is limited.

Sicherer et al. (1999b) published the first description of the clinical characteristics of allergic reactions to peanuts on commercial aircraft in subjects with peanut allergy. In a registry of 62 peanut- and tree-nut-sensitive people who had reported "airplane/airport" as a location where they had experienced allergic reactions, the authors reached 48. Reactions in 42 were thought to have begun on a plane including 34 reportedly allergic to peanuts. Among the 34, allergic responses were judged to have resulted from inhalation, ingestion, or skin contact in 14, 14, and 7 subjects, respectively.

A finding of the Sicherer et al. (1999b) investigation was that few affected passengers or their guardians had notified the flight crew or airline of their suspected allergic reactions to peanuts (James 1999). Most of the reactions identified in the study were not life-threatening, but five of the subjects received epinephrine while in flight to manage severe allergic reactions. The authors concluded that exercising caution and having emergency medication available were important, considering the number of persons who may experience allergic reactions as a consequence of ingesting peanut or tree nuts or inhaling allergen on commercial aircraft. Emergency medical kits mandated for commercial aircraft in the United States include epinephrine to treat severe anaphylaxis (14 CFR 121, Appendix A, First-Aid Kits and Emergency Medical Kits).

The Department of Transportation (DOT) proposed that airlines make "peanut-free zones" available on request from passengers with medically documented severe allergies to peanuts, as would be covered in the Air Carrier Access Act of 1986 (14 CFR 382, Nondiscrimination on the Basis of Disability in Air Travel). This proposal has not been implemented, although there is continued support for such a regulation and it could be reconsidered after submission to Congress of "a peer-reviewed scientific study that determines that there are severe reactions by passengers to peanuts as a result of contact with very small airborne peanut particles of the kind that passengers might encounter in an aircraft" (Resolution 117. Congressional Record, 105th Congress, 2nd Session, 1998). This committee was unable to locate such a study. The only studies that provide such evidence are the ones by Sicherer et al. (1999b) with self-reported symptom and exposure data, and a study in which peanut allergen was found on air filters from commercial aircraft (Jones et al. 1996), although this investigation was not published in a peer-reviewed journal. Overall, there is little evidence of responses in sensitized persons to airborne food allergens. Although rare, severe responses have been reported to cooking aerosols (e.g., from fish and hot dogs) (Crespo et al., 1995; Polasani et al., 1997), but not on aircraft. Major air carriers usually accommodate the requests of peanut-allergic travelers that peanuts not be served on a flight or in adjacent rows, and some airlines (e.g., United Air Lines, US Airways) have discontinued serving peanuts as snack foods.

Infectious Disease

The first NRC report on air quality in aircraft cabins included several recommendations for reducing the risk of transmission of infectious agents.

The recommendations were based on incidents in which transmission was clear and on the environmental factors that had contributed to the spread of infectious agents during those events, such as insufficient outdoor air, lack of air treatment to remove infectious particles, and exposure to mosquitoes (NRC 1986). Since publication of the report, transmission of other infectious agents has been reported on aircraft, awareness of the importance of the subject for national and international public health has increased, and protocols for responding to the consequences of possible exposure to infectious agents on aircraft have been developed.

Large numbers of people use air travel for business, for tourism, and for other reasons (e.g., to immigrate or seek asylum) (WHO 1998a). An estimated 50 million North Americans will cross international borders in 2001; more than 10 million of them will travel to tropical destinations that pose a serious risk of infectious disease (Weiss 2001). On aircraft, people are confined in close quarters for long periods (see Chapter 1) and then disembark to many distant places (Wilson 1995). Thus, the risk of exposure to exotic infectious agents is higher for travelers than other persons, especially on international flights. Furthermore, the consequences of exposure to infectious agents may extend beyond the travelers to all others with whom they later have contact. Those factors give substantial public-health significance to the relationship between infectious diseases and air travel (Sato et al. 2000; BRE 2001; IEH 2001; Maloney and Cetron 2001).

Before air travel became the major medium of international travel, an infectious disease commonly had time to develop to its recognized clinical form before travelers reached their destinations (Grainger et al. 1995). But air travel is rapid, and people can complete a journey in the preclinical stage of an infectious disease (Clayton et al. 1976; Maloney and Cetron 2001). Furthermore, longer nonstop flights are becoming possible, and they increase the time that travelers spend together in an aircraft and the opportunities for exposure to infectious agents.

The reservoirs for infectious agents in cabin air are the people on board; the viruses, bacteria, and fungi they carry in and on their bodies; and vectors (such as arthropods) that may be found in aircraft. The infectious agents of concern in relation to cabin air are those transmitted by person-to-person droplet contact and airborne transmission of droplet nuclei. Droplet nuclei can remain suspended in cabin air and be distributed throughout an aircraft.

Studies of potential infectious disease transmission on aircraft are summarized in Table 4-3. Primary emphasis is on infectious agents that are known

TABLE 4-3 Investigations of Potential Infectious-Disease Transmission on Aircraft

Reference	Source cases	Travel history	Number	Outcome
Influenza				
Moser et al. (1979)	21-y-old woman	Homer to Kodiak, Alaska (>3-h ground delay, cabin unventilated for 2 h), Boeing 737, 1977	49 passengers (1 not contacted) and 5 crew; 36 passengers and 2 crew members (72%) became ill; attack rate varied with time on grounded aircraft	Transmission likely to have occurred in closed, poorly ventilated, grounded aircraft and possibly in flight
Klontz et al. (1989)	11 persons actively coughing	Puerto Rico to Key West, Florida (2.5 h), DC-9, 1986	90 naval personnel; 23 (30%) of 77 susceptible persons became ill	Transmission likely to have occurred in flight; evidence of secondary transmission
Measles				
Amler et al. (1982)	3 children, 1 with early symptoms	Venezuela to Miami, 1981	No information provided	Transmission suspected; evidence of secondary transmission
CDC (1983)	27-y-old man	San Diego to Seattle to San Diego, 1982	Infection in one passenger on return flight	Transmission may have occurred in flight or at airport; evidence of secondary transmission
Slater et al. (1995)	Not identified	New York to Tel Aviv (2-h ground delay, 10-h flight), Boeing 747, 1994	350 passengers; 8 cases identified	Transmission likely to have occurred in flight, but exposure in one of two airports also possible
Tuberculosis				
Driver et al. (1994), CDC (1995)	Female flight attendant	39 international flights (U.S. to Europe or Mexico), 128 domestic flights; exposure duration of persons possibly infected, >1 h; median exposure duration of skin-tested passengers, 3.8 h; 1992	223 and 51 crew contacts; 212 exposed and 247 unexposed crew were skin-tested; 59 passengers were skin-tested	Infection of 2 crew members confirmed; evidence of transmission to passengers inconclusive

McFarland et al. (1993), CDC (1995)	Passenger, foreign-born	London to Minneapolis (9 h, seated in first class), 1992	325 passengers and 18 crew; 79 were skin-tested	No evidence of transmission
CDC (1995)	Passenger, foreign-born	Mexico to San Francisco (4.5 h) 1993	92 passengers; 22 were skin-tested	No evidence of transmission
CDC (1995), Miller et al. (1996)	Passenger, male Russian refugee	Frankfurt to New York (8.3 h), Boeing 767; New York to Cleveland (1.3 h), BAe 146; 1993	219 passengers and crew; 120 were skin-tested	Transmission to 2 persons could not be excluded, but likelihood was considered low
Moore et al. (1996)	Passenger, male	2 domestic flights (1.25 h each), Boeing 757, 1994	212 passengers and 15 crew; 100 were skin-tested	Transmission to 5 persons could not be excluded, but likelihood was considered low
CDC (1995)	Passenger, U.S.citizen, long-term resident in Asia	Taiwan to Tokyo (3 h), Tokyo to Seattle (9 h), Seattle to Minneapolis (3 h), Minneapolis to Wisconsin (0.5 h), 1994	661 passengers; 87 were skin-tested	Transmission could not be excluded, but likelihood was considered low
CDC (1995), Kenyon et al. (1996)	Passenger, 32-y-old Korean woman	Round trip: Honolulu to Chicago (8.4, and 8.75 h), Boeing 747-100, Chicago to Baltimore (1.75 and 2 h); Airbus 320-200, 1994	1,042 passengers and crew; 760 were skin-tested	Transmission to 5 passengers and 1 crew member could not be excluded
Wang (2000)	Passenger, 44-y-old Taiwanese woman	Los Angeles to Taipei (14 h) Boeing 747-400, 1997	308 passengers and crew; 225 were skin-tested	Transmission to 3 persons could not be excluded
Meningococcal Disease				
CDC (2001c)	Passenger, 62-y-old man	Sydney, Australia to Los Angeles to New York, 2001	1 of 2 adjacent passengers was identified and remained asymptomatic	No evidence of transmission, but only adjacent passengers were followed

or suspected to have been transmitted in aircraft—influenza, measles, and tuberculosis—and those whose transmission is considered possible—meningococcal disease and acute respiratory infections. An infectious diseases was not considered if the disease is transmitted in a way unrelated to the cabin air or the ventilation system, release of an agent within an aircraft cabin results from an unprecedented event (e.g., an act of bioterrorism), or exposure during air travel is highly unlikely (e.g., viral hemorrhagic fevers or pneumonic plague) (Clayton et al. 1976; Fritz et al. 1996; Wenzel 1996; Withers and Christopher 2000; Maloney and Cetron 2001; Weiss 2001). Of the communicable diseases discussed in this report, tuberculosis, meningococcal meningitis, and plague are among the diseases for which there are special requirements related to travelers and the operation of conveyances (21 CFR 1240.50, 42 CFR 70.5, Certain Communicable Diseases; Special Requirements), and for which travelers can be detained to prevent the introduction, transmission, or spread of the disease in the United States (42 CFR 70.6, 21 CFR 1240.54, Apprehension and Detention of Persons with Special Diseases; 42 CFR 71.32, Persons, Carriers, and Things).

The risk of exposure to infectious persons is highest for the passengers seated closest to a source person and for the cabin crew who work in the same section (see the following section on cabin ventilation and exposure to bioaerosols). The risk of exposure to infectious agents may be higher for the cabin crew than for other passengers because the crew fly more often, and interact (if only briefly) with more persons on any given flight.

Influenza

Influenza is a highly contagious, viral respiratory illness. Influenza A and influenza B are the major types of influenza viruses that cause disease in persons of all ages (CDC 2001b,d). Influenza viruses are spread from person to person primarily via the airborne route after coughs and sneezes (Murphy and Webster 1996). The incubation period for influenza is 1-4 d (CDC 2001b,d) and the infectious period can start the day before symptoms begin until about 5 d after illness onset; children can be infectious for longer periods.

Influenza and Air Travel

Moser et al. (1979) reported on a 1977 outbreak of influenza among passengers and crew exposed to an acutely ill passenger aboard a 56-seat commercial jet that had a greater than 3-h ground delay before takeoff. The influenza attack rate among the passengers and crew was very high (38 of 53, 72%) and four persons required hospitalization. People who were on the disabled plane for more than 3 h had the highest attack rate (25 of 29, 86%). There was no significant association between illness and sex, smoking status, history of recent influenza vaccination, activity while waiting on the airplane, or later travel. The high attack rate was attributed to the ventilation system's not operating during the ground delay and the doors' being kept closed for about 2 h. The authors concluded that proper operation of the air circulation equipment and isolation of the ill passenger might have prevented this outbreak. The investigation led the first NRC committee to the conclusion: "Because a likelihood of occurrence of epidemic disease when forced-air ventilation is not available on the ground has been demonstrated, the Committee recommends...a regulation...that requires removal of passengers from an airplane within 30 min or less after a ventilation failure or shutdown on the ground and maintenance of full ventilation whenever onboard or ground air-conditioning is available" (NRC 1986). The 30-min limit was based on the time required to return a full load of passengers to a terminal. The FAA response to the recommendation was as follows: "Because the occurrence of complete ventilation cessation on passenger-laden airplanes is extremely rare and sometimes unavoidable, we do not believe that regulatory action is necessary. However, there may be value in bringing this concern to the attention of the air carriers. The FAA will advise air carriers of the need to deplane passengers, if possible, after 30 min without ventilation" (DOT 1987).

Suspected transmission of influenza associated with air travel also has been reported in association with an aircraft that had an operating ventilation system and no ground delay (Klontz et al. 1989). Ninety squadron members traveled on two DC-9 aircraft from Puerto Rico to a naval station in Key West, Florida. Twenty-three of 77 previously well persons on these 2.5-h flights developed severe influenza-like respiratory illness within 72 h of their return. Eleven of the case patients reported actively coughing during the return flight. A significant difference in risk of acquiring influenza was observed between the two aircraft—53% (18 ill of 34 susceptible persons) vs 12% (5 of 43)—and was related to the number of symptomatic persons on board—18% (8 of 44 passengers were ill during travel) vs 7% (3 of 46)—and occupant load factor—94% (44 passengers and 47 seats) vs 69% (46 of 67). The difference in attack rate remained greater in the first aircraft even if persons who had shared sleeping quarters with an ill person before air travel were excluded—50% (10 of 20) vs 13% (4 of 30). Secondary transmission of influenza to family members and roommates was identified.

Public-Health Measures Related to Influenza and Travel

Immunization is the primary method used to prevent influenza infection and its complications (CDC 2001b,d) (see following section on control of exposure to biological agents). Annual influenza immunization has been recommended for tourism industry workers (Bodnar et al. 1999). Public-health authorities do not recommend general use of antiviral medications for all travelers to prevent influenza (in the event of immunization failure), because the drugs can have side effects and must be prescribed by a physician. However, antiviral medications might be appropriate during an influenza outbreak for unvaccinated travelers and persons who are at increased risk for influenza-related complications (CDC 2001d).

Measles

Measles is an acute, highly communicable viral disease with an average incubation period of 10-12 d (CDC 1998). It may be severe and frequently is complicated by middle ear infection or bronchopneumonia (CDC 2001b). Before widespread immunization, measles was common in childhood; 90% of people were infected by the age of 20. Since 1993, fewer than 1,000 measles cases have been reported each year (CDC 2001b), many of which are imported from outside the United States and occur among adults. Measles remains a common disease in many countries, including some developed countries in Europe and Asia, making exposure due to air travel a possibility.

Measles and Air Travel

Suspected transmission of measles at an airport and between passengers who shared a domestic flight has been reported (CDC 1983), and transmission also may have occurred in association with international air travel (Amler et al. 1982; Slater et al. 1995). Three children from Venezuela who entered the United States had measles later; one of them had early symptoms while on board. The onset of rash in the other two cases was 12 d after the flight, and secondary transmission to other children was identified (Amler et al. 1982). Eight cases of measles were associated with a New York–Tel Aviv commercial flight, but no source case was identified (Slater et al. 1995). Transmission could have occurred during a 2-h ground delay (during which the aircraft's air-conditioning system reportedly was not working) or during the 10-h flight.

Public-Health Measures Related to Measles and Travel

Immunization is the primary method used to prevent measles infection and its complications (Slater et al. 1995; CDC 1998, 1999, 2001b). Although vaccination against measles is not a requirement for entry into any country (including the United States), persons traveling abroad should ensure that they are immune. Most persons born before 1957 are likely to have had measles and generally are not considered susceptible. However, measles vaccine may be given to older persons if there is reason to believe that they may be susceptible and could be exposed during travel. Because the risk of contracting measles is greater in many countries than in the United States, infants and children should be vaccinated before leaving the United States, even if this involves vaccination at an earlier age than is recommended for infants and children remaining in the United States (CDC 2001b).

Tuberculosis

Tuberculosis (TB) is caused by *Mycobacterium tuberculosis*, a bacterium that can attack any part of the body but most frequently is associated with pulmonary infection. People typically become infected after spending a long time in a closed environment where the air is contaminated by a person with untreated tuberculosis who is coughing and has numerous organisms in secretions from the lungs (CDC 2001b). The time from exposure to detectable infection typically is 4-12 wk (Chin 2000). One-third of the world population of about 6 billion is estimated to be infected with *M. tuberculosis* (WHO 1998a), however, about 90% of otherwise healthy adults who acquire tuberculosis infection never develop active disease and therefore do not experience symptoms and are not infectious to others.

Tuberculosis and Air Travel

The ease and availability of air travel, the large number of persons traveling yearly, the emergence of *M. tuberculosis* strains resistant to one or more of the primary drugs used for treatment, and the movement of immigrants and refugees increase the possibility for all persons to be exposed to someone with infectious tuberculosis (WHO 1998a). Tuberculosis is the most thoroughly studied communicable disease possibly associated with transmission during commercial air travel (Driver et al. 1994; CDC 1995; Kenyon et al. 1996; WHO 1998a) and the disease with the most-detailed formal guidelines for recognition and prevention (Withers and Christopher 2000; BRE 2001). No case of active tuberculosis has been identified as a result of exposure on a commercial aircraft, but the number of potentially exposed people who have been successfully screened in these investigations has been very small (see Table 4-3). One additional investigation showed no evidence of tuberculosis infection in 47 commercial airline pilots who flew DC-9-series aircraft (without air recirculation) in the company of a pilot with active tuberculosis (Parmet 1999).

In 1992-1995, CDC, in conjunction with state and local health departments, conducted seven investigations involving one flight attendant and six passengers with active tuberculosis (McFarland et al. 1993; Driver et al. 1994; CDC 1995; Kenyon et al. 1996; Miller et al. 1996; Moore et al. 1996). The number of potentially exposed passengers and crew was more than 2,600 on a total of 191 flights involving nine types of aircraft (WHO 1998a). In each investigation, the index case was considered to be highly infectious. Two of the passengers knew that they had active tuberculosis at the time of their flights to the United States. Diagnosis of the other five cases occurred after travel (CDC 1995). Only two of the seven investigations produced firm evidence to suggest transmission of *M. tuberculosis* infection: the first from a flight attendant to other crew members (Driver et al. 1994), the second from a passenger to other passengers (Kenyon et al. 1996). In the first report, evidence of transmission was limited to fellow crew members who were exposed to the infectious source for at least 12 h. In the second, transmission was demonstrated only to a few passengers seated close to the passenger with active tuberculosis (in adjacent rows in the same section) and on only one flight segment that lasted longer than 8 h.

Since 1995, one investigation has been conducted after identification of active tuberculosis in a passenger who traveled from Los Angeles to Taipei (Wang 2000). Skin-test conversion (a change in skin test status from negative to positive) related to exposure during air travel could not be ruled out in three of 225 passengers. Although none of the three people was in same section of the aircraft as the index case, the author believed that exposure during the flight could not be excluded, because the cabins shared an air supply, the source person was highly infectious, and the flight was long. This investigation highlighted the value of two-step tuberculin skin testing used to differentiate between prior and recent infection. The 1-wk time interval between the first and second baseline skin tests in a two-step test is too short to identify recent infection but is sufficient to boost the weakened immune response of someone with prior infection. Eleven persons with negative initial skin tests showed positive reactions on their second baselines tests. These persons would have shown positive reactions on any later skin test (in an outbreak investigation, this typically would not be given until several months post exposure). Without the two-step baseline test, the apparent change in immune status of the 11 persons would have been mistakenly interpreted as possible travel-related infection, as happened in the investigation by Kenyon et al. (1996).

Public-Health Measures Related to Tuberculosis and Travel

A tuberculosis patient should not travel unless his or her physician judges it safe and should be instructed to cover coughs and sneezes with the hands or tissue paper at all times. Available data on the transmission of *M. tuberculosis* on aircraft indicate that the risk to passengers is no greater than is posed by other activities in which contact with potentially infectious individuals may occur (e.g., train travel, bus travel, and attending conferences) (Rogers 1962; Sacks et al. 1985; Lodi 1994; CDC 1995, 2001b; Moore et al. 1999; Witt 1999). There also is no evidence of increased risk to flight attendants, and routine and periodic tuberculin screening of flight crew is not justified or indicated for otherwise asymptomatic employees (Driver et al. 1994; WHO 1998a). However, air travelers who make frequent and regular stops in countries and areas with a high tuberculosis burden may be advised to have a baseline skin test to determine their infection status (CDC 2001b) and, if it is negative, to have periodic tests to identify subsequent exposure. The World Health Organization (WHO) and the Centers for Disease Control and Prevention (CDC) jointly developed guidelines for the prevention and control of tuberculosis during air travel and criteria for determining when followup of potential exposure is warranted (Table 4-4) (WHO 1998a).

TABLE 4-4 Criteria for Deciding to Inform Passengers and Crew Members of Possible Exposure to *M. tuberculosis*

Consideration	Criterion
Infectiousness of person identified as having had tuberculosis at time of air travel	Index case must be judged to have been capable of transmitting infection at time of travel on basis of clinical evidence and appropriate medical testing
Duration of exposure	At least 8 h
Interval between flight and notification of health authorities	No longer than 3 mo
Proximity of exposed persons to index case	Only passengers seated close to person with active tuberculosis and crew members working in same cabin need be informed initially

Source: Adapted from WHO (1998a).

Meningococcal Disease

Meningococcal disease is an acute bacterial infection characterized by sudden onset with fever, intense headache, nausea (often with vomiting), and stiff neck. The incubation period typically is 3 to 4 d but may range from 2 to 10 d (Chin 2000). *Neisseria meningitidis* is a leading cause of bacterial meningitis and sepsis in children and young adults in the United States and is spread through direct contact with respiratory secretions (CDC 2000). Case-fatality rates of meningococcal disease used to exceed 50%, but with early diagnosis, modern therapy, and supportive measures, they are now 5-15% (WHO 1998b; CDC 2001b). Up to 10% of populations in countries with endemic disease may be asymptomatic carriers of *N. meningitidis*.

Meningococcal Disease and Air Travel

Persons with meningococcal disease are known to have traveled on commercial aircraft (Duffy 1993; CDC 2001c), and passengers next to an infected person on long flights may be at higher risk than other passengers for developing meningococcal disease (Maloney and Cetron 2001; CDC 2001c,e). However, no cases of transmission of *N. meningitidis* to fellow air travelers have been identified.

Public-Health Measures Related to Meningococcal Disease and Air Travel

Prompt chemoprophylaxis of persons in close contact with an index-case patient is the primary means of preventing cases of meningococcal disease. CDC uses a passive-surveillance system by which local health departments report suspected cases of air-travel-associated meningococcal disease (CDC 2001c). CDC annually receives reports of about 12 cases of confirmed disease in which the index patient likely was contagious aboard an international conveyance (ship or aircraft) (CDC 2001c). In collaboration with the Council of State and Territorial Epidemiologists, CDC has developed procedures for the management of suspected exposure to *N. meningitidis* associated with air travel (CDC 2001e; CSTE 2001). In the absence of data on increased risk to other passengers, antimicrobial chemoprophylaxis is recommended only for passengers in seats next to an index case-patient (i.e., on either side of the potentially infectious person) (CDC 2001c).

CONTROL OF EXPOSURES TO BIOLOGICAL AGENTS

Hypersensitivity Diseases, Toxins, and Nuisance Agents

As discussed earlier, extensive microbiological growth in aircraft cabins appears to be unlikely. And there is no reason to think that the design, maintenance, or operation of the ventilation systems on commercial aircraft increase exposures to biological agents. Other than the identification of legionellae in cockpit humidifiers, bacterial or fungal growth in aircraft ventilation systems has not been reported.

Except for cat and peanut allergens, passengers appear to be the primary means of allergen entry into aircraft cabins. Therefore, exposure to allergens can be controlled only insofar as they are allowed to accumulate in dust that settles into upholstered furniture and onto carpeted floors and other surfaces and to the degree that allergenic particles are removed by filters in the return air system. High-efficiency filtration of cabin return air would remove essentially all potentially irritating, inflammatory, or toxic microorganisms and particles carrying allergens. Even lower-efficiency filters (e.g., 80-90% efficient) will remove most of the particles.

The presence of a cat in a cabin may be a more important source of cat allergen than what people carry into an aircraft. Persons who are hypersensitive to cat allergen generally do not have cats in their homes, and those persons who are extremely sensitive restrict visits to places where they know that cats may be present. Therefore, limiting the number of cats or other pets that travel in aircraft cabins may reduce exposures that could have serious consequences.

Insufficient evidence is available to recommend eliminating peanut-containing food from being served to passengers on aircraft. But not serving peanuts to passengers adjacent to peanut-allergic travelers upon request would allay these person's concerns and would reduce the chances for inhalation, contact, or accidental ingestion of peanut allergens that could result in severe adverse reactions.

Infectious Agents

Crew members and passengers on commercial aircraft can protect themselves and others from infectious diseases by adhering to current recommendations for immunization, practicing good personal hygiene, and not traveling when unwell (Slater et al. 1995; Rayman 1997; IEH 2001). Persons with communicable diseases that do not pose a direct threat to the health or safety of others cannot be denied access to air travel (14 CFR 382, Nondiscrimination on the Basis of Disability in Air Travel), although air carriers can impose restrictions (e.g., the wearing of a face mask) on persons with communicable diseases. Ill and well passengers have worn respiratory protection during air travel to avoid transmission of infectious agents, but the efficacy of this practice has not been demonstrated (Hendley 1987; Withers and Christopher 2000). Properly fitting respirators are likely to be uncomfortable, would stigmatize the wearer, and could impede communication; all these effects could reduce compliance.

Immunization against Infectious Diseases

The best means to protect travelers against vaccine-preventable diseases is immunization (Slater et al. 1995; Rayman 1997; Maloney and Cetron 2001). Table 4-5 lists the minimal, universally recommended immunizations for young children, adolescents, and adults. Influenza immunization also is recommended for persons younger than 65 who are at high risk of exposure and of exposing

TABLE 4-5 Universally Recommended Vaccinations for Children, Adolescents, and Adults

Population	Vaccination
All young children	Measles, mumps, and rubella
	Diphtheria-tetanus toxoid and pertussis vaccine
	Poliomyelitis
	Haemophilus influenzae type B
	Hepatitis B
	Rotavirus
	Varicella
Previously unvaccinated or partially vaccinated adolescents	Hepatitis B
	Varicella (if no previous history of varicella)
	Measles, mumps, and rubella
	Tetanus-diphtheria toxoid (if not vaccinated during previous 5 years)
All adults	Tetanus-diphtheria toxoid
All adults aged 65 years and older	Influenza
	Pneumococcal

Source: CDC (1999)

high-risk persons (CDC 2001d; Buxton et al. 2001). Selection of additional immunizations for travelers should be based on the requirements of the local health authorities at the travel destination and individual travelers' risk of infection. Potential exposures to infectious agents during the time spent on an aircraft to reach a destination generally are not considered in these decisions, because the exposure time during flight relative to other travel-related exposures typically is brief.

Recognition of Potential Infection During or Shortly After Flight

Passengers and crew members occasionally become ill in flight. Persons with hypersensitivity diseases generally recognize the onset of allergic responses and asthma and carry appropriate medication to manage symptoms. Cabin crew usually can recognize passengers suffering from, for example, air sickness or excessive alcohol consumption. Persons suffering from diarrhea or vomiting and fever, with or without a skin rash, should be considered to be infectious (Grainger et al. 1995). Flight crew must notify the local health authority or quarantine officer at the airport where they are scheduled to land of arriving passengers who appear ill (e.g., with fever, rash, unusually flushed or pale complexion, jaundice, shivering, profuse sweating, diarrhea, or inability to walk without assistance) (Weiss 2001; 21CFR1240.45 Report of Disease; 42CFR70 Interstate Quarantine; 42CFR71.21(b) Foreign Quarantine). Recommendations have been formulated for responding to notification of the arrival of one or more ill travelers (Maloney and Cetron 2001; CDC 2001c,e). Those responses often can be initiated before an aircraft reaches an airport.

A WHO guideline (1998a) outlines basic actions that flight attendants should take when a person reports or is suspected of having tuberculosis during a flight. Many of the recommendations also would apply to other communicable diseases that are transmitted from person to person. For example, a symptomatic person should be isolated from other passengers, made comfortable, provided with tissues and waste containers, advised to move around the cabin as little as possible, and instructed to cover the nose and mouth when coughing.

State and local health departments and private physicians should ask all persons with infections that are transmitted by droplets and droplet nuclei about recent travel (WHO 1998a; CDC 2001c). The case report forms that healthcare providers and laboratories are required to submit to local public health authorities for notifiable diseases should include information on recent travel.

In almost all of the investigations outlined in Table 4-3, the person's infection was not detected until after the flight was completed and the passengers had dispersed. Notification of fellow travelers frequently has been hindered by difficulty in obtaining contact information for passengers. Airlines typically maintain passenger manifests and other records for 2-7 d, after which they are archived or destroyed. However, this period may be too short to allow follow up of infectious diseases with incubation periods longer than one wk (e.g., tuberculosis, measles, and possibly meningococcal disease). To facilitate timely identification and public health notification and management of at-risk passengers, airlines should ensure that electronic passenger mainfests and contact information are preserved and readily available for a period of at least one month following disembarkation (CDC 2001c,e). WHO (1998a) recommends that airlines preserve for at least three years passenger records for flights involved in investigations of possible exposure to *M. tuberculosis*.

Postexposure Prophylaxis (PEP)

Chemoprophylaxis may be recommended to prevent infection if exposure to some infectious agents (e.g., influenza viruses, *N. meningitidis*, and *M. tuberculosis*) is recognized during a flight or shortly thereafter. PEP is appropriate if exposure is judged to have been certain or highly likely, appropriate therapy is available and can be administered promptly, and the consequences of infection are sufficiently severe or the risk of complications from an infection is high.

Cabin Ventilation and Exposure to Bioaerosols

Ventilation Rate and Air Movement

Microorganisms can remain suspended in cabin air for long periods of time if there is little air movement or the exhaust rate is very low, as happened in an influenza outbreak on a grounded aircraft (Moser et al. 1979). Higher concentrations of airborne microorganisms have been measured on aircraft with higher passenger loads (Nagda et al. 1989, 1992) and in coach or economy class relative to business or first class (ATA 1994; Dechow et al. 1997; Dechow 1996; M. Dechow, Airbus, personal communication, January 3, 2001) (Table 4-2). Elevated concentrations of microorganisms also have been observed during periods of passenger activity, such as boarding and deplaning (Spengler et al. 1997; Wick and Irvine 1995; Dechow et al. 1997; ASHRAE/ CSS 1999; Pierce et al. 1999; Dumyahn et al. 2000; Dechow 1996, personal communication, January 3, 2001; Lee et al. 1999, 2000; Janczewski 2001). Nagda et al. (1989, 1992) observed higher concentrations of airborne bacteria, but lower concentrations of fungi, on flights with lower nominal air change rates. Wick and Irvine (1995) measured higher concentrations of bacteria near the floor vents than in the breathing zone of passengers. This observation, in conjunction with the available evidence that only persons seated near a source person are exposed to a sufficient number of bacteria or viruses to become infected, indicate that bioaerosols as small as droplet nuclei are readily entrained in moving air streams (see Figure 2-3). Therefore, the movement of bioaerosols along the length of aircraft cabins is likely restricted, and particles, including bioaerosols, are removed from the cabin with the exhaust or return air (see section on ventilation practices in aircraft cabins in Chapter 2).

On the issue of aircraft ventilation with regard to airborne infectious agents, the 1986 NRC report recommended "that maximal airflow be used with full passenger complements to decrease the potential for microbial exposure and that recirculated air be filtered (to remove particles larger than 2-3 μm) to reduce microbial aerosol concentrations" (NRC 1986). The FAA does not require airlines to provide maximal air flow or filtration of recirculated air, but the latter has become common practice. Control of exposure to biological agents, except for bioeffluents as discussed in Chapter 2, is not a primary function of cabin ventilation. Nevertheless, increasing outside air ventilation or the amount of filtered recirculated air would decrease the mathematical probability of disease transmission in aircraft cabins by diluting airborne viruses or bacteria in a larger volume of microorganism-free air (see Equation 2-1). Although increasing cabin ventilation rates or changing the air mixing patterns are unlikely to prevent the transmission of infectious agents entirely, aircraft cabins should be provided with at least the minimal recommended supply of outside air whenever passengers are on board. If adequate ventilation cannot be provided, passengers should be deplaned and moved to a better-ventilated location, such as an airport terminal (NRC 1986; IEH 2001).

Recirculation of Cabin Air and Exposure to Infectious Agents

The practice of recirculating some cabin air has been questioned with regard to the transmission of infectious agents. Several studies measured higher concentrations of bacteria on aircraft that recirculated air (Nagda et al. 1989, 1992; Spengler et al. 1997), but no significant difference was observed in another study (ATA 1994). The higher bacterial concentrations could be due to a lower supply of outside air or low-efficiency or no filtration of return air on the aircraft that recirculated air. The seven investigations of possible transmission of *M. tuberculosis* on aircraft (Table 4-3) found no evidence that air recirculation facilitated transmission of the bacterium (CDC 1995; WHO 1998a). Similar rates of post-flight upper respiratory tract infections have been seen in travelers on flights with and without air recirculation, 18.5% (108 of 584 passengers) and 20.7% (106 of 516 passengers), respectively (J. Nutik Zitter, personal communication). Therefore, as has been seen in other public transportation vehicles, spread of airborne infectious agents in aircraft cabins appears to be limited to droplet and droplet nuclei transmission within close proximity (Rogers 1962; Sacks et al. 1985; Lodi 1994; Moore et al. 1999; Witt 1999).

Treatment of Recirculated Air and Exposure to Infectious Agents

Some particles as small as single viruses pass through even HEPA filters because some viruses are near 0.3 μm in size, but the fraction of the total number of such particles that may penetrate a filter is negligible (see the discussion of recirculation in Chapter 2) (ASHRAE 2000). The use of ultraviolet germicidal irradiation (UVGI) for air disinfection has been suggested for aircraft cabins (Hendley 1987; Hall et al. 2000). Irradiation of return air to inactivate infectious agents in theory may be beneficial if an environmental control system (ECS) cannot accommodate a filter to treat recirculated air, but that method of air disinfection has not been demonstrated in aircraft (Slater et al. 1995). Use of UVGI in place of high-efficiency filtration is not recommended because irradiation would not remove the allergenic or toxic properties of biological particles. Although the potential benefit of the combined use of UVGI and HEPA filtration is not known (CDC 1994), any reduction in the concentration of viable airborne infectious agents beyond that provided by a HEPA filter would likely be insignificant and the survival time of microorganisms on filters is so short that killing them before or after collection on a filter is unnecessary. Use of germicidal lamps in the occupied space of an aircraft directly to irradiate cabin air, as is done in some high-risk health-care environments (CDC 1994), is not practical, because of the small volume of air that could be irradiated and necessary constraints on lamp placement to avoid direct human exposure to UVGI.

CONCLUSIONS

- A person's risk of acquiring an infection on an aircraft depends on several factors, such as the presence of an infectious person and release of infectious agents by that person, the ventilation rate and mixing of cabin air, the amount of air that is recirculated and how it is treated, proximity to the source person, duration of exposure, and susceptibility to the specific infectious agents. These factors could also increase inhalation exposure to allergens and other potentially hazardous biological materials generated by passengers and activities within aircraft cabins.
- The proper design, operation, and maintenance of an aircraft ventilation system can limit but not eliminate the transmission of infectious agents and exposure to other biological agents on aircraft. Exposure to biological agents is increased when people are confined in an aircraft cabin without adequate ventilation.

Aeroallergens, Toxins, and Biological Irritants

- Exposure of passengers to aeroallergens, toxins, and biological irritants from outdoor air on grounded aircraft likely would be similar to what they would encounter in an airport terminal or while traveling to and from an airport.
- During flight, biological agents in aircraft cabins arise almost exclusively from inside sources rather than from outside air that is introduced through the ventilation system. People are a primary source of skin, scalp, nasal, and oral bacteria on aircraft, whereas most of the fungi and environmental bacteria in the cabin environment enter with outside air on the ground or are carried in by the occupants (e.g., on their shoes, clothing, or hand luggage).
- Available bioaerosol data are of little value for the assessment of the quality of cabin air and the estimation of the magnitude of the health risks that may be associated with bioaerosol exposures (acute hypersensitivity and infectious disease). The need for measurements of inflammatory, irritant, or toxic biological agents (e.g., bacterial endotoxin or fungal toxins or glucans) can be determined better when the results of current studies become available.
- Dust samples from commercial aircraft cabins have been analyzed for allergens and cat and dust mite allergens have been detected. There is limited evidence that individuals allergic to peanuts will respond to inhalation of, contact with, or accidental ingestion of airborne peanut allergens in the cabin environment.

Infectious Agents

- Infectious agents can be transmitted from person to person aboard aircraft on the ground and during flight.
- It is known that passengers and crew members with common respiratory infections occasionally travel while infectious. Passengers and cabin crew are exposed fairly often to the agents of common viral and bacterial infections and less often to the agents of more serious respiratory infections.

RECOMMENDATIONS

Aeroallergens, Toxins, and Biological Irritants

- FAA and the airlines should work with the medical community to evaluate whether the presence of animals (other than service animals) in passenger cabins on commercial aircraft can lead to allergic reactions in hypersensitive individuals.

• FAA should conduct research to determine whether there is a potential for severe reactions by passengers with peanut allergies exposed to airborne peanut particles on aircraft. Until such research can be completed, airlines should consider complying with requests that peanuts not be served to passengers next to those passengers with peanut allergies.

• Cabin crew should be trained to recognize and respond to the severe, potentially life-threatening responses (e.g., anaphylaxis or severe asthma attacks) that hypersensitive people may experience from exposures to airborne allergens.

Infectious Agents

• Physicians treating persons with communicable diseases should discuss the advisability of air travel with them and should instruct their patients to take appropriate precautions that will protect other persons from infectious agents. In addition, physicians treating persons who are unusually susceptible to infectious disease should inform their patients of the potential risks associated with air travel and of the appropriate precautions that may protect susceptible persons from infection.

• Health-care providers should obtain travel histories from persons with reportable infectious diseases and should notify public health authorities if a patient with a communicable disease recently has traveled. Public health authorities and air carriers should cooperate to notify air travelers as soon as possible if they may have been exposed to an infectious agent for which postexposure prophylaxis may reduce the risk or the severity of an infection. Airlines should ensure that electronic passenger manifests and contact information is preserved and readily available for a period of at least one month following disembarkation to facilitate timely identification and notification of passengers who may have been exposed to an infectious agent during air travel.

• Increased efforts should be made to provide cabin crew, passengers, and health professionals with information on health issues related to air travel. To that end, FAA and the airlines should work with federal and international agencies (such as the U.S. Public Health Service and WHO) and medical organizations (such as the American Medical Association, the Aerospace Medical Association, and the International Society of Travel Medicine) to improve the awareness of health professionals of the need to advise patients of the risks posed by air travel.

REFERENCES

Ahlbom, A., A. Backman, J. Bakke, T. Foucard, S. Halken, I.M. Kjellman, L. Malm, S. Skerfving, J. Sundell, and O. Zetterström. 1998. "NORDPET" pets indoors—A risk factor for or protection against sensitisation/allergy. Indoor Air 8(4):219-235.

Aintablian, N., P. Walpita, and M.H. Sawyer. 1998. Detection of Bordetella pertussis and respiratory synctial virus in air samples from hospital rooms. Infect. Control Hospital Epidemiol. 19(12):918-923.

Amler, R.W., A.B. Bloch, W.A. Orenstein, K.J. Bart, P.M. Turner Jr, and A.R. Hinman. 1982. Imported measles in the United States. JAMA 248(17):2219-2133.

Ammann, H.M. 1999. Microbial volatile organic compounds. Pp. 26.1-26.17 in Bioaerosols: Assessment and Control, J.M. Macher, H.M. Ammann, H.A. Burge, D.K. Milton, and P.R. Morey, eds. Cincinnati, OH: American Conference of Government Industrial Hygienists.

Arlian, L.G. 1999. House dust mites. Pp. 22.1-22.9 in Bioaerosols: Assessment and Control, J.M. Macher, H.M. Ammann, H.A. Burge, D.K. Milton, and P.R. Morey, eds. Cincinnati, OH: American Conference of Government Industrial Hygienists.

ASHRAE(American Society of Heating, Refrigerating and Air-Conditioning Engineers). 2000. Air cleaners for particulate contaminants in 2000 ASHRAE Handbook: Heating, Ventilating, and Air-Conditioning Systems and Equipment. Atlanta, GA:　American Society of Heating, Refrigerating and Air-Conditioning Engineers, Inc.

ASHRAE/CSS (American Society of Heating Refrigerating and Air-conditioning Engineers and /Consolidated Safety Services). 1999. Relate Air Quality and Other Factors to Symptoms Reported by Passengers and Crew on Commercial Transport Category Aircraft. Final Report. ASHRAE Research Project 957-RP. Results of Cooperative Research Between the American Society of Heating, Refrigerating and Air-Conditioning Engineers, Inc., and Consolidated Services, Inc. February 1999.

ATA (Air Transport Association of America). 1994. Airline Cabin Air Quality Study. Submitted to: Air Transport Association of America, Washington, DC. April 1994.

Batterman, S.A. 1995. Sampling and analysis of biological volatile organic compounds. Pp. 249-268 in Bioaerosols, H.A. Burge, ed. Boca Raton: Lewis.

Bellanti, J.A., and D.B. Wallerstedt. 2000. Allergic rhinitis update: epidemiology and natural history. Allergy Asthma Proc. 21(6):367-370.

Bodnar, U.R., K.L. Fielding, A.W. Navin, S.A. Maloney, M.S. Cetron, C.B. Bridges, K. Fukuda, and J.C. Butler. 1999. Preliminary Guidelines for the Prevention and Control of Influenza-Like Illness Among Passengers and Crew Members on Cruise

Ships. Division of Quarantine, National Center for Infectious Diseases, Centers for Disease Control and Prevention, Atlanta, GA. August.

BRE (British Research Establishment, Environment Division). 2001. Study of Possible Effects on Health of Aircraft Cabin Environments- Stage 2. British Research Establishment, Environment Division, Garston, Watford, UK. [Online]. Available: http://www.aviation.dtlr.gov.uk/healthcab/aircab/index.htm [posted September 2001].

Brook, I. 1985. Bacterial flora of airline headset devices. Am. J. Otolaryngol. 6(2):111-114.

Brook, I., and W.E. Jackson. 1992. Changes in the microbial flora of airline headset devices after their use. Laryngoscope. 102(1):88-89.

Burge, H.A., W.R. Solomon, and J.R. Boise. 1980. Microbial prevalence in domestic humidifiers. Appl. Environ. Microbiol. 39(4):840-844.

Burton, N.C., and R.E. McCleery. 2000. Exposure potentials during cleaning, overhauling and repairing of aircraft lavatory tanks and hardware. Appl. Occup. Environ. Hyg. 15(11):803-808.

Buxton, J.A., D.M. Skowronski, H. Ng, S.A. Marion, Y. Li, A. King, and J. Hockin. 2001. Influenza revaccination of elderly travelers: antibody response to single influenza vaccination and revaccination at 12 weeks. J. Infect. Dis. 184(2):188-191.

CDC (Centers for Disease Control and Prevention). 1983. Epidemiological notes and reports. Interstate importation of measles following transmission in an airport - California, Washington, 1982. MMWR 32(16):210, 215-216.

CDC (Centers for Disease Control and Prevention). 1994. Emerging infectious diseases. Detection of notifiable diseases through surveillance for imported plague-New York, September-October 1994. MMWR 43(44):805-807.

CDC (Centers for Disease Control and Prevention). 1995. Exposure of passengers and flight crew to Mycobacterium tuberculosis on commercial aircraft, 1992-1995. MMWR 44(08):137-140.

CDC (Centers for Disease Control and Prevention). 1998. Measles, mumps, and rubella - vaccine use and strategies for elimination of measles, rubella, and congenital rubella syndrome and control of mumps: recommendations of the Advisory Committee on Immunization Practices (ACIP). MMWR. 47(RR-8):1-59.

CDC (Centers for Disease Control and Prevention). 1999. Vaccine-preventable diseases: improving vaccination coverage in children, adolescents, and adults. A report on recommendations of the Task Force on Community Preventive Services. MMWR 48(RR-8):1-15.

CDC (Centers for Disease Control and Prevention). 2000. Prevention and control of meningococcal disease: recommendations of the Advisory Committee on Immunization Practices (ACIP). MMWR. 49(RR07):1-10.

CDC (Centers for Disease Control and Prevention). 2001a. Public health dispatch: Update: assessment of risk for meningococcal disease associated with the Hajj 2001. MMWR 50(12):221-222.

CDC (Centers for Disease Control and Prevention). 2001b. Health Information for International Travel 2001-2002. Atlanta, GA: U.S. Dept. of Health and Human services, Public Health Services, Centers for Disease Control and Prevention, National Center for Infectious Disease, Division of Quarantine.

CDC (Centers for Disease Control and Prevention). 2001c. Exposure to patients with meningococcal disease on aircrafts - United States, 1999-2001. MMWR. 50:485-488.

CDC (Centers for Disease Control and Prevention). 2001d. Prevention and control of influenza. MMWR 50(RR04):1-46.

CDC (Centers for Disease Control and Prevention). 2001e. Guidelines for the Management of Airline Passengers Exposed to Meningococcal Disease. [Online]. Available: http://www.cdc.gov/travel/menin-guidelines.htm. [June 2001].

Cetron, M., J. Keystone, D. Shlim, and R. Steffen. 1998. Travelers' health. Emerg. Infect. Dis. 4(3):405-407.

Chapman, M.D. 1995. Analytical methods: immunoassays. Pp. 235-248 in Bioaerosols, H.A. Burge, ed. Boca Raton: Lewis.

Chew, G.L., H.A. Burge, D.W. Dockery, M.L. Muilenberg, S.T. Weiss, and D.R. Gold. 1998. Limitations of a home characteristics questionnaire as a predictor of indoor allergen levels. Am. J. Respir. Crit. Care Med. 157(5 Pt1):1536-1541.

Chin, J. 2000. Control of Communicable Diseases Manual: An Official Report of the American Public Health Association, 17th Ed. Washington, DC: American Public Health Association.

Clayton, A.J., D.C. O'Connell, R.A. Gaunt, and R.E. Clarke. 1976. Study of the microbiological environment within long- and medium-range Canadian Forces aircraft. Aviat. Space Environ. Med. 47(5):471-482.

Cooper, R.C., and R.E. Danielson. 1992. The Occurrence of Legionellaceae in Aircraft Water Supply and Humidifiers. UCB/SEEHRL No. 92-1. Sanitary Engineering and Environmental Health Research Laboratory (SEEHRL), College of Engineering, School of Public Health, University of California, Berkeley.

Crespo, J.F., C. Pascual, C. Dominguez, I. Ojeda, F.M. Munoz, and M.M. Esteban. 1995. Allergic reactions associated with airborne fish particles in IgE-mediated fish hypersensitive patients.
Allergy 50(3):257-261.

CSTE (Council of State and Territorial Epidemiologists). 2001. Guidelines for management of contacts of a patient with meningococcal disease who has recently traveled by airline. [Online]. Available: http://www.cste.org/ps/ 2000/2000-id-02.htm. [June 2001].

Custovic, A., H. Woodcock, M. Craven, R. Hassall, E. Hadley, A. Simpson, and A. Woodcock. 1999a. Dust mite allergens are carried on not only large particles. Pediatr. Allergy Immunol. 10(4):258-260.

Custovic, A., B. Simpson, A. Simpson, C. Hallam, M. Craven, and A. Woodcock. 1999b. Relationship between mite, cat, and dog allergens in reservoir dust and ambient air. Allergy 54(6):612-616.

D'Amato, G., R. Liccardi, M. Russo, D. Barber, M. D'Amato, and J. Carreira. 1997. Clothing is a carrier of cat allergens. J. Allergy Clin. Immunol. 99(4):577-578.

Danielson, R.E., and R.C. Cooper. 1992. Distribution of Legionellaceae in Aircraft Water and Humidified Air. University of California, Environmental Engineering and Health Sciences Laboratory, College of Engineering and School of Public Health, Berkeley, CA.

Dechow, M. 1996. Airbus Cabin Air Quality—Only the Best! FAST Airbus Technical Digest, December 1996.

Dechow, M. 2001. Response from Airbus to NAS questions following 3 January 2001 public session in Washington, DC. [unpublished].

Dechow, M., H. Sohn, and J. Steinhaus. 1997. Concentrations of selected contaminants in cabin air of airbus aircraft. Chemosphere 35(1-2):21-31.

De Lucca, S.D., T.J.O'Meara, and E.R. Tovey. 2000. Exposure to mite and cat allergens on a range of clothing items at home and the transfer of cat allergen in the workplace. J. Allergy Clin. Immunol. 106(5):874-879.

Dharmage, S., M. Bailey, J. Raven, T. Mitakakis, A. Cheng, D. Guest, J. Rolland, A. Forbes, F. Thien, M. Abramson, and E.H. Walters. 2001. Current indoor allergen levels of fungi and cats, but not house dust mites, influence allergy and asthma in adults with high dust mite exposure. Am. J. Respir. Crit. Care Med. 164(1):65-71.

DOT (Department of Transportation). 1987. Airline Cabin Air Quality. Report to Congress. Washington, DC: Department of Transportation, Federal Aviation Administration. February.

Driver, C.R., S.E. Valway, W.M. Morgan, I.M. Onorato, and K.G. Castro. 1994. Transmission of Mycobacterium tuberculosis associated with air travel. JAMA. 272(13):1031-1035.

Duffy, T.P. 1993. Clinical problem-solving. The sooner the better. N. Eng. J. Med. 329(1):710-713.

Dumyahn, T.S., J.D. Spengler, H.A. Burge, and M. Muilenburg. 2000. Comparison of the environments of transportation vehicles: results of two surveys. Pp. 3-25 in Air Quality and Comfort in Airliner Cabins, N. Nagda, ed. West Conshohocken, PA: American Society for Testing and Materials.

Echavarria, M., S.A. Kolavic, S. Cersovsky, F. Mitchell, J.L. Sanchez, C. Polyak, B.L. Innis, and L.N. Binn. 2000. Detection of adenoviruses (AdV) in culture-negative environmental samples by PCR during an AdV-associated respiratory disease outbreak. J. Clin. Microbiol. 38(8):2982-2984.

Enberg, R.N., S.M. Shamie, J. McCullough, and D.R. Ownby. 1993. Ubiquitous presence of cat allergen in cat-free buildings: probable dispersal from human clothing. Ann. Allergy 70(6): 471-474.

Fischer, G., T. Muller, M. Moller, R. Ostrowski, and W. Dott. 1999. MVOC of fungiuse as an indicator for exposure level. [in German]. Schriftenr. Ver. Wasser Boden Lufthyg. 104:183-192.

Fritz, C.L., D.T. Dennis, M.A. Tipple, G.L. Campbell, C.R. McCance, and D.J. Gubler. 1996. Surveillance for pneumonic plague in the United States during an international emergency: a model for control of imported epidemic diseases. Emerg. Infect. Dis. 2(1):30-36.

Gelber, L.E., L.H. Seltzer, J.K. Bouzoukis, S.M. Pollart, M.D. Chapman, and T.A. Platts-Mills. 1993. Sensitization and exposure to indoor allergens as risk factors for asthma among patients presenting to hospital. Am. Rev. Respir. Dis. 147(3):573-578.

Glines, C.V. 1991. Should long-range aircraft be humidified? Air Line Pilot. 60(3):30-32, 50.

Grainger, C.R., M.J. Young, and H.H. John. 1995. A code of practice on dealing with infectious diseases on aircraft. J. R. Soc. Health 115(3):175-177.

Hall, R.J., J.J. Sangiovanni, H.H. Hollick, T.N. Obee, and S.O. Hay. 2000. Design of air purifiers for aircraft passenger cabins based on photocatalytic oxidation technology. Pp. 135-160 in Air Quality and Comfort in Airliner Cabins, N.L. Nagda, ed. West Conshohocken, PA: American Society for Testing and Materials.

Hendley, J.O. 1987. Risk of acquiring respiratory tract infection during air travel. JAMA. 258(19):2764.

Hernandez, M., and M. Swartz. 2000. A Review of Sampling and Analysis Methods for Assessing Airborne Microbiological Contamination on Commercial Aircraft. A Literature Survey and Review. Draft Report to NIOSH per Order #0009936697, Requisition 9938VQC.

Hourihane, J.O'B., S.A. Kilburn, J.A. Nordlee, S.L. Hefle, S.L. Taylor, and J.O. Warner. 1997. An evaluation of the sensitivity of subjects with peanut allergy to very low doses of peanut protein: a randomized, double-blind, placebo-controlled food challenge study. J. Allergy Clin. Immunol. 100(5):596-600.

Hung, L.L., C.S. Yang, F.J. Dougherty, F.A. Lewis, F.A. Zampiello, and L. Mangiaracina. 1993. Dust mite and cat dander allergens in office buildings in the mid-Atlantic region. Pp. 163-170 in Environments for People. Proceedings of IAQ92 Conference, San Francisco, California. Atlanta, GA: American Society of Heating, Refrigerating and Air-Conditioning Engineers, Inc.

IEH (Institute for Environment and Health). 2001. Consultation on the Possible Effects on Health, Comfort and Safety on Aircraft Cabin Environments. IEH Web Report W5. Leicester, UK: Institute for Environment and Health. [Online]. Available: http://www.le.ac.uk/ieh/webpub.html [posted March 2001]

IOM (Institute of Medicine). 1993a. Assessing exposure and risk. Pp. 185-205 in Indoor Allergens: Assessing and Controlling Adverse Health Effects, A.M. Pope, R. Patterson, and H. Burge, eds. Washington, DC: National Academy Press.

IOM (Institute of Medicine). 1993b. Magnitude and dimensions of sensitization and disease caused by indoor allergens. Pp. 44-85 in Indoor Allergens: Assessing and Controlling Adverse Health Effects, A.M. Pope, R. Patterson, and H. Burge, eds. Washington, DC: National Academy Press.

IOM (Institute of Medicine). 2000. Executive summary. Pp. 1-18 in Clearing the Air: Asthma and Indoor Air Exposures. Washington, DC: National Academy Press.

James, J.M. 1999. Airline snack foods: tension in the peanut gallery. J. Allergy Clin. Immunol. 104(1):25-27.

Janczewski, J. 2001. Airline Cabin Air Quality Study. Consolidated Safety Services, Inc. Presentation to the NRC Committee on Air Quality in Passenger Cabins on Commercial Aircraft, January 3-4, 2001, Washington, DC.

Janko, M., D.C. Gould, L. Vance, C.C. Stengel, and J. Flack. 1995. Dust mite allergens in the office environment. Am. Ind. Hyg. Assoc. J. 56(11):1133-1140.

Jones, R.T., D.B. Stark, G.L. Sussman, S. Waserman, and J.W. Yunginger. 1996. Recovery of peanut allergens from ventilation filters of commercial airliners. J. Allergy Clin. Immunol. 97(1 Part 3):423.

Kenyon, T.A., S.E. Valway, W.W. Ihle, I.M. Onorato, and K.G. Castro. 1996. Transmission of multidrug-resistant Mycobacterium tuberculosis during a long airplane flight. N. Engl. J. Med. 334(15):933-938.

Kim, J. 1994. Atmospheric environment of bioaerosols. Pp. 28-67 in Atmospheric Microbial Aerosols: Theory and Applications, B. Lighthart, and A.J. Mohr, eds. New York: Chapman & Hall.

Klontz, K.C., N.A. Hynes, R.A. Gunn, M.H. Wilder, M.W. Harmon, and A.P. Kendal. 1989. An outbreak of influenza A/Taiwan/1/86 (H1N1) infections at a naval base and association with airplane travel. Am. J. Epidemiol. 129(2):341-348.

Lacey, J., and J. Venette. 1995. Outdoor air sampling techniques. Pp. 407-471 in Bioaerosols Handbook, C.S. Cox, and C.M. Wathes, eds. Boca Raton, FA: Lewis.

Lee, S.C., C.S. Poon, X.D. Li, and F. Luk. 1999. Indoor air quality investigation on commercial aircraft. Indoor Air 9(3):180-187.

Lee, S.C., C.S. Poon, X.D. Li, F. Luk, M. Chang, and S. Lam. 2000. Air quality measurements on sixteen commercial aircraft. Pp. 45-58 in Air Quality and Comfort in Airliner Cabins, N.L. Nagda, ed. West Conshohocken, PA: American Society for Testing and Materials.

Lighthart, B., and L.D. Stetzenbach. 1994. Distribution of microbial bioaerosols. Pp. 68-98 in Atmospheric Microbial Aerosols: Theory and Applications, B. Lighthart, and A.J. Mohr, eds. New York: Chapman & Hall.

Lodi Tuberculosis Working Group. 1994. A school- and community-based outbreak of Mycobacterium tuberculosis in Northern Italy, 1992-3. Epidemiol. Infect. 113(1):83-93.

Logan, N.A., and C.B. Turnbull. 1999. Bacillus and recently derived genera. Pp. 357-369 in Manual of Clinical Microbiology, 7th Ed., P.R. Murray, E.J. Baron, M.A. Pfaller, F.C. Tenover, and R.H. Yolken, eds. Washington, DC: ASM Press.

Lundblad, F.P. 1991. House dust mite allergy in an office building. Appl. Occup. Environ. Hyg. 6(2):94-96.

MacNeil, L., T. Kauri, and W. Robertson. 1995. Molecular techniques and their potential application in monitoring the microbiological quality of indoor air. Can. J. Microbiol. 41(8):657-665.

Maloney, S.A., and M.S. Cetron. 2001. Investigation and management of infectious diseases on international conveyances (airplanes and cruise ships). Pp. 519-530 in Textbook of Travel Medicine and Health, 2nd Ed., H.L. DuPont, and R. Steffen, eds. Hamilton, Ontario: BC Decker.

Masterton, R.G., and A.D. Green. 1991. Dissemination of human pathogens by airline travel. Soc. Appl. Bacteriol. Symp Ser. 20:31S-38S.

Mastorides, S.M. R.L. Oehler, J.N. Greene, J.T. Sinnott 4th, M. Kranik, and R.L. Sandin. 1999. The detection of airborne Mycobacterium tuberculosis using micropore membrane air sampling and polymerase chain reaction. Chest 115(1):19-25. Comment in Chest 116(4):1143-1145.

McFarland, J.W., C. Hickman, M. Osterholm, and K.L. MacDonald. 1993. Exposure to Mycobacterium tuberculosis during air travel. Lancet 342(8863):112-113.

Menzies, R, R. Tamblyn, P. Comtois, C. Reed, J. Pasztor, Y. St. Germaine, and F. Nunes. 1993. Case-control study of microenvironmental exposures to aero-allergens as a cause of respiratory symptoms-Part of the sick building syndrome (SBS) symptom complex. Pp. 201-210 in Environments for People, Proceedings of IAQ92 Conference, San Francisco, California. Atlanta, GA: American Society of Heating, Refrigerating and Air-Conditioning Engineers, Inc.

Miller, M.A., S.E. Valway, and I.M. Onorato. 1996. Tuberculosis risk after exposure on airplanes. Tuber. Lung Dis. 77(5):414-419.

Milton, D.K. 1996. Bacterial endotoxins: a review of health effects and potential impact in the indoor environment. Pp. 179-195 in Indoor Air and Human Health, 2nd Ed., R.B. Gammage, and B.A. Berven, eds. Boca Raton: CRC.

Mohr, A.J. 1997. Fate and transport of microorganisms in air. Pp. 641-650 in Manual of Environmental Microbiology, C.J. Hurst, G.R. Knudson, M.J. McInerney, L.D. Stetzenbach, M.V. Walter, eds Washington, DC: ASM Press.

Montealegre, F., and M. Bayona. 1996. An estimate of the prevalence, severity and seasonality of asthma in visitors to a Ponce shopping center. P. R. Health Sci. J.15(2):113-117.

Moore, M., K.S. Fleming, and L. Sands. 1996. A passenger with pulmonary/laryngeal tuberculosis: no evidence of transmission on two short flights. Aviat. Space Environ. Med. 67(11):1097-1100.

Moore, M., S.E. Valway, W. Ihle, and I.M. Onorato. 1999. A train passenger with pulmonary tuberculosis: evidence of limited transmission during travel. Clin. Infect. Dis. 28(1):52-56.

Moritz, M., H. Schleibinger, and H. Ruden. 1998. Investigations on the survival time of outdoor microorganisms on air filters. Zentralbl. Hyg. Umweltmed. 201(2):125-133.

Moser, R.M., T.R. Bender, H.S. Margolis, G.R. Noble, A.P. Kendal, and D.G. Ritter. 1979. An outbreak of influenza aboard a commercial airliner. Am. J. Epidemiol. 110(1):1-6.

Muilenberg, M.L. 1995. The outdoor aerosol. Pp. 163-204 in Bioaerosols, H.A. Burge, ed. Boca Raton: Lewis.

Murphy, B.R., and R.G. Webster. 1996. Orthomyxoviruses. Pp. 1397-1445 in Fields Virology, 3rd Ed., B.N. Fields, D.M. Knipe, P.M. Howley, R.M. Chanock, T.P. Monath, J.L. Melnick, B. Roizman, and S.E. Straus, eds. Philadelphia, PA: Lippincott-Raven.

Nagda, N.L., M.D. Fortmann, M.D. Koontz, S.R. Baker, and M.E. Ginevan. 1989. Airliner Cabin Environment: Contaminant Measurements, Health Risks, and Mitigation Options. DOT-P-15-89-5. NTIS/PB91-159384. Prepared by GEOMET Technologies, Germantown, MD, for the U.S. Department of Transportation, Washington DC.

Nagda, N.L., M.D. Koontz, A.R. Konheim, and S.K. Hammond. 1992. Measurement of cabin air quality aboard commercial airliners. Atmos. Environ. Part A Gen. Top.26(12):2203-2210.

NRC (National Research Council). 1986. Pp. 1-12, 152-160 in The Airliner Cabin Environment: Air Quality and Safety. Washington, DC: National Academy Press.

Ormstad, H., E. Namork, P.I. Gaarder, and B.V. Johansen. 1995. Scanning electron microscopy of immunogold labeled cat allergens (Fel d 1) on the surface of airborne house dust particles. J. Immunol. Methods 187(2):245-251.

Parmet, A.J. 1999. Tuberculosis on the flight deck. Aviat. Space. Environ. Med. 70(8):817-818.

Pasanen, A.L., J. Keinanen, P. Kalliokoski, P.I. Martikainen, and J. Ruuskanen. 1993. Microbial growth on respirator filters from improper storage. Scand. J. Work Environ. Health 19(6):421-425.

Pierce, W., J. Janczewski, B. Roethlisberger, and M. Janczewski. 1999. Air quality on commercial aircraft. ASHRAE J. (Sept.):26-34.

Platts-Mills, T.A., and M.C. Carter. 1997. Asthma and indoor exposure to allergens. N. Engl. J. Med. 336(19):1382-1384.

Platts-Mills, T.A., W.R. Thomas, R.C. Aalberse, D. Vervloet, and M.D. Chapman. 1992. Dust mite allergens and asthma: report of a second international workshop. J. Allergy Clin. Immunol. 89(5):1046-1060.

Polasani, R., L. Melgar, R.E. Reisman, and M. Ballow. 1997. Hot dog vapor-induced status asthmaticus. Ann. Allergy Asthma Immunol.78(1):35-36.

Powell, G.S. 1994. Allergens. Pp. 458-475 in Physical and Biological Hazards of the Workplace, P.H. Wald, and G.M. Stave, eds. New York: Van Nostrand Reinhold.

Rayman, R.B. 1997. Passenger safety, health, and comfort: a review. Aviat. Space Environ. Med. 68(5):432-440.

Rogers, E.F.H. 1962. Epidemiology of an outbreak of tuberculosis among school children. Public Health Rep. 77(5):401-409.

Rylander, R. 1994. Endotoxins. Pp. 73-79 in Organic Dusts: Exposure, Effects, and Prevention, R. Rylander, and R.R. Jacobs, eds. Boca Raton: Lewis.

Sacks, J.J., E.R. Brenner, D.C. Breeden, H.M. Anders, and R.L. Parker. 1985. Epidemiology of a tuberculosis outbreak in a South Carolina junior high school. Am. J. Public Health. 75(4):361-365.

Sato, K., T. Morishita, E. Nobusawa, Y. Suzuki, Y. Miyazaki, Y. Fukui, S. Suzuki, and K. Nakajima. 2000. Surveillance of influenza viruses isolated from travellers at Nagoya International Airport. Epidemiol Infect. 124(3):507-514.

Schafer, M.P., J.E. Fernback, and P.A. Jensen. 1998. Sampling and analytical method development for qualitative assessment of airborne mycobacterial species of the Mycobacterium tuberculosis complex. Am. Ind. Hyg. Assoc. J. 59(8):540-546.

Shieh, Y.S., R.S. Baric, and M.D. Sobsey. 1997. Detection of low levels of enteric viruses in metropolitan and airplane sewage. Appl. Environ. Microbiol. 63(11):4401-4407.

Sicherer, S.H., A. Muñoz-Furlong, A.W. Burks, and H.A. Sampson. 1999a. Prevalence of peanut and tree nut allergy in the US determined by a random digit dial telephone survey. J. Allergy Clin. Immunol. 103(4):559-562.

Sicherer, S.H., T.J. Furlong, J. DeSimone, and H.A. Sampson. 1999b. Self-reported allergic reactions to peanut on commercial airliners. J. Allergy Clin. Immunol. 104(1):186-189.

Silvestri, M., G.A. Rossi, S. Cozzani, G. Pulvirenti, and L. Fasce. 1999. Age-dependent tendency to become sensitized to other classes of aeroallergens in atopic asthmatic children. Ann. Allergy Asthma Immunol. 83(4):335-340.

Slater, P.E., E. Anis, and A. Bashary. 1995. An outbreak of measles associated with a New York/Tel Aviv flight. Travel Med. Int. 13(3):92-95.

Spengler, J, H. Burge, T. Dumyahn, C. Dalhstrom, M. Muilenberg, and D. Milton. 1994. Aircraft Cabin Environmental Survey - Executive Summary. Department of Environmental Health, Harvard University School of Public Health, Boston, MA. May 16, 1994.

Spengler, J., H. Burge, T. Dumyahn, M. Muilenberg, and D. Forester. 1997. Environmental Survey on Aircraft and Ground-Based Commercial Transportation Vehicles. Prepared for the Commercial Airplane Group, The Boeing Company, by Harvard School of Public Health, Harvard University, Cambridge, MA. May 31, 1997.

Squillace, S.P. 1995. Allergens of arthropods and birds. Pp. 133-148 in Bioaerosols, H. Burge, ed. Boca Raton, FL: Lewis.

Suda, T., A. Sato, M. Ida, H. Gemma, H. Hayakawa, and K. Chida. 1995. Hypersensitivity pneumonitis associated with home ultrasonic humidifiers. Chest 107(3):711-717.

Tovey, E.R. A. Mahmic, and L.G. McDonald. 1995. Clothing-an important source of mite allergen exposure. J. Allergy Clin. Immunol. 96(6 Pt 1):999-1001.

Trudeau, W.L., and E. Fernández-Caldas. 1994. Identifying and measuring indoor biologic agents. J. Allergy Clin. Immunol. 94(2 Pt 2):393-400.

Wang, P.D. 2000. Two-step tuberculin testing of passengers and crew on a commercial airplane. Am. J. Infect. Control 28(3):233-238.

Waters, M .,T. Bloom, and B. Grajewski. 2001. Cabin Air Quality Exposure Assessment. National Institute for Occupational Safety and Health, Cincinnati, OH., Federal Aviation Administration Civil Aeromedical Institute. Presented to the

NRC Committee on Air Quality in Passenger Cabins of Commercial Aircraft, January 3, 2001, Washington, DC.

Weiss, E.L. 2001. Epidemiologic alert at international airports. Pp. 530-533 in Textbook of Travel Medicine and Health, 2nd Ed., H.L. DuPont, and R. Steffen, eds. Hamilton, Ontario: BC Decker.

Wenzel, R.P. 1996. Airline travel and infection. N. Engl. J. Med. 334(15):981-982.

WHO (World Health Organization). 1998a. Tuberculosis and Air Travel: Guidelines for Prevention and Control. WHO/TB98.256. Geneva: World Health Organization.

WHO (World Health Organization). 1998b. Control of Epidemic Meningococcal Disease. WHO Practical Guidelines, 2nd Ed. WHO/EMC/BAC/98.3. Geneva: World Health Organization.

Wick, R.L., and L.A. Irvine. 1995. The microbiological composition of airliner cabin air. Aviat. Space Environ. Med. 66(3):220-224.

Wickens, K., I. Martin, N. Pearce, P. Fitzharris, R. Kent, N. Holbrook, R. Siebers, S. Smith, H. Trethowen, S. Lewis, I. Town, and J. Crane. 1997. House dust mite allergen levels in public places in New Zealand. J. Allergy Clin. Immunol. 99(5):587-593.

Wilson, M.E. 1995. Travel and the emergence of infectious diseases. Emerg. Infect. Dis. 1(2):39-46.

Withers, M.R., and G.W. Christopher. 2000. Aeromedical evacuation of biological warfare casualties: a treatise on infectious diseases on aircraft. Mil. Med. 165(11 Suppl):1-21.

Witt, M.D. 1999. Trains, travel, and the tubercle. Clin. Infect. Dis. 28(1):57-58.

Wood, R.A., A.N. Laheri, and P.A. Eggleston. 1993. The aerodynamic characteristics of cat allergen. Clin. Exp. Allergy. 23(9):733-739.

5

Health Considerations Related to Chemical Contaminants and Physical Factors

The 1986 National Research Council (NRC) report on commercial airliner cabin air quality notes that information regarding the environmental characteristics (e.g., relative humidity and air pressure) and contaminants identified in surveys of airline cabin air "suggests a diverse set of adverse health effects that could arise from exposure to the cabin environment—from acute effects . . . to long-term effects."

Any consideration of health effects in the context of airline cabin air must distinguish between effects of exposures that result from the ambient environment encountered during boarding, waiting at the gate with the aircraft door open, and normal operation of the aircraft and effects of exposures that result from incidents during flight. Examples of the two categories of exposures are listed in Table 5-1. The myriad health complaints registered by flight crews and passengers are broad and nonspecific, and that makes it difficult to define or discern a precise illness or syndrome.

Among the many plausible explanations of the complaints are the flight environment (e.g., partial pressure of oxygen (PO_2) and relative humidity), chemical or biological contaminants, psychological and physiological stressors, and exacerbation of pre-existing medical conditions. (Biological agents are discussed in Chapter 4).

TABLE 5-1 Exposure Sources Relevant to Aircraft Cabin Air Quality

Exposures Related to Normal Operations of the Aircraft	Exposures Related to Incidents
Ozone	Carbon monoxide
Carbon dioxide	Smoke, fumes, mists, vapors from
Temperature	leaks of engine oils, hydraulic
Relative humidity	fluids, and deicing fluids and
Off-gassing from interior material and cleaning agents	their combustion products
Bioeffluents	
Personal-care products	
Allergens	
Infectious or inflammatory agents	
Ambient airport air	
Cabin pressure/partial pressure of oxygen	
Pesticides	
Jet exhaust fumes (runway)	
Alcohol	

FLIGHT ENVIRONMENT

Cabin Pressure

As discussed in Chapter 2, at cruise altitude, the aircraft cabin is typically pressurized to the equivalent of an altitude of 6,000-8,000 ft (1,829-2,438 m), with a corresponding barometric pressure of 609-564 mm Hg and an ambient PO_2 of 128-118 mm Hg (see Table 5-2). As specified in the Federal Aviation Regulation (FAR) 25.841, aircraft "cabin pressure altitude" must not exceed 8,000 ft at the aircraft's highest operating altitude (14 CFR 1986).

The reduced pressure in the cabin environment results in several physiological changes in the passengers and crew. Specifically, the reduced ambient air pressure will cause the gas in body cavities (e.g., middle ear, sinuses, and gastrointestinal tract) to expand in volume by as much as 25%. In the lungs, the lower PO_2 in ambient air will reduce the oxygen (O_2) pressure in the alveoli from the normal value of 105 mm Hg. That decrease will lower systemic arterial PO_2. In healthy people, the arterial PO_2 is usually about 5-10 mm Hg lower than the alveolar PO_2 (see Table 5-2). It is the arterial PO_2 that determines the amount of O_2 that is carried by the hemoglobin in the blood, expressed as the percent hemoglobin saturation.

TABLE 5-2 Barometric Pressure and PO_2

Altitude, ft (m) above sea level	Barometric Pressure, mm Hg (kPa)	Ambient PO_2, mm Hg (kPa)	Alveolar PO_2 (for healthy person at rest), mm Hg	Arterial PO_2 (for healthy person at rest), mm Hg
0	760 (101.3)	160 (21.3)	105	95-100
6,000 (1,828)	609 (81.2)	128 (17.1)	76	66-71
8,000 (2,438)	564 (75.2)	118 (15.7)	72	62-67

Source: Adapted from Slonim and Hamilton (1971).

Table 5-3 shows the arterial O_2 saturation at various altitudes without supplemental O_2. At an altitude of 7,500 ft (2,287 m), the PO_2 begins to approach the steep-slope portion of the O_2-hemoglobin dissociation curve (Figure 5-1). Further small changes in PO_2 in the cabin can lead to large changes in the O_2 content of the blood.

The relationship between arterial PO_2 and hemoglobin saturation is an S-shaped curve. As shown in Figure 5-1, at an arterial PO_2 greater than 60 mm Hg, hemoglobin is more than 90% saturated with O_2.

For healthy adults at sea level, hemoglobin is 95-97% saturated. Even at an altitude of 8,000 ft (2,439 m), the O_2 saturation of hemoglobin remains at least 90% for healthy adults at rest because their arterial PO_2 is above 60 mm Hg. Below arterial PO_2 of 60 mm Hg, there is a steep decline in the curve, over which a slight change in arterial PO_2 can lead to large changes in hemoglobin saturation. The significance of this curve with respect to the aircraft

TABLE 5-3 Hypobaric Pressure and Arterial O_2 Saturation

Pressure Altitude, ft	Atmospheric Pressure, mm Hg	PO_2, mm Hg[a]	Arterial O_2 Saturation Without Supplemental O_2, %
0	760	160	96
2,500	694	147	95
5,000	632	133	95
7,500	575	121	93
10,000	523	110	89

[a] 21% of atmospheric pressure.
Source: NRC (1986).

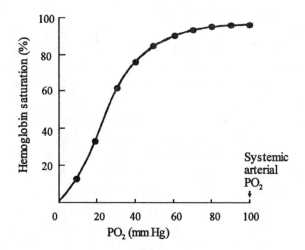

FIGURE 5-1 O_2-hemoglobin dissociation curve shows relationship between PO_2 in blood (X-axis) and amount of O_2 held by blood hemoglobin (Y-axis). Important feature of curve is sharp drop in O_2 content of hemoglobin when PO_2 falls below 60 mm Hg.

environment is that at cruise altitudes (reduced air pressure), alveolar and arterial PO_2 are reduced. However, the hemoglobin saturation decreases only slightly unless arterial PO_2 falls below 60 mm Hg. A drop in arterial PO_2 from 100 to 60 mm Hg will result in only a 10% drop in hemoglobin saturation.

In persons with chronic obstructive pulmonary disease (COPD) or asthma, there may be inadequate O_2 exchange between the air in the lungs and the blood. In this case, even at sea level, arterial PO_2 may be considerably lower than alveolar PO_2 (i.e., some affected persons may have less than 90% hemoglobin saturation when at rest; many more will experience this effect on exertion). The situation is worsened at higher altitudes or with exercise; under these conditions, hemoglobin saturation will be substantially lower than in healthy persons at sea level or in sedentary persons, respectively. For these people, lowering alveolar PO_2 by lowering ambient air pressure may also decrease the arterial PO_2 and substantially lower the amount of O_2 carried by the blood

The 1986 NRC report on airliner cabin environment summarized the effects of altitude on PO_2 and recommended that passengers with heart or lung disease be educated about the risks posed by flight. It also recommended that passengers with middle ear problems be told about the effects of cabin pressure in general. However, that committee had few direct data on the hemoglobin saturation that might be expected in passengers and cabin crew under normal flight conditions of commercial aircraft. Some studies have since examined the effects of cabin pressurization on PO_2 in humans. Real-time continuous monitoring of hemoglobin saturation in flight has indicated that considerable changes can occur in a given person and that there are considerable differences among people.

The effects of hypoxia have been studied in a number of situations, including combat flights and passenger transport (Ernsting 1978). Studies conducted during the 1940s suggested that the maximal acceptable degree of hypoxia in passenger aircraft corresponded to a cabin altitude of 8,000 ft, but it was recommended that under routine operating conditions cabin pressure altitude should not exceed 5,000-6,000 ft. The altitude of 8,000 ft was a compromise between the aircraft design and operation requirements and the human performance impairments. Studies by McFarland and Evans (1939), McFarland (1946), and others (Ernsting et al. 1962; Denison et al. 1966; Ledwith 1970) showed that mild hypoxia, as is found in subjects at 8,000 ft, might impair the learning of new tasks and the performance of complex tasks. Most of the studies were conducted on young, healthy men, primarily in the military, who were engaged in vigilance tasks. McFarland and Evans (1939) found an increase in the absolute brightness threshold of the dark-adapted eye at hypoxia equivalent to 7,400 ft but concluded that the change was so small that it was of no practical significance. Ernsting (1978) noted that a "reduction of the cabin altitude from 8,000 to 6,000 ft is associated with a lower incidence of otiotic barotrauma and disturbances in passengers with cardiorespiratory disease." Studies on performance deficit under hypoxic conditions—86% arterial oxyhemoglobin equivalent to 8,900 ft—showed no effect on performance, including night vision; the threshold for effects appeared to be 82% oxyhemoglobin (9,750 ft) (Fowler et al. 1987).

Cottrell et al. (1995) used continuously reading pulse oximeters to measure O_2 saturation in 38 pilots on 21 flights of about 4 h each. Pressure altitudes in the aircraft cockpits during the cruise portion of the flight were 6,000-9,000 ft (average, 7,610 ft). Maximal and minimal O_2 saturations were 95-99% (mean, 97% ± 1.1%) and 80-93% (mean, 88.6% ± 2.9%), respectively. Of the 38 subjects, 20 (53%) developed an O_2 saturation of less than 90% at some time during the flight (duration of time below 90% not given). Baseline O_2 saturation was 95-99%; at 6,000-7,000 ft, saturation was 87-92%; and at 8,000 ft or more, it was 80-91%. No symptoms were reported, and no attempt was made to measure performance decrement. There was no correlation between the minimal oxyhemoglobin saturation and age, height, weight, length of flight, maximal saturation, smoking history, peak cabin altitude, or whether the flight was during the day or night.

In a study of the effects of hypoxia on healthy infants, 34 babies (1-6 mo old; average 3.1 mo) were exposed to 15% O_2 in nitrogen for a mean duration of 6.3 h ± 2.9 h (Parkins et al. 1998). During exposure, there was a significant increase in the infants' heart rate and time spent in periodic apnea and a decrease in the amount of time spent in regular breathing; respiratory rates did not change significantly. Baseline O_2 saturation decreased from a median of 97.6% (range, 94-100%) in ambient air to 92.8% (range, 84.7-100%) in 15% O_2. Four of the infants had to be removed from the hypoxic conditions early because their O_2 saturation fell below 80% for more than 1 min. The authors suggest that exposure to reduced PO_2, like that encountered on high-altitude flights can result in hypoxia in some infants.

A newborn infant's red blood cells contain up to 77% fetal hemoglobin (Delivoria-Papadopoulos and Wagerle 1990). At 6 mo, the red cells still contain up to 4.7% fetal hemoglobin. Fetal hemoglobin binds O_2 with greater affinity than does adult hemoglobin (see Figure 5-2). That results in greater saturation for any given partial pressure, but it also means that at any given partial pressure fetal hemoglobin holds O_2 more than adult hemoglobin. The net effect is that less O_2 is available for tissue metabolic needs. Thus, infants who experience substantial falls in arterial O_2 saturation may be at greater risk for tissue hypoxia than adults at the same O_2 saturation.

With decreasing cabin pressure, air in the middle ear and sinuses expands and escapes via nostrils and ostia to the outside via the nasopharynx. On descent, increasing ambient pressure requires gas to re-enter the middle ear and sinuses through the same pathways. Both movements can happen only if air can move freely in either direction. If there is blockage due, for example, to an upper respiratory infection, allergy, or tumor, the free flow of air may be impeded, and earache, a feeling of sinus fullness, or dizziness can ensue, particularly on descent. In severe cases, the pressure differential can cause extreme pain, bleeding, or rupture of a tympanic membrane. Likewise, gas in the respiratory tract and gastrointestinal tract expands and contracts with decreasing and increasing cabin pressure. Although gas readily escapes from those parts of the body, minor stomach cramping or bloating might still occur.

At cabin altitudes during cruise, all crew and passengers are in a decreased-O_2 environment. It is clear that under ordinary conditions of commercial flight, PO_2 can be reduced substantially during rest or in situations of minimal exertion. As stated above, this relatively small decrement usually causes

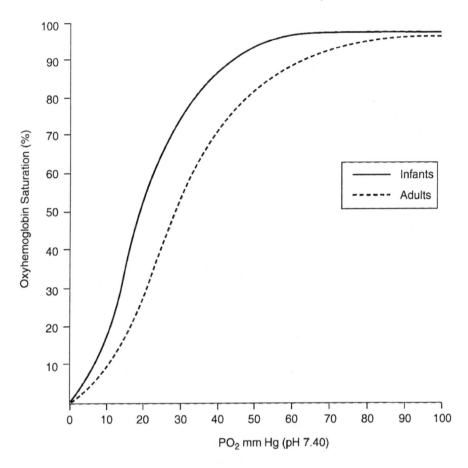

FIGURE 5-2 Oxyhemoglobin equilibrium curves of blood from infants and adults. Source: Adapted from Delivoria-Papadopoulos and Wagerle (1990).

no symptoms in healthy people because hemoglobin remains well saturated with O_2 at altitude. Nevertheless, some people might be sensitive to a lowering of PO_2 and experience various symptoms, including headache, dizziness, fatigue, numbness, and tingling (Sheffield and Heimbach 1996). The array of possible symptoms varies from person to person or even in the same person on different days. The PO_2 values are such that substantial reductions in arterial O_2 content occur and could pose a definite health risk for persons with underlying pulmonary or cardiac disease or untreated or partially treated anemia.

Relative Humidity

As noted in Chapters 1 and 2, the relative humidity in most aircraft cabins is low, typically from 10 to 20% (see Table 1-2) with an average of 15-19%, depending on the aircraft (Nagda et al. 2000). Although, low relative humidity is not an air contaminant, it can affect passenger and crew comfort and health (Rayman 1997). Low relative humidity may cause drying of the skin, mucous membranes, and conjunctivae in the latter case adding a risk for conjunctivitis with its symptoms of tearing and pain, especially in those wearing contact lenses (Eng 1979). Studies have indicated that passengers and cabin crews find the air in the aircraft cabin to be too dry and to lead to such symptoms as dry, itchy, or irritated eyes; dry or stuffy noses; and skin dryness or irritation (Lee et al. 2000). Stuffy, dry nose was the primary complaint in another study of 3,630 passengers in all cabin classes on standard and wide-body aircraft. Although symptoms could not be correlated with aircraft type (comfort rating, 5.34 of 7 for first-class versus 4.46 for coach and no difference in passenger comfort ratings for relative humidity between standard-body and wide-body aircraft), they did reflect flight length: longer flights resulted in more symptoms (Rankin et al. 2000). Low relative humidity has also been associated with fatigue, headaches, and nosebleeds (Space et al. 2000). Low humidity can have a greater effect on passengers who have respiratory infections, asthma, or tracheotomy (Rayman 1997). It has been found that in dry environments, mucus can concentrate; this can reduce ciliary clearance and phagocytic activities in the respiratory tract (Berglund 1998).

In a study of the effects of low humidity on the human eye, Laviana et al. (1988, as cited in Nagda and Hodgson 2001) found that at a relative humidity of either 10% or 30%, eye pain (described as scratchiness, pain, or burning) increased over time up to the fourth hour of exposure (total exposure duration of 10 h) for both a naked eye and an eye covered with a soft contact lens. However, there was no difference in the severity of responses at either 10% or 30% relative humidity and the humidity level did not significantly affect acuity, refractive error or cornea curvature in either eye. In aircraft cabins, symptoms of low humidity, such as eye and nasal irritation, seem to occur within 2 h after exposure begins (Eng 1979); skin symptoms may require at least 4 h to occur (Carleton and Welch 1971, as cited in Nagda and Hodgson 2001). Furthermore, all symptoms increase in severity with time, and an adaptive response to low-humidity environments is not evident (Nagda and Hodgson 2001). At reduced barometric pressures (259-700 mm Hg), symptoms associated with low humidity develop more quickly and are more severe (Carleton and Welch 1971, as cited in Nagda and Hodgson 2001). In a review of the studies that examined potential health effects of exposure to low relative humidity, Nagda and Hodgson (2001) indicated that the study subjects in general were relatively young and that an older study population might have been more likely to perceive changes in relative humidity and also might be more susceptible to health effects of such exposures. They concluded that a modest increase in the relative humidity in an aircraft cabin (e.g., from the current average of 14-19% to about 22-24%) might have beneficial effects similar to those seen in building studies in which a 10% increase in humidity alleviated many of the symptoms of "sick-building syndrome." Such humidities are below the values that may affect the safety of an aircraft—that is, that might cause condensation in and corrosion of the aircraft shell—or that would result in increased microbial growth.

Although cabin relative humidity is well below the preferred values of 30-60% suggested in American Society of Heating, Refrigerating and Air-Conditioning Engineers Standard 62-1999, it is questionable whether low humidity has substantial short- or long-term health effects. One investigator reported that in a room with 0% humidity, the degree of dehydration in undressed subjects was insignificant over a period of 7 h (Nicholson 1996). It might be expected that the adverse effects of low relative humidity experienced by crew and passengers will be temporary and will be alleviated when they leave the aircraft. The time required for rehydration will depend on individual physiology and the ambient environment.

Humidity influences the perception of air quality. In one study, as relative humidity increased from 24% to 79%, the air was considered to be less fresh (Berglund 1998). Acceptability of air-quality decreases with increasing temperature and relative humidity (Fang et al. 1998a,b).

CHEMICAL CONTAMINANTS OF CONCERN

During flight, passengers and crew can be exposed to a variety of air contaminants. Some of them are naturally occurring chemicals, such as ozone (O_3), which is found in greater concentrations in the upper troposphere and lower stratosphere; others result from incidents that suggest equipment failure; and still others result from preventive measures, such as the use of pesticides. The toxicity of these contaminants is discussed below.

Ozone

Although ground-level O_3 is a major contributor to photochemical air pollution in urban air, the presence of O_3 in the upper troposphere and lower stratosphere provides a necessary health benefit to humans by screening out harmful ultraviolet radiation. As discussed in Chapter 3, many commercial flight paths are at altitudes where O_3 concentration might be greater than those typically found at ground level. O_3 in the cabin is required not to exceed 0.25 ppm (250 ppb) during a flight (FAR 25.832[1]). Mean O_3 measured on aircraft has ranged from 22 ppb (Nagda et al. 1989) to 200 ppb (Waters 2001). The health effects of ground-level O_3 have been well studied (EPA 2001). Those effects are relevant to travelers who might be exposed to increased O_3 during high-altitude flights if the aircraft is not equipped with an O_3 converter or the equipment is not operating properly. The ground-level national ambient air-quality standard for O_3, established by the Environmental Protection Agency (EPA) in 1997, is 0.12 ppm by volume for a 1-h exposure and 0.08 ppm by volume for 8-h exposures (EPA 1997).

O_3 can cause acute respiratory problems, aggravate asthma, and impair the body's immune system making people more susceptible to respiratory illnesses, including bronchitis and pneumonia (EPA 1996). Exposures to O_3 as low as approximately 0.08-0.10 ppm for about 6 h have resulted in impairments of the immune system (EPA 1996). Inflammatory responses occurred within 1 h after a 1-h exposure to O_3 at 0.3 ppm or more. Exposures for up to 7 h to O_3 as low as 0.08 ppm caused small decrements in lung function and increases in respiratory symptoms (EPA 1996). Those effects are exacerbated by exercise. In a study of healthy adults exposed during exercise to ambient O_3 at 21-124 ppb, there was a statistically significant decrement in lung function (Spektor et al. 1988a). Similar lung function decrements were seen in healthy children (at a summer camp) exposed to O_3 at the ambient air standard of 120 ppb, where average forced vital capacity (FEV), forced expiratory volume in the first second (FEV$_1$), peak expiratory flow rate (PEFR), and forced expiratory flow (FEF$_{25-75}$) decrements were 4.9%, 7.7%, 17%, and 11%, respectively (Spektor et al. 1988b).

---

[1]FAR Section 25.832 Cabin ozone concentration: "(a) The airplane cabin ozone concentration during flight must be shown not to exceed: (1) 0.25 parts per million by volume, sea level equivalent, at any time above flight level 320, and (2) 0.1 parts per million by volume, sea level equivalent, time-weighted average during any 3-hour interval above flight level 270."

Asthmatics, particularly children, show an increase in respiratory symptoms and decrements in peak expiratory flow rate with increasing O_3 (EPA 1996). Studies on children with moderate-to-severe asthma found that increasing O_3 from 84 to 160 ppb for 1 h resulted in increased medication use and an increase in the number of chest symptoms (Thurston et al. 1997). The long-term sequelae of inhalation exposure are unknown, but continued exposure O_3 at low concentrations could result in chronic effects in humans (EPA 1996).

The committee was able to identify only one study that examined the possible effects of O_3 on flight attendants (Tashkin et al. 1983). The authors attempted to determine, on the basis of symptoms reported on a questionnaire distributed to flight attendants on high-altitude flights of Boeing 747SP aircraft, whether the symptoms were consistent with possible exposure to increased O_3. Although the authors concluded that the reported symptoms were consistent with exposure to toxic concentrations of O_3 (0.4-1.09 ppm, measured onboard other 747SP aircraft), no O_3 measurements were made on flights where the questionnaire was used. There were substantial limitations in the study, as discussed in Chapter 6.

Engine Oils and Hydraulic Fluids

Engine lubricating oils and hydraulic fluids are complex mixtures of primarily organic compounds (see Table 3-12 for major components). Among the major additives to the petroleum base of these mixtures are organophosphate compounds, of which those organophosphates, of greatest toxicological concern are tricresyl phosphate (TCP) derivatives and the generation of trimethylolpropane (TMPP). Other organophosphates that can be present in the oils and fluids are tributyl phosphate, dibutyl phenyl phosphate, and triphenyl phosphate. The general toxicity of these agents is discussed below.

Engine lubricating oils and hydraulic fluids have been reported to enter the passenger cabin of aircraft through the environmental control system (ECS) as discussed in Chapter 3. The fluids (and their possible pyrolysis products) can result in mists, fumes, vapors, and smoke in the cabin. The major pyrolysis products of jet engine oils and hydraulic fluids, at less than very high temperatures, will probably be volatile organic compounds (VOCs) and carbon monoxide (CO). Although there is considerable toxicological information on the major constituents of engine oils and hydraulic fluids, there are few data on the toxicity of the formulated oils and fluids themselves. Only a couple of studies were found in the published literature, and they are discussed below.

In an early study to determine whether the introduction of engine oils or hydraulic fluids or their combustion products into the ECS would impair aircraft crew, rats were exposed via inhalation to synthetic engine lubricating oils (Exxon turbo oil 2380 or Mobil II jet oil) or one synthetic hydraulic fluid (Skydrol 550B) (Crane et al. 1983). The authors concluded that for combustion of the engine oils at 400°C, the major toxic component was CO, whether or not combustion produced flames or only smoke. The hydraulic fluid Skydrol, under flaming conditions, was more toxic (the animals became incapacitated and died more quickly) than the engine oils, but it was not toxic under non-flaming (smoking) conditions.

In another study, the hydraulic fluid Skydrol 500B-4 (manufactured by Monsanto), containing primarily tributyl phosphate and dibutyl phenyl phosphate (80-90% phosphate esters) was tested with rats via inhalation. The animals were exposed for 6 h/d, 5 d/wk, to very high Skydrol concentrations (up to 300 mg/m^3, with about 95% of particles less than 10 μm in diameter) for up to 13 wk. Rats exhibited reddish nasal discharge and salivation, suggesting irritation. No pathological changes were seen in animals exposed at 100 mg/m^3 (the no-observed-adverse-effect level). Pathological changes at 300 mg/m^3 included increased liver weight with hepatocellular hypertrophy, decreased hematocrit in both males and females, and decreased plasma cholinesterase in females only (Healy et al. 1992). In other animal studies with organophosphate ester hydraulic fluids, rabbits exposed for less than 4 h/d, 5 d/wk, for 11 or 22 d, to aerosols of Cellulube 200 hydraulic fluid at 2,000 mg/m^3 died with severe dyspnea and mild diarrhea (Carpenter et al. 1959, as cited in ATSDR 1997).

Tricresyl Phosphate Esters

As noted in Table 3-12, engine oils and hydraulic fluids contain several phosphate esters as antiwear additives. Hydraulic fluids contain tributyl phosphate (at up to 80%), dibutyl phenyl phosphate, and butyl diphenyl phosphate and can also contain up to 1% TCP (Hewstone 1994). Engine lubricating oils contain TCP at 1-5% (typically 3%). The commercial TCP found in engine oils and hydraulic fluids is actually a mixture of aryl phosphates, including the ortho, meta, and para isomers and other *o*-cresol compounds. Although older TCP contained up to about 25-40% *o*-cresyl residues, currently manufactured TCP contains less than 0.3% *o*-cresyl residues (some reports indicate that *o*-cresyl isomers are present in TCP at 0.05-0.13%) (Goode 2000), and some

TCP mixtures contain almost no *o*-cresyls or *o*-xylenyls (Craig and Barth 1999).

TCP is a toxic mixture that can cause a wide array of transitory or permanent neurological dysfunction—including convulsions, flaccid paralysis, and polyneuropathy—although these serious sequelae are almost always secondary to ingestion. The committee did not identify any reports of human neuropathy following inhalation of TCP. Episodes of penetration of engine oil mists and fumes into the aircraft cabin might lead to dermal and inhalation exposure of passengers and crew members. Although the committee was unable to find any objective information that would substantiate and document frank neurotoxicity in persons exposed to such mists and fumes on aircraft, such a possibility cannot be a priori discounted if airborne concentrations of TCP are substantial and exposures long enough.

One of the most toxic components of TCP is tri-*o*-cresyl phosphate (TOCP). The primary consequence of TOCP ingestion is delayed neuropathy, characterized by degenerative changes in the axons ("dying-back neuropathy"), also known as OPIDN (organophosphate-induced delayed neurotoxicity or neuropathy), that generally occurs 1-2 wk after exposure (Baron 1981). Acute toxicity may be manifest immediately as nausea, vomiting, diarrhea, and abdominal pain and be followed by a long asymptomatic period (8-35 d), after which there is a bilaterally symmetrical degeneration of sensory and motor axons in peripheral nerves and spinal-cord tracts; fibers that are longest, and have the largest diameter tend to be the most affected. Signs and symptoms are paresthesias in the extremities (numbing and tingling), pain in the calves and legs, and absence of reflexes, mostly in the feet and legs; in severe cases of poisoning, all four limbs can be affected in the form of a "glove and stocking" distribution. Development of uncoordinated movements (ataxia) may occur at the same time and progress to flaccid paralysis. Recovery is usually poor; in time, flaccidity may be replaced with spasticity, reflecting some regeneration of the peripheral nerves, but with residual damage in the spinal cord.

Studies with experimental animals (primarily hens and cats) have shown that modern lubricating oils have comparatively low toxicity via ingestion and inhalation. An acute oral 5-g/kg dose of engine lubricating oil containing 3% TCP did not result in any clinical signs of delayed neurotoxicity in hens, the animals most sensitive to the neurotoxic action of TCP and related phosphate esters, after a 3-wk followup period; a repeat oral dose also produced no toxic or histopathological changes (Daughtrey et al 1990). Groups of 17-20 hens received a daily oral 1-g/kg body weight dose of engine lubricating oils containing either 3% TCP (the commercial TCP contained less than 1% TOCP), triphenyl phosphorothionate, or butylated triphenyl phosphate for 5 d/wk for 13 wk. The animals did not exhibit any inhibition of neuropathic target esterase (NTE) activity in the brain and spinal cord at 6 wk; at 13 wk, NTE was inhibited by 23-34% compared with the result in saline-treated hens; no clinical or neuropathological signs of OPIDN were evident at either 6 or 13 wk. Hens that received TOCP daily at 7.5 mg/kg (plus an oral dose of 500 mg/kg 12 d before the end of treatment) exhibited clinical impairment and lesions indicative of OPIDN at 6 and 13 wk (Daughtrey et al. 1996). The authors concluded that TCP should not pose a hazard to humans under exposure conditions that can realistically be expected in the handling of these materials.

Only one set of animal studies of the inhalation toxicity of TOCP was found by the committee (Siegel et al. 1965). One of four hens with continuous (23 h/d), whole-body exposure to hydraulic fluid mist containing not more than 1.5% TOCP (with trixylenyl phosphates and other trialkylphenyl esters) at to 23 mg/m^3 exhibited signs of neurotoxicity at day 58. At hydraulic fluid concentrations of 102-110 mg/m^3, early signs of neurotoxicity were evident in 19 of 20 hens between days 22 and 29. Four of 12 squirrel monkeys exposed continuously to hydraulic fluid mists at 4.3 or 4.4 mg/m^3 (six animals per concentration) died within the 108-day exposure period although no signs of neurotoxicity were seen in the surviving animals (four of the 10 control animals also died) (Siegel et al. 1965). For intermittent exposures, groups of six chickens were exposed to hydraulic fluid mist containing 1.5% TOCP at 25 or 50 mg/m^3 for 8 h/d, 5 d/wk, for a total of 30 exposures. Hens in the lower concentration group showed no signs of neurotoxicity although there was an increase in weight; at the higher concentration, three of the hens developed signs of paralysis by day 29 (Siegel et al. 1965).

In humans, exposures to TOCP that are not likely to produce toxic effects are estimated to be 2.5 mg/kg for a single oral dose (175 mg for a 70-kg person) and 0.13 mg/kg per day for repeated exposures (9 mg/d for a 70-kg person) on the basis of the assumption that there was no significant difference in TOCP sensitivity among cats, hens, and humans. A safety factor of 10 was used to extrapolate from animals to humans (Craig and Barth 1999). Because of the lack of data on TOCP toxicity in humans, the committee is unable to assess whether the assumption of equivalent sensitivity in humans, cats, and hens is appropriate but notes that a safety factor of 100 is typically used, which includes a safety factor of 10 for extrapolation from animals to humans and a second safety factor of 10 to account for interhuman variability (Faustman and Omenn 2001).

A further assessment of the inhalation toxicity of TOCP was conducted by comparing the amount of TOCP in an oil mist with worst-case occupational-exposure scenarios (Craig and Barth 1999). For the purposes of comparing risks, the authors assumed that inhaled and ingested doses of TOCP were of equivalent toxic potential (but do not justify the assumption). If the TOCP concentration in the oil mist were 3%, a sedentary worker (ventilation rate, 0.5 m^3/h) with an 8-h workday would inhale TOCP at 0.6 mg/d, and a worker engaged in heavy activity (ventilation rate, 3.6 m^3/h) with a 12-h workday would inhale TOCP at 6.5 mg/d. Those values do not exceed the estimated safe dose for a 70-kg person of 9 mg/d. The authors note that although a lubricating oil or hydraulic fluid is unlikely to contain 3% pure TOCP, if the other *o*-aryl phosphates present approach this concentration, it is the atmospheric concentrations of the neurotoxic components that might be of concern because they could exceed the American Conference of Governmental Industrial Hygienists (ACGIH) 8-h threshold limit value of 0.1 mg/m^3 time-weighted average for pure TOCP (Craig and Barth 1999; ACGIH 2001).

Some reports suggest that although the "conventional" type of jet engine oils contains substantial concentrations of TOCP (2-3% *o*-cresyl isomers in the TCP), a newer "low-toxicity" jet engine oil contains very little of the ortho isomer and instead contains primarily the less-toxic meta and para isomers. It was estimated that oral doses of jet engine oil of 5.7-33.3 g/kg containing 3% "conventional" TCP would be required to inhibit brain NTE in the hen (the most sensitive species) by 70%; an oral dose of 87-330 g/kg would be required to elicit the same effect if jet engine oil containing 3% "low-toxicity" TCP were used (Craig and Barth 1999; Mackerer et al. 1999).

A recent risk assessment to estimate the potential of conventional and low-toxicity jet engine oils containing TCP to cause OPIDN found that ingestion of jet engine oil containing 3% TCP at approximately 9-10 g/d would be a minimal toxic dose for a 70-kg person; for conventional jet engine oils, the minimal toxic dose was estimated to be 280 mg/kg per day (Mackerer et al. 1999). Studies by Henschler (1958a,b) on older TCP that contained up to 30% *o*-cresyl isomers showed that the asymmetrical mono-*o*-cresyl-substituted mixed esters were about 10 times as toxic as pure TOCP, with the di-*o*-cresyl isomers in between (Henschler 1959). It has been estimated that the conventional TCP can contain the mono-*o*-cresyl isomers at 3,070 ppm and the di-*o*-cresyl isomers at 6 ppm, compared with less than 0.005 ppm for TOCP; however, the newer TCP has been calculated to contain TOCP at less than 0.001 ppm, mono-*o*-cresyl isomers at approximately 1,760 ppm, and di-*o*-cresyl isomers at 1.1 ppm (Mackerer and Ladov 2000).

Other Phosphate Esters

Hydraulic fluids contain phosphate esters in addition to the TCP mixture. Among them are tributyl phosphate, dibutyl phenyl phosphate, butyl diphenyl phosphate, and triisobutyl phosphate. Toxicity information on those phosphate esters in the published literature is relatively sparse and is summarized below.

The use of synthetic engine lubricants raises the possibility that at extremely high temperatures (250-700°C) a very potent neurotoxicant can be formed (Centers 1992; Wright 1996). If the lubricant contains trimethylolpropane esters as well as TCP, they may thermally degrade to form TMPP at rates of 7-10 mg/mL of lubricant at those very high temperatures (Centers 1992). Detection of TMPP after a fire (on a ship that used a lubricant containing the compounds) confirmed that the generation of this extremely toxic compound can occur under actual fire conditions (Wyman et al. 1993). The intraperitoneal dose of TMPP required to kill 50% of mice was 1.0 mg/kg; 50% mortality was achieved with dermal exposures of 50-100 mg/kg (Centers 1992).

Dibutyl phenyl phosphate and di-*tert*-butylphenyl phosphate manifest many of the health effects typical of organophosphate compounds in that they are expected to produce neurotoxic symptoms after ingestion. Irritation of the mucous membranes of the eyes, nose, and upper respiratory tract with coughing and wheezing follow inhalation exposure to their vapors or aerosolized formulations. Repeated dermal contact with dibutyl phenyl phosphate has resulted in drying and cracking of exposed skin (Hazardous Substances Data Bank [HSDB] 2000). Specific toxicity data were not identified for di-*tert*-butylphenyl phosphate.

Tributyl phosphate (TBP) appears to be much less toxic than TCP. Vapors of TBP can cause irritation of the mucous membranes in the eyes, nose, and throat. Workers exposed to TBP at 15 mg/m^3 complained of headache and nausea. Prolonged exposure may result in paralysis; however, neurotoxic effects are seen only at very high doses and then primarily via ingestion. TBP has only weak cholinesterase-inhibition activity and only via ingestion of large quantities (e.g., 100 mL). It has been estimated that the probable oral lethal dose for a 70-kg person is 0.5-5 g/kg (HSDB 2000). Skin contact may result in irritation.

In summary, the presence of phosphate esters in engine lubricating oils and hydraulic fluids may constitute a potential neurotoxic hazard. Practically all known cases of human toxicity of these compounds involve ingestion of substantial amounts. It should be noted that no TCP isomers, including TOCP, and no TMPP were detected on several flights of different aircraft (Nagda et al. 2001); however, TCP vapors were detected in air when engine lubricating oils and hydraulic fluids were pyrolized under laboratory conditions (van Netten 2000, van Netten and Leung 2000, van Netten and Leung 2001). Measurement of the airborne concentrations of these compounds or modeling of possible exposure scenarios in aircraft cabins, with objective demonstration of neurotoxicity in experimental animals exposed to these compounds by inhalation, is needed for a proper assessment of risk.

Carbon Monoxide

CO is a colorless, odorless gas produced by incomplete combustion of carbonaceous material. EPA has established standards for ambient air concentrations of CO of 9 ppm (10 mg/m^3) for an 8-h exposure and 35 ppm (40 mg/m^3) for a 1-h exposure (EPA 2000). It has also established "significant harm" concentrations of 50 ppm (8-h average), 75 ppm (4-h average) and 125 ppm (1-h average) as conditions in which exposure could result in carboxyhemoglobin (COHb) concentrations of 5-10%, which could cause significant health effects in sensitive people (EPA 2000). CO in cabin air is regulated by FAR 25.831.[2] Concentrations measured in aircraft under normal operating conditions are typically below 1.0 ppm (Nagda et al. 2001) and in a recent survey did not exceed 0.87 ppm (Waters 2001). The presence of CO in the aircraft cabin is important for health because it has an affinity for hemoglobin about 250 times that of O_2, forming COHb, reducing the O_2-carrying capacity of the blood, and possibly causing hypoxia. Formation of COHb depends on CO concentrations in the ambient air and on respiratory minute volume. Under normal physiological conditions, the brain can increase the blood flow or tissue O_2 extraction to compensate for the hypoxia. When ambient CO is high—for example, in a fire or in the presence of incomplete combustion—blood CO can reach lethal concentrations within minutes and are not necessarily preceded by such symptoms as headaches and dizziness. At

---

[2] FAR Section 25.831 "Ventilation: 1. Carbon monoxide concentrations in excess of 1 part in 20,000 parts of air (50 ppm) are considered hazardous. For test purposes, any acceptable carbon monoxide detection method may be used."

COHb of at least 2.3 and at least 4.3%, maximal exercise duration and perfor-mance, respectively, are slightly reduced in healthy people; no effects on submaximal exercise were observed when COHb levels was increased to 15-20% (EPA 2000). A large clinical study of the cardiovascular effects of CO on patients with angina was sponsored by the Health Effects Institute (Allred et al. 1991). Exposure to CO concentrations during exercise sufficient to raise the blood COHb concentration to 2%, resulted in a reduced time to angina symptoms, 4% COHb resulted in a further reduction. These CO exposures also produced depressions of one temporal segment of the cardiograms.

Smokers are more susceptible to the effects of CO because they have COHb at about 5%; nonsmokers typically have COHb at close to zero. Early symptoms of CO poisoning in humans are headaches and lightheadedness. At COHb of 15%, hypoxia begins. When blood COHb reaches 30-40%, severe headache, weakness, dimness of vision, confusion, dizziness, nausea, and vomiting may occur; consciousness can be lost sometimes for hours before death. At higher concentrations—usually more than 40-50% blood saturation with CO—severe ataxia, increasing confusion, hallucinations, and accelerated respiration occur. At 50-60% saturation, convulsions, tachycardia, and coma are evident; death occurs at COHb of greater than 70% (Gosselin et al. 1984, as cited in HSDB 2000). Fatal COHb concentrations would probably be lower at higher altitudes, with reduced PO_2, than at sea level.

Although CO is known to have fetotoxic effects in laboratory animals, effects on human fetuses are less evident, although of concern. In animals, COHb concentrations of less than 10% do not appear to affect fetal develop-ment adversely until possibly later in gestation, and it has been postulated that the same response may be expected in human pregnancy (Robkin 1997; as cited in EPA 2000). One prospective study of 40 cases of acute CO poisoning of pregnant women (Koren et al. 1991, as cited in EPA 2000) found that even hypoxemia resulting in mild to moderate COHb saturations, up to 18%, did not impair fetal growth; this suggests that most pregnant women are not increasing fetal risk through ordinary ambient exposure. In another study, maternal expo-sures to CO at 150 to 200 ppm, resulting in 15-25% COHb, did result in lower birth weights, heart abnormalities, delays in behavioral development, and dis-ruption of cognitive function in several laboratory animal species (EPA 2000). Exposures to CO at levels as low as 60-65 ppm CO (6-11% COHb) through-out gestation produced many of the same developmental effects (EPA 2000).

Healthy people have shown central nervous effects after 1-h peak CO exposure which resulted in 5-20% COHb. Effects included decrements in hand-eye coordination (driving or tracking), attention, and vigilance (detection of infrequent events) (EPA 2000).

People who have suffered and recovered from acute CO poisoning occasionally develop neuronal loss in the cortex and necrosis in the basal ganglia, which result in parkinsonism-like signs (Smith, 1996; Steinmetz 1998).

Formaldehyde

Another cabin air contaminant that may have adverse effects is formaldehyde. Formaldehyde is a colorless gas with a distinct odor that is noticeable at approximately 0.5-1.0 ppm. It is a ubiquitous chemical in ambient air at very low concentrations. However, excessive concentrations could be found in aircraft cabins as a result of the thermal decomposition of engine oils or hydraulic fluids or the reaction of O_3 with cabin surfaces. Chapter 3 contains more information on the generation of formaldehyde in cabin air. Under normal operating conditions, measured concentrations on aircraft have not exceeded 26 ppb (13 $\mu g/m^3$) (Nagda et al. 2001)

Formaldehyde is a known irritant of the mucous membranes of the eyes, nose, and respiratory tract at concentrations of approximately 0.4-3 ppm (ATSDR, 1999). Paustenbach et al. (1997) reported that for most people, eye irritation does not occur until at least 1 ppm; more severe eye, throat, and lung irritation occurs at 2-3 ppm; and acclimation occurs (Paustenbach et al. 1997). On the basis of its evaluation of the literature, the Industrial Health Panel recommended an 8-h time-weighted average occupational exposure limit of 0.3 ppm for formaldehyde with a ceiling value of 1.0 ppm. The panel did not identify any hypersensitive subgroups, such as asthmatics, and found no evidence of sensitization (Paustenbach et al. 1997).

Chronic inhalation exposures (for 6.8 yr; range, 2-19 yr) to formaldehyde at 0.39 ppm resulted in abnormalities of the nasal epithelium in plywood-factory workers (Ballarin et al., 1992, as cited in ATSDR 1999). In 70 chemical workers who produced formaldehyde for an average of 7.3 yr (range, 1-36 yr), expousre to it at approximately 0.24 ppm also resulted in nasal epithelium abnormalities (Holmstrom et al. 1989, as cited in ATSDR 1999); the chronic daily human exposure that is estimated to be without risk of adverse effects is 0.008 ppm, on the basis of this study. Concentrations of greater than 20 ppm are life-threatening. Formaldehyde has been designated by EPA as a probable human carcinogen (Integrated Risk Information System [IRIS] 2001).

Deicing Fluids

Glycols (ethylene glycol and propylene glycol) are the main constituents of deicing fluids used for aircraft. Exposure to deicing fluids is expected to be season- and climate-dependent. Deicing fluid is typically applied hot. As noted in Chapter 3, there are no published studies on the potential for deicing fluids to enter the cabin air via bleed air.

Ethylene glycol can be extremely toxic, particularly if ingested at high doses or if high concentrations of heated vapors are inhaled. Most known cases of ethylene glycol toxicity resulted from ingestion; ethylene glycol has a low vapor pressure. Ingestion of small amounts can cause drowsiness and slurred speech. Ingestion of antifreeze fluids containing ethylene glycol, often accidentally or suicidally, causes serious metabolic acidosis and renal toxicity (tubular necrosis) in humans—because of the toxic metabolite oxalic acid—and CNS depression, cardiopulmonary disturbances, pulmonary edema, and congestive heart failure (Snyder and Andrews 1996). The minimal lethal oral dose of ethylene glycol for adults is 1.4 mL/kg (1,330 mg/kg of body weight) (ATSDR 1997). Because of its poor absorption through the skin, systemic effects through skin contact are unlikely.

There are few data on inhalation exposures to ethylene or propylene glycol and no reports of death following exposure by this route (ATSDR 1997). After inhalation exposure to ethylene glycol aerosol at 3-67 mg/m^3 for 20-22 h/d for about 4 wk, human volunteers complained of upper respiratory tract irritation and occasionally of slight headache and low backache. Ethylene glycol at above 140 mg/m^3 was highly irritating and above 200 mg/m^3 was intolerable. No ethylene glycol or its metabolites were found in the blood or urine of the subjects (Wills et al. 1974).

Propylene glycol has almost no intrinsic oral toxicity and is widely used in foods, cosmetics, and pharmaceuticals. Its low toxicity is explained by its metabolic conversion to lactic acid and pyruvic acid, normal body constituents (Snyder and Andrews 1996). Both propylene glycol and ethylene glycol are rapidly metabolized by humans and cannot be detected in tissues 48 h after exposure. Repeated, short-term exposures to propylene glycol may result in some irritation of the eyes, skin, and nasal and oral mucosa; some skin sensitization is possible (ATSDR 1997). Over the last 5 yr, the market share of propylene glycol used in deicing fluids has grown from 10% (124 million pounds) to 70% (140 million pounds), primarily at the expense of ethylene glycol (Ritter 2001). The increased use of propylene glycol in deicing fluids has reduced the potential for toxic effects after accidental exposure to these compounds.

Pyrethroid Pesticides

Today, only a few foreign countries require the use of insecticides on aircraft coming from other countries, including aircraft from the United States. Consequently, the interiors of U.S. aircraft destined for those countries must be disinsected, a practice the United States discontinued in 1979 (see Chapter 3). The pesticides are usually synthetic pyrethroids, primarily phenothrin and permethrin. Chapter 3 describes the application methods for these pesticides.

The disinsection practices of some countries are expected to result in the exposure of cabin crews and passengers to pesticides. Although pyrethroid pesticides have very low toxicity in humans, they can cause adverse effects in some people and are recognized as neurotoxicants at very high doses. The acute oral dose that is lethal to 50% of rats is 0.5-5 g/kg; the dose that is lethal to 50% of animals in 4-h is 2,280 mg/kg; and the concentration that is lethal to 50% of animals via inhalation with intermittent exposure is 500 mg/m^3 (NRC 1994). The pesticides act by slowing inactivation of the neuronal sodium channels, thus causing hyperexcitability of the nerves. That hyperexcitability produces signs of CNS and peripheral nerve disturbances. CNS toxicity is manifested as coarse tremors, involuntary muscle movements, and excessive lacrimation. Although their lethality is low, the pyrethroids are recognized as dermal irritants and, when they are in fume or vapor form, as respiratory and ocular irritants (Ecobichon and Joy 1994).

Systemic poisoning is less likely from dermal exposure than from oral exposure oral because absorption through the skin is considerably less than absorption through the gastrointestinal tract. In general, dermal exposure to pyrethroids results in a localized tingling sensation of the skin (paresthesia) without loss of normal sensation within 0.5-5 h and persists for up to 3 d (WHO 1995). Because skin appears to bind pyrethroids, localized effects can be expected to last for long periods, and the skin might act as a reservoir; this could result in symptoms that last for several weeks in acute cases of pyrethroid poisoning (He et al. 1989), although thorough cleaning of the skin may prevent or minimize this effect. In a study with 184 human volunteers, a 21-d repeat patch test with a 40% solution of permethrin did not result in any skin sensitization, although there were some reports of transient burning, stinging, and itching (NRC 1994). Patch tests with permethrin have shown no allergic response (Naumann and McLachlan 1999).

Areas on the body where skin is thin are affected first, because the agent can interact more readily with nerve endings. The average threshold dose for producing paresthesia is 0.2 mg/cm^2 of skin in humans (WHO 1995). About 2% of permethrin applied directly to human skin is expected to be absorbed (NRC 1994). In a scabies-control program involving about 1,000 people, skin application of a cream containing 5% pyrethroids resulted in no complaints of paresthesia (Yonkosky et al. 1990). There appear to be individual susceptibilities to the effect; this may be related to the variation in the concentrations of carboxyesterase enzymes in the skin (Leng et al. 1999). Thus, lymphocyte carboxyesterase activity might serve as a biomarker of the effect (Leng et al. 1999). It has been suggested that persons with pre-existing disease (including skin disease and lowered immunity), infants, and children can be more sensitive than healthy adults to the effects of permethrin (Naumann and McLachlan 1999). Phenothrin is used to control headlice in children and has been reported to have relatively few adverse health effects (Naumann and McLachlan 1999).

Occupational exposure to or inappropriate handling of synthetic pyrethroids has been found to produce dizziness, burning, itching, or tingling of the skin. More severe contact, such as spilling of the pesticides onto the skin, resulted in lacrimation, photophobia, and eye irritation. Accidental swallowing may cause headaches, dizziness, vomiting, fatigue, anorexia, chest tightness, blurred vision, paresthesias, and, in severe cases, convulsive attacks. However, the signs and symptoms of pyrethrin toxicity are usually fully reversible, and no long-term human toxicity has been reported after single or repeated exposure (He et al. 1989; Ecobichon and Joy 1994; Harp 1998; Leng et al. 1999; Vandenplas et al. 2000). It is possible that after exposure sensitive people will develop a rash or allergy-like symptoms such as wheezing, cough, and shortness of breath. That might happen particularly in people who have atopy or bronchial asthma, although the relationship between asthma and pyrethroid exposure is not clear (WHO 1995). Aerosolized pesticides can trigger a "nonspecific" asthmatic response—bronchoconstriction and respiratory symptoms—although it is has been difficult to link sensitization in nonasthmatics to exposure to aerosolized pyrethroid insecticides (WHO 1995).

Aside from the obvious inhalation and dermal exposures that occur when passengers and crew are actively sprayed, typically with phenothrin, residual exposures can occur when permethrin is used to treat unoccupied cabins.

Residual exposures are likely to be dermal and oral. See Chapter 3 for a more detailed description of possible exposures resulting from disinsection practices.

Pyrethroid pesticides are rapidly destroyed by carboxyesterase enzymes in the liver and to a lesser extent in skin, muscle, kidney, brain, and serum (Leng et al.1999). Organophosphates are also destroyed by carboxyesterases and so can compete with pyrethroids for the enzymes, the competition can result in a synergism that increases overall toxicity. The enzyme cholinesterase, which is also inhibited by organophosphates, does not appear to be important in pyrethroid metabolism (Leng et al.1999). There is little information on long-term exposure to permethrin and none such on exposure of humans (Naumann and McLachlan 1999).

Most flight attendants' reports of pesticide incidents describe lung, eye, throat, or skin irritation as the primary adverse effect. Sensitization, or enhanced responsiveness with successive pesticide exposures, may be a problem faced by cabin attendants because they are intermittently exposed; such an exposure regimen is ideal for inducing sensitization or magnified responses to the same exposures (Weiss and Santelli 1978; Gilbert 1995; Jones et al. 1996).

If people are exposed to both pyrethroid and some enzyme inhibitors, there is a potential for synergistic effects. Pyrethroids are lipophilic and are rapidly absorbed and metabolized after ingestion by mammals. However, they do not accumulate in mammalian tissues. In mammals and insects, pyrethroids are generally metabolized by carboxylases and the mixed-function oxidase (MFO) system of the microsomes (Casida et al. 1983). Piperonyl butoxide, a potent inhibitor of the MFO system, is often added to the pesticide formulations to increase the toxicity by a factor of 10-300 (Casida et al. 1983). Carboxyesterase inhibitors may also act synergistically to enhance the toxicity of pyrethroids. The synergism is an asset for insect control, but it can have potentially serious implications for humans in the rare event that a leak of engine oils or hydraulic fluids causes these compounds to enter the cabin and causes simultaneous exposure to carboxyesterase inhibitors. No incidents of such interactions have been documented in humans, but engine-oil seal failures and hydraulic-fluid leaks into the ECS of aircraft have been reported (Parliament of Commonwealth of Australia 2000; van Netten 2000; van Netten and Leung 2000, 2001). See Chapter 3 for a more detailed discussion of engine-oil and hydraulic fluid leaks.

Two other features of possible pyrethroid toxicity may warrant further research and discussion: their potential as developmental neurotoxicants (Ericsson, 1997; Landrigan et al, 1999) and their potential as endocrine disruptors in vivo, which so far has been demonstrated only in vitro (Go et al. 1999).

Aldehydes

Aldehydes, such as acetaldehyde and acrolein (an unsaturated aldehyde), could be found in cabin air in the case of leaking or pyrolysis of engine oil or hydraulic fluids. Concentrations of aldehydes (and ketones) measured in aircraft cabins are "lower than those encountered in ground level buildings" (Nagda et al. 2001).

Maximal concentrations of acetaldehyde were 26.4-30.7 $\mu g/m^3$ in bleed air and 20.8-70.2 $\mu g/m^3$ in cabin air (see Chapter 3). Some sensitive people might suffer adverse effects, and there could be repeated exposures. Exposure to other aldehydes results in decreasing toxicity as the chain length of the aldehyde increases in the order of acetaldehyde, propionaldehyde, isobutyraldehyde, n-butyraldehyde, valeraldehyde, and isovaleraldehyde (HSDB 2000). Effects of inhalation exposure to acetaldehyde at high concentrations (about 100-200) ppm include irritation of the mucous membranes (200 ppm for 15 min), irritation of the respiratory tract (134 ppm for 30 min), and lung edema. Prolonged inhalation exposure to acetaldehyde results in symptoms that mimic alcohol intoxication. Repeated dermal or ocular exposures may cause dermatitis or conjuncitivitis, respectively (HSDB 2000). Irritation of the eye may occur at acetaldehyde vapor concentrations as low as 25 ppm after 15 min. Acetaldehyde may facilitate the uptake of other air contaminants by humans because of its ciliotoxic and mucus-coagulating effects (HSDB 2000). Acetaldehyde has been classified as a possible human carcinogen.

The unsaturated aldehyde acrolein may also cause adverse effects when inhaled. However, acrolein was not detected in bleed air or cabin air of several aircraft under normal operating conditions (Nagda et al. 2001). It is an irritant of the eyes and respiratory tract. The odor of acrolein can be perceived at 0.7 mg/m^3. The threshold concentrations of acrolein for irritation and health effects are 0.13 mg/m^3 for eye irritation, 0.3 mg/m^3 for nasal irritation and blinking, and 0.7 mg/m^3 for decreased respiratory rate. Exposure to acrolein vapor in air at 1 ppm (2.3 mg/m^3) causes lacrimation and marked eye, nose, and throat irritation within 5 min. It irritates the conjunctiva and mucous membranes of the upper respiratory tract. Above 3 mg/m^3, it causes injury to the lungs, and respiratory insufficiency may persist for 18 mo or more. A 10-min exposure at 350 mg/m^3 was lethal (HSDB 2000). Acrolein has been classified as a possible human carcinogen on the basis of an increased incidence of adrenal cortical adenomas in female rats (IRIS 2001). EPA established a reference concentration[3] of 0.0002 mg/m^3 for acrolein on the basis of squamous metaplasia and neutrophilic infiltration of nasal epithelium in subchronic inhalation studies in rats (Kutzman 1981, as cited in IRIS 2001).

Carbon Dioxide

Carbon dioxide (CO_2) in the aircraft cabin is primarily a "bioeffluent" in that it is released in the exhaled breath of the occupants and is generated by normal metabolism (see also the section on sources inside the cabin in Chapter 3). CO_2 inhaled at 2% in air increases pulmonary ventilation by 50% (Thienes and Haley 1972, as cited in HSDB 2000). Asphyxiant effects may occur as the concentration exceeds 6%, causing shortness of breath, dizziness, tingling, and slowed mentation (Maresh et al. 1997). Adding 1% CO_2 to air increased pulmonary ventilation rate by 37% at sea level. Under a pressure that simulated an altitude of 5,000 m (16,400 ft), pulmonary ventilation rate increased by 7%. CO_2 concentrations of 0.5% or 1% stimulated hyperventilation to a degree that prevented a decrease in psychomotor performance at a simulated altitude of 5,800 m (19,000 ft) but not at 5,000 m (16,400 ft) (Vieillefond et al. 1981, as cited in HSDB 2000). Repeated daily exposures to 0.5-1.5% inspired CO_2 at 1 atm are well tolerated by healthy people (ACGIH 1980).

FAR 25.831 has a set limit for CO_2 in the cabin at 0.5% (5,000 ppm).[4] Cabin pressure altitude is limited to 2,440 m (8,000 ft) and that limit appears to protect occupants from adverse effects of exposure to CO_2. The highest CO_2 concentrations measured in aircraft cabins and reported in the literature are below the limit (see Table 1-2, Chapter 1). Typical values reported are in the range of 1,100-1,700 ppm, although concentrations as high as 4,200 ppm have been reported (Waters 2001). Backman and Highighat (2000) reported CO_2 concentrations of 455 ppm, 586 ppm, and 706 ppm on an Airbus 320 (altitude, 39,000 ft; 86 passengers), a Boeing 767 (39,000 ft; 70 passengers), and a DC-9 (35,000 ft; 75 passengers), respectively.

---

[3] The reference concentration is an estimate of a daily exposure to the human population (including sensitive subgroups) that is likely to be without an appreciable risk of deleterious effects during a lifetime.

[4] FAR Section 25.831: "Ventilation: 2. CO_2 concentration during flight must be shown not to exceed 0.5 percent by volume (sea level equivalent) in compartments normally occupied by passengers or crew members."

CO_2 is used to assess the adequacy of ventilation. Normally, CO_2 produced as the metabolic byproduct of occupants—respiration—is the primary source of CO_2 generation in an occupied space. Although the CO_2 generated by respiration is unlikely to reach the point at which adverse effects would be expected, other bioeffluents are generated more or less in proportion to CO_2. These other bioeffluents may be responsible for complaints about odors, stale air, lack of fresh air, and stuffiness.

Particulate Matter

Airborne particles in the passenger cabin may comprise coarse particles (such as powders, dusts, dirt, and hair with aerodynamic diameters over 2.5 μm) and fine and ultrafine particles. Fine particles (aerodynamic diameter, less than 2.5 μm) and ultrafine particles (aerodynamic diameter, less than 0.1 μm) are the products of combustion of materials such as fuels, engine oil, and hydraulic and deicing fluids. Particulate matter (PM), generally with an aerodynamic diameter of less than 10 μm (PM_{10}) has been measured aboard aircraft during normal operations with concentration means ranging from less than 10 $\mu g/m^3$ (Nagda et al. 2001) to 176 $\mu g/m^3$ (CSS 1994). Typical indoor air PM concentrations are 3-35 $\mu g/m^3$ (Burton et al. 2000). Coarse particles, because of their size, are deposited in the upper respiratory airways and can lead to coughing, sneezing, and nasal irritation. Chapter 4 discusses exposure to and health effects of particles of biological origin.

PM_{10} or smaller may be deposited in the lower (thoracic) regions of the human respiratory tract, including $PM_{2.5}$, which deposit primarily in deep lung airways. Exposure to fine and ultrafine PM has been associated with heart rate variability (Peters et al. 2000, as cited in EPA 2001) and may be related to cough. Asthmatics have small decrements in lung function and increases in coughing, phlegm, breathing difficulty, and bronchodilator use after exposure to PM_{10} and $PM_{2.5}$ (EPA 2001). Fine and ultrafine particles, which are deposited in the small airways, may contain toxic constituents that have been shown to cause cardiopulmonary effects. However, few data are available on the acute effects of particle-associated organic carbon constituents (EPA 2001). See Chapter 4 for a more detailed discussion of the possible health effects associated with exposure to biological agents.

OTHER HEALTH CONSIDERATIONS

Physiological Stressors

In addition to the environmental and chemical stressors discussed above, other stressors may play a role in the perception of and susceptibility to air quality or changes in it.

Fatigue is a common complaint among flight attendants. Work schedules can be erratic, with unplanned layovers and long hours. Furthermore, jet lag can pose a problem. Sleeping accommodations are not always conducive to good rest, and meals can be erratic. Fatigue can also be accentuated by increased CO_2 in the air: other factors contributing to fatigue are the effects of vibration, turbulence, and noise; however, these are beyond the scope of this report and will not be discussed further here.

Multiple and Interactive Factors

As discussed earlier in this chapter, cabin occupants (whether flight crew or passengers) may be subjected to numerous air contaminants and other physiological stessors in some situations. Space et al. (2000) argue that cabin air quality generally is better than that in most indoor environments, according to standard criteria. Although that assessment may describe the experience of passengers under most circumstances, problems do occur. Engine oil and hydraulic fluids, and their pyrolysis products, as noted in other chapters, may contaminate the cabin environment, releasing organic compounds, including cresyl phosphates, some of which are neurotoxic. Because cabin pressurization is limited to a virtual altitude of 8,000 ft, PO_2 declines; this a potential hazard for passengers with cardiovascular impairment. Elevated CO_2 concentrations in that cabin may not present a serious health threat, but in office buildings they are reported to be associated with respiratory symptoms (Apte et al. 2000). Some countries expose passengers to insecticides as part of a program designed to prevent the introduction of nonnative insects.

Although individual factors in isolation might each contribute only a small diminution in well-being, in combination they might be additive, synergistic, or even antagonistic, with their combined influence on the health status of crew and passengers virtually unknown.

Such factors as cramped seating also influence passenger comfort. However, physical activity is more likely to be a potentiating than an additive factor. Unlike the relatively sedentary passengers and cockpit crew, cabin attendants tend to have sustained periods of moderate activity during the course of a flight. As shown for O_3 earlier, exercise could enhance the effects of exposure to many possible cabin air contaminants, including CO and CO_2.

The interactions of CO and high altitude have been examined, and results suggest that the effects are additive (EPA 2000). By binding to hemoglobin, CO may increase the hypoxic effects of high altitude in the elderly and those with coronary arterial disease. At high altitudes on earth (e.g., 15,000 ft), the body adjusts to these hypoxic effects within a few days, but the adjustments do not occur over the duration of even a long-distance flight. In a review of the long-term effects of CO at high altitude, EPA found few effects at CO concentrations of less than 100 ppm and altitudes below 15,000 ft (4,573 m) (EPA 2000).

Passengers and crew members frequently report fatigue, dizziness, headaches, sinus and ear problems, dry eyes, and sore throats during and after air travel. Such symptoms are common to many conditions and can be mistaken for infections (IEH 2001). Other characteristics of air travel, however, can make people more susceptible to infection. For example, travelers are subject to stresses (unrelated to their time on an aircraft) that may reduce their resistance to infectious diseases, such as long hours of waiting, disruption of eating habits, and changes in climate and time zone (Rayman 1997; CDC 2001). Psychological stress has been associated with increased susceptibility to the common cold (Cohen et al., 1991) and may play a role in other infectious diseases. Drying of the nasal mucosa during air travel (due to the low moisture content of the air) may result in nosebleeds and is thought by some to increase susceptibility to respiratory infections, although this is uncertain (BRE 2001). Rose et al. (1999) hypothesized that the reduced-O_2 environment in aircraft leads to mild hypoxia and suppression of the cellular immune system, which in turn contributes to increased susceptibility to infection. If that is borne out, this pathway could explain in part why some frequent flyers complain of infections after air travel.

Other gases, such as those associated with human body odor, are present in the cabin air and may contribute to the discomfort of passengers. Emissions vary widely and depend on individual health, diet, activity, and personal hygiene, such as bathing habits, frequency of clothing change, and use of deodorant (Berg-Munch et al. 1986). Personal-hygiene products and fragrances contain various plant and animal components, such as aldehydes and sterols, that may be in the cabin air. Additional sources of disagreeable biological odors on aircraft are the toilets, waste-storage areas in galleys, and bacterial and fungal growth in the cabin. Evidence of adverse health effects (as opposed to annoyance) of body odors, cosmetics and colognes, or VOCs of microbial origin remains controversial. Respiratory irritation is one condition that could result from exposure to odors on aircraft (Batterman 1995; Ammann 1999; Korpi et al. 1999; Rose et al. 2000).

Susceptibility Factors

Although most members of the traveling public are healthy, given the large number of people who fly each year it is to be expected that many of them will have some level of compromised health. At cabin altitudes during cruise, all crew and passengers are in an environment with decreased O_2. Under ordinary conditions of commercial flight, relatively small O_2 decrements usually cause no symptoms in healthy people, because hemoglobin remains well saturated with O_2 at altitude. Nevertheless, some persons with underlying pulmonary or cardiac disease or significant untreated or partially treated anemia might be sensitive to a lowered PO_2 and might experience any number of symptoms, including headache, lightheadedness, dizziness, fatigue, numbness, and tingling (Sheffield and Heimbach 1996). The health implications of commercial flight for passengers with pre-existing illness are described in greater detail in *Medical Guidelines for Airline Travel* (Aerospace Medical Association 1997). Air crew and passengers with particular preexisting medical conditions may not tolerate the cabin environment described above as well as would a healthy traveler. People with underlying cardiovascular or respiratory diseases, particularly adults and children with asthma, may also be at increased risk because of the increased O_3 exposures that occur in high-altitude flights. Data are lacking on the relationship between asthma and O_3 in aircraft cabins at high altitudes. People with COPD or emphysema have lung-tissue damage with an abnormal capability for transporting O_2 across the lung tissue into the bloodstream. Consequently, they have less O_2 in their blood than a normal person, whether at sea level or, worse, at altitude. Such passengers may become symptomatic and have shortness of breath, wheezing, and coughing. Schwartz et al. (1984) studied subjects with severe COPD (average resting arterial PO_2, 68.0 ± 7.3 mm Hg) during flights at altitudes of 1,650-2,250 m (about 5,400-7,400 ft) in unpressurized aircraft cabins. PO_2 decreased to an average of 51.0 ± 9.1 mm Hg at an altitude of 1,650 m; there was little further change at 2,250 m.

Several simulation studies have been carried out in hypobaric chambers with COPD patients and healthy subjects (Dillard et al. 1989; Naughton et al. 1995). Declines in PO_2 were observed in all patients at rest; values fell to below 50 mm Hg in many healthy and COPD subjects and the declines were made worse by light exercise (Figure 5-3). Of the 100 patients with COPD, 44 reported traveling by air in the 2 years before the interview (Dillard et al. 1991). Eight of the 44 patients reported increased symptoms during flight (no direct physiological measurements were available). Five of the 8 patients experienced shortness of breath when walking in the cabin, and two requested supplemental O_2 for their symptoms. Variables to consider are clinical evaluation, pulmonary-function tests, and blood-gas analysis. The measurement of arterial blood gas is most useful because it is considered the best predictor of arterial PO_2 at altitude. A ground-level value greater than 9.3 kPa (70 mm Hg) is considered adequate in most cases (Gong et al. 1984; Cottrell 1998). If it is lower, consideration must be given to prescribing inflight medical O_2 or postponing air travel depending on the clinical circumstances. It cannot be overemphasized that the clinical judgment of a physician is essential for a person with COPD in deciding whether to fly. The committee was unable to locate any data to suggest that most people with COPD cannot fly safely.

People with significant coronary arterial disease may also be at increased risk for cardiac symptoms because of reduced PO_2 at altitude. The committee was unable to locate any published studies of air-travel experiences of passengers with coronary arterial disease. Because of the lack of data, the broad spectrum of disease severity, and individual variability, the committee cannot put forward any recommendations other than to note that, as with COPD, a decision on a traveler's fitness to fly should be made by a physician.

A number of characteristics may affect a person's risk of becoming infected if exposed and affect the severity of disease that could result from infection (see Chapter 4 for more information on infectious agents on aircraft). Those characteristics include age, nutritional condition, alcoholism, coexisting (particularly chronic) diseases (such as those discussed above), immune status, and abnormalities in the skin or respiratory tract that allow the entry of infectious agents (Macher and Rosenberg 1999). Many persons with altered immunocompetence are able to travel, but they should consult their physicians before traveling.

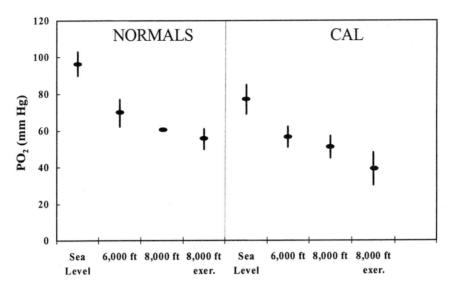

FIGURE 5-3 PO_2 in six healthy subjects and nine subjects with chronic air-flow limitation (CAL; $FEV_1/FVC \leq 70\%$) during hypobaric high-attitude simulation. Data at 8,000 ft are at rest and after 2 min of "light exercise" (approximate work rate, 200 kJ/min) to approximate a short walk on aircraft. Source: Naughton et al. 1995. Redrawn from the *American Journal of Respiratory and Critical Care Medicine*; copyright 1995, American Thoracic Society.

People with medical conditions that may require treatment during travel (e.g., COPD, asthma, and autoimmune diseases) should carry letters describing their condition and treatment. Various types of medical alert tags and bracelets are available and should be worn to inform emergency-care providers of appropriate treatments or precautions for various medical conditions and how to reach travelers' health-care providers.

Populations that may be particularly sensitive to CO poisoning are anemic persons who have low hemoglobin concentrations, children who have higher metabolic rates that would exacerbate the adverse effects of CO, persons with a history of coronary heart disease or respiratory disease, and the elderly. Smokers typically have increased COHb and may have an adaptive response to elevated COHb (EPA 2000). People with coronary arterial disease and reproducible exercise-induced angina have decreased exercise tolerance at COHb concentrations of 3-6% (EPA 2000). Pregnant women may be at particular risk. The fetus may be very susceptible to the effects of CO because it readily crosses the placenta and might result in neurological damage to the infant (EPA 2000). Asthmatics may be especially sensitive to formaldehyde.

Cabin crews as a cohort population are relatively healthy. Air-transport pilots are required by FAA to undergo a flight physical examination every 6 mo, and must be granted an exception for significant illness or some classes of medications. Air-transport pilots can no longer fly as pilots once they are 60 yr old. However, flight attendants are required to undergo only an entry physical examination. There is no requirement for periodic medical examinations as there is for pilots. There are no health requirements for passengers.

CONCLUSIONS

• The flight environment, with its lowered barometric pressure, may result in passenger and crew discomfort and in susceptible people, in health effects. Infants may also be at greater risk for hypoxia under conditions of reduced PO_2.

• Although low relative humidity in the aircraft cabin can result in temporary discomfort as a result of the drying of mucous membranes and eye, nose, and respiratory tract irritation, symptoms are expected to subside after exposure is discontinued. There is no information on the potential for long- or short-term adverse effects associated with exposure to low relative humidity.

• High-altitude flights might result in increased O_3 levels in an aircraft. Elevated O_3 concentrations have been associated with increased respiratory symptoms, such as coughing, wheezing, and asthma.

• Phosphate esters, formaldehyde, other aldehydes, and CO—found in engine oil, hydraulic fluids, and their pyrolysis products—may cause respiratory and neurological effects, particularly at high concentrations. However, more data are needed to establish an association between the presence and concentrations of cabin contaminants and potential health effects in passengers and crew.

• Although pyrethroid pesticides, used for disinsection on some aircraft, have very low toxicity in humans, they can cause adverse effects and are recognized as neurotoxins.

• In passengers and cabin crew who have pre-existing illness—such as anemia, asthma, COPD, and coronary arterial disease—the stresses of flight could exacerbate symptoms.

RECOMMENDATIONS

● Potential synergistic and interactive effects of exposure in the aircraft cabin to reduced barometric pressure, low humidity, O_3, other chemical contaminants, and pesticides should be examined.

● If future research, such as that described in Chapter 8, indicates that some cabin air contaminants or other environmental characteristics, such as relative humidity, pose hazards to the health of passengers or crew, FAA should work with other organizations—such as the Occupational Safety and Health Administration, EPA, and ACGIH—to establish standards or guidelines to regulate them.

REFERENCES

ACGIH (American Conference of Governmental Industrial Hygienists). 1980. Documentation of the Threshold Limit Values, 4th Ed. Cincinnati, OH: ACGIH.

ACGIH (American Conference of Governmental Industrial Hygienists). 2001. 2001 TLVs and BEIs: Threshold Limit Values for Chemical Substances and Physical Agents: Biological Exposure Indices. Cincinnati, OH: ACGIH.

Aerospace Medical Association. 1997. Medical Guidelines for Airline Travel. Alexandria, VA: Aerospace Medical Association.

Allred, E.N., E.R. Bleecker, B.R. Chaitman, T.E. Dahms, S.O. Gottlieb, J.D. Hackney, M. Pagano, R.H. Selvester, S.M. Walden, and J. Warren. 1991. Effects of carbon monoxide on myocardial ischemia. Environ. Health Perspect. 91:89-132.

Ammann, H.M. 1999. Microbial volatile organic compounds. Pp. 26.1-26.17 in Bioaerosols: Assessment and Control, J.M. Macher, H.M. Ammann, H.A. Burge, D.K. Milton, and P.R. Morey, eds. Cincinnati, OH: American Conference of Government Industrial Hygienists.

Apte, M.G., W.J. Fisk, and J.M. Daisey. 2000. Associations between indoor CO2 concentrations and sick building syndrome symptoms in U.S. office buildings: An analysis of the 1994-1996 BASE study data. Indoor Air 10(4):246-257.

ATSDR (Agency for Toxic Substances and Disease Registry). 1997. Toxicological Profile for Ethylene Glycol and Propylene Glycol. U.S. Department of Health and Human Services, Public Health Service, Atlanta, GA.

ATSDR (Agency for Toxic Substances and Disease Registry). 1999. Toxicological Profile for Formaldehyde. U.S. Department of Health and Human Services, Public Health Service, Atlanta, GA.

Backman, H., and F. Haghighat. 2000. Air quality and ocular discomfort aboard commercial aircraft. Optometry 71(10):653-656.

Ballarin, C., F. Sarto, L. Giacomelli, G.B. Bartolucci, and E. Clonfero. 1992. Micronucleated cells in nasal mucosa of formaldehyde-exposed workers. Mutat. Res. 280(1):1-7.

Baron, R.L. 1981. Delayed neurotoxicity and other consequences of organophosphate esters. Annu. Rev. Entomol. 26:29-48.

Batterman, S.A. 1995. Sampling and analysis of biological volatile organic compounds. Pp. 249-268 in Bioaerosols, H.A. Burge, ed. Boca Raton: Lewis.

Berglund, L.G. 1998. Comfort and humidity. ASHRAE J. 40(10):35-41.

Berg-Munch, B., G. Clausen, and P.O. Fanger. 1986. Ventilation requirements for the control of body odor inspaces occupied by women. Environ. Int. 12(1-4):195-199.

BRE (British Research Establishment). 2001. Study of Possible Effects on Health of Aircraft Cabin Environments- Stage 2. British Research Establishment, Environment Division, Garston, Watford, UK. [Online]. Available: http://www.aviation. dtlr.gov.uk/healthcab/aircab/index.htm [September 2001].

Burton, L.E., J.G. Girman, and S.E. Womble. 2000. Airborne particulate matter within 100 randomly selected office buildings in the United States (BASE). Pp. 157-162 in Healthy Buildings 2000: Exposure, Human Responses, and Building Investigations, Proceedings, Vol. 1, O. Seppänen, and J. Säteri, eds. Helsinki, Finland: SIY Indoor Air Information.

Carleton, W.M., and B. Welch. 1971. Fluid Balance in Artificial Environment. U.S. Air Force School of Aerospace Medicine. NASA Report CR-114977.

Carpenter, H.M., D.J. Jenden, N.R. Shulman, and J.R. Tureman. 1959. Toxicology of a triaryl phosphate oil: I. Experimental toxicology. AMA Arch. Ind. Health. 20(3):234-252.

Casida, J.E., D.W. Gammon, A.H. Glickman, and L.J. Lawrence. 1983. Mechanisms of selective action of pyrethroid insecticides. Annu. Rev. Pharmacol. Toxicol. 23:413-438.

CDC (Centers for Disease Control and Prevention). 2001. Health Information for International Travel 2001-2002. Atlanta, GA: National Center for Infectious Disease, Division of Quarantine, Centers for Disease Control and Prevention.

Centers, P. 1992. Potential neurotoxin formation in thermally degraded synthetic ester turbine lubricants. Arch. Toxicol. 66(9):679-680.

Cohen, S., D.A. Tyrrell, and A.P. Smith. 1991. Psychological stress and susceptibility to the common cold. N. Engl. J. Med. 325(9):606-612.

Commonwealth of Australia Senate. 1999. Air Safety-BAe 146 Air Quality. The Proof and Official Hansard transcripts of Senate Rural and Regional Affairs and Transport References Committee, Canberra, Australia. November 1, 1999.

Cottrell, J.J. 1998. Altitude exposures during aircraft flight. Flying higher. Chest 93(1):81-84.

Cottrell, J.J., B.L. Lebovitz, R.G. Fennell, and G.M. Kohn. 1995. Inflight arterial saturation: Continuous monitoring by pulse oximetry. Aviat. Space Environ. Med. 66(2):126-130.

Craig, P.H., and M.L. Barth. 1999. Evaluation of the hazards of industrial exposure to tricresyl phosphate: A review and interpretation of the literature. J. Toxicol. Environ. Health B Crit. Rev. 2(4):281-300.

Crane, C.R., D.C. Sanders, B.R. Endecott, and J.K. Abbot. 1983. Inhalation Toxicology: III. Evaluation of Thermal Degradation Products From Aircraft and Automobile Engine Oils, Aircraft Hydraulic Fluid, and Mineral Oil. FAA-AM-83-12. Washington, DC: Federal Aviation Administration.

CSS (Consolidated Safety Services). 1994. Airline Cabin Air Quality Study. Prepared for Air Transport Association of America, Washington, DC. April 1994.

Daughtrey, W.C., R.A. Scala, L.N. Curcio, and J.O. Kuhn. 1990. Evaluation of the Acute Delayed Neurotoxic Potential of a Commercial Tricresylphosphate-Containing Turbo Oil. Toxicologist 10: Abstracts of the 29th Annual Meeting No. 343.

Daughtrey, W., R. Biles, B. Jortner, and M. Ehrich. 1996. Subchronic delayed neurotoxicity evaluation of jet engine lubricants containing phosphorus additives. Fundam. Appl. Toxicol. 32(2):244-249.

Delivoria-Papadopoulos, M., and L.S. Wagerle. 1990. Oxygen diffusion and transport. Pp. 281-305 in Pulmonary Physiology: Fetus, Newborn, Child and Adolescent, 2nd Ed., E.M. Scarpelli, ed. Philadelphia: Lea and Febiger.

Denison, D.M., F. Ledwith, and E.C. Poulton. 1966. Complex reaction times at simulated cabin altitudes of 5,000 feet and 8,000 feet. Aerosp. Med. 37(10):1010-1013.

Dillard, T.A., B.W. Berg, K.R. Rajagopal, J.W. Dooley, and W.J. Mehm. 1989. Hypoxemia during air travel in patients with chronic obstructive pulmonary disease. Ann. Inter. Med. 111(5):362-367.

Dillard, T.A., W.A. Beninati, and B.W. Berg. 1991. Air travel in patients with chronic obstructive pulmonary disease. Arch. Intern. Med. 151(9):1793-1795.

Ecobichon, D.J., and R.M. Joy. 1994. Pesticides and Neurological Diseases, 2nd Ed. Boca Raton, FL: CRC.

Eng, W.G. 1979. Survey on eye comfort in aircraft: I. Flight attendants. Aviat. Space Environ. Med. 50(4):401-404.

EPA (U.S. Environmental Protection Agency). 1996. Air Quality Criteria for Ozone and Related Photochemical Oxidants. EPA/600/P-93/004aF, Vol. 1; EPA/600/P-93/004bF, Vol. 2; EPA/600/P-93/004cF, Vol. 3. National Center for Environmental Assessment, Office of Research and Development, U.S. Environmental Protection Agency, Research Triangle Park, NC. [Online]. Available: www.epa.gov/ncea/ozone.htm. [July 11, 2001].

EPA (U.S. Environmental Protection Agency). 1997. National Ambient Air Quality Standards (NAAQS). Office of Air Quality Planning and Standards, U.S. Environmental Protection Agency. [Online]. Available: http://www.epa.gov/airs/criteria.html [Oct. 24, 2001].

EPA (U.S. Environmental Protection Agency). 2000. Air Quality Criteria for Carbon Monoxide. EPA/600/P-99/001F. National Center for Environmental Assessment, Office of Research and Development, U.S. Environmental Protection Agency, Research Triangle Park, NC.

EPA (U.S. Environmental Protection Agency). 2001. Air Quality Criteria for Particulate Matter, Vol. 2, 2nd External Review Draft. EPA/600/P-99/002bB. National Center for Environmental Assessment, Office of Research and Development, U.S. Environmental Protection Agency, Research Triangle Park, NC. [Online]. Available:

http://www.epa.gov/NCEA/pdfs/partmatt/VOL_II_AQCD_PM_2nd_Review_D raft.pdf [July 12, 2001].

Eriksson, P. 1997. Developmental neurotoxicity of environmental agents in the neonate. Neurotoxicology 18(3):719-726.

Ernsting, J. 1978. Prevention of hypoxia—Acceptable compromises. Aviat. Space Environ. Med. 49(3):495-502.

Ernsting, J., J.L. Gedye, and G.J.R. McHardy. 1962. Anoxia subsequent to rapid decompression. Pp. 359-368 in Human Problems of Supersonic and Hypersonic Flight, A. Buchanan-Barbour, and H.E. Whittingham, eds. Oxford, England: Pergamon Press.

Fang, L., G. Clausen, and P.O. Fanger. 1998a. Impact of temperature and humidity on the perception of indoor air quality. Indoor Air 8(2):80-90.

Fang, L., G. Clausen, and P.O. Fanger. 1998b. Impact of temperature and humidity on perception of indoor air quality during immediate and longer whole-body exposures. Indoor Air 8(4):276-284.

Faustman, E.M., and G.S. Omenn. 2001. Risk assessment. Pp. 83-104 in Casarett and Doull's Toxicology, The Basis Science of Poisons, 6th Ed., C.D. Klaassen, ed. New York: McGraw-Hill.

Fowler, B., D.D. Elcombe, B. Kelso, and G. Porlier. 1987. The threshold for hypoxia effects on perceptual-motor performance. Human Factors 29(1):61-66.

Gilbert, M.E. 1995. Repeated exposure to lindane leads to behavioral sensitization and facilitates electrical kindling. Neurotoxicol. Teratol. 17(2):131-141.

Go, V., J. Garey, M.S. Wolff, and B.G. Pogo. 1999. Estrogenic potential of certain pyrethroid compounds in the MCF-7 human breast carcinoma cell line. Environ. Health Perspect. 107(3):173-177.

Gong Jr., H., D.P. Tashkin, E.Y. Lee, and M.S. Simmons. 1984. Hypoxia-altitude simulation test. Evaluation of patients with chronic airway obstruction. Am. Rev. Respir. Dis. 130(6):980-986.

Goode, M. 2000. Chemistry and History of TCP Usage in Aviation Lubricating. Health Safety and Environmental Overview. Presentation at Chemical Exxon Mobile Conference, Great Lakes, December 2000.

Gosselin, R.E., R.P. Smith, and H.C. Hodge. 1984. Pp. III-98 in Clinical Toxicology of Commercial Products, 5th Ed. Baltimore: Williams and Wilkins.

Harp, P.R. 1998. Pyrethrin/pyrethroids. Pp. 610-611 in Encyclopedia of Toxicology, Vol. 2., P. Wexler and S.C. Gad, eds. San Diego: Academic Press.

He, F., S. Wang, L. Liu, S. Chen, Z. Zhang, and J. Sun. 1989. Clinical manifestations and diagnosis of acute pyrethroid poisoning. Arch. Toxicol. 63(1):54-58.

Healy, C.E., R.S. Nair, W.E. Ribelin, and C.L. Bechtel. 1992. Subchronic rat inhalation study with Skydrol 500B-4 fire resistant hydraulic fluid. Am. Ind. Hyg. Assoc. J. 53(3):175-180.

Henschler, D. 1958a. Hazards in use of modern tricresyl phosphates [in German]. Zbl. Arbeitsmed. 8(11):265-267.

Henschler, D. 1958b. Tricresyl phosphate poisoning. Experimental clarification of problems of etiology and pathogenesis [in German]. Klinische Wochenschrift 36(14):663-674.

Henschler, D. 1959. Relationships between the chemical structure and the paralyzing action of triaryl phosphates [in German]. Naunyn-Schmiedeberg's Arch. Exp. Path. U. Pharmak. 237:459-472.

Hewstone, R.K. 1994. Environmental health aspects of lubricant additives. Sci. Total Environ. 156(3):243-254.

Holmstrom, M., B. Wilhelmsson, H. Hellquist, and G. Rosen. 1989. Histological changes in the nasal mucosa in persons occupationally exposed to formaldehyde alone and in combination with wood dust. Acta Otolaryngol. 107(1-2):120-129.

HSDB [Hazardous Substances Data Bank]. 2000. Toxicology Data Network, National Library of Medicine's (NLM). [Online]. Available: http://toxnet.nlm.nih.gov/

IEH (Institute for Environment and Health). 2001. Consultation on the Possible Effects on Health, Comfort and Safety on Aircraft Cabin Environments. IEH Web Report W5. Leicester, UK: Institute for Environment and Health. [Online]. Available: http://www.le.ac.uk/ieh/webpub.html [March 2001].

IPCS (International Programme on Chemical Safety). 1999. Carbon monoxide. Environmental Health Criteria 213, 2nd Ed. Geneva: World Health Organization.

IRIS. 2001. Integrated Risk Information System. U.S. Environmental Protection Agency. [Online]. Available: http://www.epa.gov/iriswebp/iris/subst/0419.htm [October 19, 2001].

Jones, S.R., T.H. Lee, R.M. Wightman, and E.H. Ellinwood. 1996. Effects of intermittent and continuous cocaine administration on dopamine release and uptake regulation in the striatum: In vitro voltammetric assessment. Psychopharmacology 126(4):331-338.

Koren, G., T. Sharay, A. Pastuszak, L.K. Garrettson, K. Hill, I. Samson, M. Rorem, A. King, and J.E. Dolgin. 1991. A multicenter, prospective study of fetal outcome following accidental carbon monoxide poisoning in pregnancy. Reprod. Toxicol. 5(5):397-403.

Korpi, A., J.P. Kasanen, Y. Alarie, V.M. Kosma, and A.L. Pasanen. 1999. Sensory irritating potency of some microbial volatile organic compounds (MVOCs) and a mixture of five MVOCs. Arch. Environ. Health 54(5):347-352.

Kutzman, R.S. 1981. A Subchronic Inhalation Study of Fischer 344 Rats Exposed to 0, 0.4, 1.4, or 4.0 ppm Acrolein. Brookhaven National Laboratory, Upton, NY. National Toxicology Program: Interagency Agreement No. 222-Y01-ES-9-0043.

Landrigan, P.J., L. Claudio, S.B. Markowitz, G.S. Berkowitz, B.L. Brenner, H. Romero, J.G. Wetmur, T.D. Matte, A.C. Gore, J.H. Godbold, and M.S. Wolff. 1999. Pesticides and inner-city children: Exposures, risks, and prevention. Environ. Health Perspect. 107 (Suppl. 3):431-437.

Laviana, J.E., F.H. Rohles Jr., and P.E. Bullock. 1988. Humidity, comfort and contact lenses. ASHRAE Transactions 94:3-11.

Ledwith, F. 1970. The effects of hypoxia on choice reaction time and movement time. Ergonomics 13(4):465-482.

Lee, S.C., C.S. Poon, X.D. Li, F. Luk, M. Chang, and S. Lam. 2000. Air quality measurements on sixteen commercial aircraft. Pp. 45-58 in Air Quality and Comfort in Airliner Cabins, N.L. Nagda, ed. West Conshohocken, PA: American Society for Testing and Materials.

Leng, G., J. Lewalter, B. Roehrig, and H. Idel. 1999. The influence of individual susceptibility in pyrethroid exposure. Toxicol. Lett. 107(1-3):123-130.

Macher, J.M., and J. Rosenberg. 1999. Evaluation and management of exposure to infectious agents. Pp. 287-371 in Handbook of Occupational Safety and Health, 2nd Ed., L.J. DiBerardinis, ed. New York: John Wiley & Sons.

Mackerer, C.R, and E.N. Ladov. 2000. Submission 13A to the Senate References Committee Rural and Regional Affairs and Transport on The Inquiry into Air Safety- BAe 146 Cabin Air Quality. Pp. 44-54 in Air Safety and Cabin Air Quality in the BAe 146 Aircraft, Vol. 3. Report by the Senate Rural and Regional Affairs and Transport References Committee, Parliament of the Commonwealth of Australia, Parliament House, Canberra. October 2000.

Mackerer, C.R., M.L. Barth, A.J. Krueger, B. Chawla, and T.A. Roy. 1999. Comparison of neurotoxic effects and potential risks from oral administration or ingestion of tricresyl phosphate and jet engine oil containing tricresyl phosphate. J. Toxicol. Environ. Health A. 57(5):293-328.

Maresh, C.M., L.E. Armstrong, S.A. Kavouras, G.J. Allen, D.J. Casa, M. Whittlesey, and K.E. LaGasse. 1997. Physiological and psychological effects associated with high carbon dioxide levels in healthy men. Aviat. Space Environ. Med. 68(1): 41-45.

McFarland, R.A. 1946. Pp. 69-101 in Human Factors in Air Transport Design, 1st Ed. New York: McGraw-Hill.

McFarland, R.A., and J.N. Evans. 1939. Alterations to dark adaptions under reduced oxygen tensions. Am. J. Physiol. 127:37-50.

Nagda, N.L, ed. 2000. Air Quality and Comfort in Airliner Cabins. West Conshohocken, PA: American Society for Testing and Materials.

Nagda, N.L., and M. Hodgson. 2001. Low relative humidity and aircraft cabin air quality. [Review]. Indoor Air 11(3):200-214.

Nagda, N.L., M.D. Fortmann, M.D. Koontz, S.R. Baker, and M.E. Ginevan. 1989. Airliner Cabin Environment: Contaminant Measurements, Health Risks, and Mitigation Options. DOT-P-15-89-5. NTIS/PB91-159384. Prepared by GEOMET Technologies, Germantown, MD, for the U.S. Department of Transportation, Washington DC.

Nagda, N.L., H.E. Rector, Z. Li, and D.R. Space. 2000. Aircraft cabin air quality: A critical review of past monitoring studies. Pp. 215-235 in Air Quality and Comfort in Airliner Cabins, N.L. Nagda, ed. West Conshohocken, PA: American Society for Testing and Materials.

Nagda, N.L., H.E. Rector, Z. Li, and E.H. Hunt. 2001. Determine Aircraft Supply Air Contaminants in the Engine Bleed Air Supply System on Commercial Aircraft. ENERGEN Report AS20151. Prepared for American Society of Heating, Refrigerating, and Air-Conditioning Engineers, Atlanta, GA, by ENERGEN Consulting, Inc., Germantown, MD. March 2001.

Naughton, M.T., P.D. Rochford, J.J. Pretto, R.J. Pierce, N.F. Cain, and L.B. Irving. 1995. Is normobaric simulation of hypobaric hypoxia accurate in chronic airflow limitation? Am. J. Respir. Crit. Care Med. 152(6 Pt 1):1956-1960.

Naumann, I.D., and K. McLachlan. 1999. Aircraft Disinsection. Report commissioned by the Australian Quarantine and Inspection Service. June 1999.

Nicholson, A.N.   1996.   Low Humidity: Dehydration, Dipsosis or Just Dryness? Royal Air Force School of Aviation Medicine Report No 01/96, Hampshire, UK. May 1996.

NRC (National Research Council). 1986. The Airliner Cabin Environment: Air Quality and Safety. Washington, DC: National Academy Press.

NRC (National Research Council. 1994. Health Effects of Permethrin-Impregnated Army Battle-Dress Uniforms. Washington, DC: National Academy Press.

Parkins, K.J., C.F. Poets, L.M. O'Brien, V.A. Stebbens, and D.P. Southall. 1998. Effect of exposure to 15% oxygen on breathing patterns and oxygen saturation in infants: Interventional study. BMJ. 316(7135):887-894.

Parliament of the Commonwealth of Australia. 2000. Air Safety and Cabin Air Quality in the BAe 146 Aircraft. Report by the Senate Rural and Regional Affairs and Transport References Committee, Parliament House, Canberra. October 2000.

Paustenbach, D., Y. Alarie, T. Kulle, N. Schachter, R. Smith, J. Swenberg, H. Witschi, and S.B. Horowitz. 1997. A recommended occupational exposure limit for formaldehyde based on irritation. J. Toxicol. Environ. Health 50(3):217-263.

Peters, A., E. Liu, R.L. Verrier, J. Schwartz, D.R. Gold, M. Mittleman, J. Baliff, J.A. Oh, G. Allen, K. Monahan, and D.W. Dockery. 2000. Air pollution and incidence of cardiac arrhythmia. Epidemiology 11(1):11-17.

Rankin, W. L., D. R. Space, and N.L. Nagda. 2000. Passenger comfort and the effect of air quality. Pp. 269-290 in Air Quality and Comfort in Airline Cabins, N.L. Nagda, ed. West Conshohocken: American Society for Testing and Materials.

Rayman, R.B. 1997. Passenger safety, health, and comfort: A review. Aviat. Space Environ. Med. 68(5):432-440.

Ritter, S. 2001. Aircraft deicers. Chem. Eng. News 79(1):30.

Rose, D.M., D. Jung, D. Parera, and J. Konietzko. 1999. Time zone shift and the immune system during long-distance flights. [in German]. Z. Arztl. Fortbild. Qualitatssich. 93(7):481-484.

Rose, L.J., R.B. Simmons, S.A. Crow, and D.G. Ahearn. 2000. Volatile organic compounds associated with microbial growth in automobile air conditioning systems. Curr. Microbiol. 41(3):206-209.

Robkin, M.A. 1997. Carbon monoxide and the embryo. Int. J. Dev. Biol. 41(2):283-289.

Schwartz, J.S., H.Z. Bencowitz, and K.M. Moser. 1984. Air travel hypoxemia with chronic obstructive pulmonary disease. Ann. Intern. Med. 100(4):473-477.

Sheffield, P.J., and R.D. Heimbach. 1996. Respiratory physiology. Pp. 68-108 in Fundamentals of Aerospace Medicine, 2nd Ed., R. DeHart, ed. Baltimore: Williams and Wilkins.

Siegel, J., H.S. Rudolph, A.J. Getzkin, and R.A. Jones. 1965. Effects on experimental animals of long-term continuous inhalation of a triaryl phosphate hydraulic fluid. Toxicol. Appl. Pharmacol. 7(4):543-549.

Slonim, N.B., and L.H. Hamilton. 1971. Respiratory Physiology, 2nd Ed. St. Louis, MO: Mosby.

Smith, R.P. 1996. Toxic responses of the blood. Pp. 335-354 in Casarett and Doull's Toxicology, The Basis Science of Poisons, 5th Ed., C.D. Klaassen, ed. New York: McGraw-Hill.

Snyder, R., and L.S. Andrews. 1996. Toxic effects of solvents and vapors. Pp. 737-772 in Casarett and Doull's Toxicology, The Basis Science of Poisons, 5th Ed., C.D. Klaassen, ed. New York: McGraw-Hill.

Space, D.R., R.A. Johnson, W.L. Rankin, and N.L. Nagda. 2000. The airplane cabin environment: Past, present and future research. Pp. 189-210 in Air Quality and Comfort in Airliner Cabins, N.L. Nagda, ed. West Conshohocken, PA: American Society for Testing and Materials.

Spektor, D.M., M. Lippmann, G.D. Thurston, P.J. Lioy, J. Stecko, G. O'Connor, E. Garshick, F.E. Speizer, and C. Hayes. 1988a. Effects of ambient ozone on respiratory function in healthy adults exercising outdoors. Am. Rev. Respir. Dis. 136(4):821-828.

Spektor, D.M., M. Lippmann, G.D. Thurston, P.J. Lioy, K. Citak, D.J. James, N. Bock, F. E. Speizer, and C. Hayes. 1988b. Effects of ambient ozone on respiratory function in active, normal children. Am. Rev. Respir. Dis. 137(2):313-320.

Steinmetz, D. 1998. Carbon monoxide. Pp. 224-226 in Encyclopedia of Toxicology, Vol.1., P. Wexler, and S.C. Gad, eds. San Diego: Academic Press.

Tashkin, D.P., A.H. Coulson, M.S. Simmons, and G.H. Spivey. 1983. Respiratory symptoms of flight attendants during high-altitude flight: Possible relation to cabin ozone exposure. Int. Arch. Occup. Environ. Health. 52(2):117-137.

Thienes, C., and T.J. Haley. 1972. Pp. 57 in Clinical Toxicology, 5th Ed. Philadelphia: Lea and Febiger.

Thurston, G.D., M. Lippmann, M.B. Scott, and J.M. Fine. 1997. Summertime haze air pollution and children with asthma. Am. J. Respir. Crit. Care Med. 155(2):654-660.

Vandenplas, O., J.P. Dewiche, J. Auverdin, J.M. Caroyer, and F. Binard-Van Cangh. 2000. Asthma to tetramethrin. Allergy 55(4):418-419.

Van Netten, C. 1998. Air quality and health effects associated with the operation of BAe 246-200 aircraft. Appl. Occup. Environ. Hyg. 13(10):733-739.

Van Netten, C. 2000. Analysis of two jet engine lubricating oils and a hydraulic fluid: Their pyrolytic breakdown products and their implication on aircraft air quality. Pp. 61-75 in Air Quality and Comfort in Airliner Cabins, N.L. Nagda, ed. West Conshohocken, PA: American Society for Testing and Materials.

Van Netten, C., and V. Leung. 2000. Comparison of the constituents of two jet engine lubricating oils and their volatile pyrolytic degradation products. Appl. Occup. Environ. Hyg. 15(3):277-283.

Van Netten, C., and V. Leung. 2001. Hydraulic fluids and jet engine oil: Pyrolysis and aircraft air quality. Arch. Environ. Health 56(2):181-186.

Vieillefond, H., J.L. Poirier, and H. Marotte. 1981. Influence de l' altitude sur la toxicite des oxydes de carbone. Pp. B11.1-B11.4 in Toxic Hazards in Aviation: Papers presented at the Aerospace Medical Panel Specialists' Meeting held in Toronto, Canada, 15-19 September 1980. AGARD-Conference Proceedings No. 309. Neuilly sur Seine, France: AGARD, Advisory Group for Aerospace Research & Development.

Waters, M., T. Bloom, and B. Grajewski. 2001. Cabin Air Quality Exposure Assessment. National Institute for Occupational Safety and Health, Cincinnati, OH. Federal Aviation Administration Civil Aeromedical Institute. Presented to the NRC Committee on Air Quality in Passenger Cabins of Commercial Aircraft, January 3, 2001, Washington, DC.

Weiss, B., and S. Santelli. 1978. Dyskinesias evoked in monkeys by weekly administration of haloperidol. Science 200(4343):799-801.

WHO (World Health Organization). 1995. Report of the Informal Consultation on Aircraft Disinsection, WHO/HQ, Geneva, 6-10 November 1995, International Programme on Chemical Safety. Geneva, Switzerland: World Health Organization.

Wills, J.H., F. Coulston, E.S. Harris, E.W. McChesney, J.C. Russell, and D.M. Serrone. 1974. Inhalation of aerosolized ethylene glycol by man. Clin. Toxicol. 7(5):463-476.

Wright, R.L. 1996. Formation of the neurotoxin TMPP from TMPE-Phosphate Formulations. Tribology Trans. 39(4):827-834.

Wyman, J., E. Pitzer, F. Williams, J. Rivera, A. Durkin, J. Gehringer, P. Serve, D. von Minden, and D. Macys. 1993. Evaluation of shipboard formation of a neurotoxicant (trimethyolpropane phosphate) from thermal decomposition of synthetic aircraft engine lubricant. Am. Ind. Hyg. Assoc. J. 54(10):584-592.

Yonkosky, D., L. Ladia, L. Gackenheimer, and M.W. Schultz. 1990. Scabies in nursing homes: An eradication program with permethrin 5% cream. J. Am. Acad. Dermatol. 23(6 Pt.1):1133-1136.

6

Health Surveillance

At the end of its review of health data in the 1986 report *The Airliner Cabin Environment: Air Quality and Safety*, the National Research Council (NRC) committee concluded that "available information on the health of crews and passengers stems largely from ad hoc epidemiologic studies or case reports of specific health outcomes [and] conclusions that can be drawn from the available data are limited to a great extent by self-selection . . . and lack of exposure information" (NRC 1986). This chapter reviews data on possible health effects of exposure to aircraft cabin air that have emerged since the 1986 report and the emergence of data resources (e.g., surveillance systems) and studies that have particular relevance for the evaluation of potential health effects related to aircraft cabin air quality. Selected earlier sources are also reviewed. The decision to ban tobacco-smoking on domestic airline flights in 1987 and on flights into and out of the United States in 1999 reduces the relevance of some studies of exposures and reported signs and symptoms that clearly could have been related to the products of tobacco smoke.

A wide array of symptoms have been attributed to various exposures to cabin air as a result of normal aircraft operations or incidents (Table 6-1). The symptoms or health effects are grouped into four categories that are intended to be descriptive and do not imply mechanisms. The column "Health Outcomes" identifies outcomes related to chronic exposure, to cabin air, or to physiological responses in groups of people who may be at particular risk in the cabin of a commercial airliner, such as passengers with underlying cardiac or pulmonary disease. Most of the symptoms listed in the first three categories have been reported primarily by cabin crews. Very few data are available on passengers, and their symptoms are primarily drying of mucous membranes.

TABLE 6-1 Signs and Symptoms Reported to Be Related to Aircraft Cabin Air

Reported Signs and Symptoms

Respiratory and Mucosal Surfaces	Neurological and Neurobehavioral	Syndrome-Symptom Complexes	Other Symptoms	Health Outcomes
Irritation, pain (eyes, nose, sinuses, throat)	Neurotoxicity (altered vision and coordination, loss of balance, slurred speech, paresthesias)	*Aerotoxic syndrome* Irritability, neurotoxicity, chemical sensitivity	Dry skin	Reproductive effects
Difficulty breathing		*Acute intoxications*	Rapid heart rate and palpitations	Cancer (unrelated to cosmic radiation)
Breathing discomfort	Neurobehavioral (impaired memory and ability to concentrate, trouble counting, general cognitive problems)	Neurological signs and symptoms, neurobehavioral effects, cardiovascular effects, gastrointestinal symptoms		Lung function effects
Pain in chest				Signs and symptoms related to underlying disease or episodes of worsening underlying diseases (cardiovascular, chronic respiratory)
Coughing		*Chemical sensitivity*		
Dry, stuffy nose	Headache	*Ozone-related* Cough, chest discomfort, irritation of mucous membranes		Signs and symptoms related to cabin pressure and oxygen content
	Disorientation and confusion			
	Lightheadedness			Acute infections (1979 influenza outbreak, tuberculosis)
	Dizziness	*Organophosphate-induced delayed neuropathy syndrome*		Immunosuppression
	Weakness and fatigue			Hair loss
	Feelings of intoxication	Delayed-onset weakness, ataxia, paralysis		

The sources of the various reports or symptoms can be grouped into two broad categories (Table 6-2): various types of systematic presentations, only a few of which represent formal studies, and selected written documents that have not been peer-reviewed and published. Most of the reports of symptoms come from collections of case reports abstracted from various reporting systems for cabin and cockpit crew members, usually in relation to known or suspected incidents. This chapter begins with a discussion of the systematic studies that have been conducted, including those on specific exposure conditions (e.g., pesticides, infectious agents, and cabin pressure). That is followed by a brief review of other sources of health information. Finally, the current health-related data collection systems are reviewed.

STUDIES OF AIRCRAFT CABIN AIR QUALITY

Relatively few formal studies have evaluated the effects on passengers and cabin crew of exposure to aircraft cabin air during routine flights or during flight-related air-quality incidents.

General Air-Quality Surveys

Although the data were available to the 1986 NRC committee, the report of Tashkin et al. (1983) is reviewed briefly here because it has been quoted often in more recent publications and testimony (e.g., Australian Senate Rural and Regional Affairs and Transport References Committee 1999). The authors analyzed survey data collected by a flight attendants union in response to complaints of various respiratory symptoms among attendants who flew on the Boeing 747SP. Attendants were concerned that the aircraft flew at high altitudes and that increased exposure to ozone (O_3) was therefore possible. O_3 concentrations as high as 1.09 ppm had been measured in the cabins of 747SP aircraft on other occasions (Tashkin et al. 1983). The authors acknowledged that they did not have a role in the survey design and that considerable methodological problems existed, including the following:

1. The response rate of the surveys was inadequate. Only 55.1% (248 of 450) of the original surveys distributed to attendants who flew on the 747SPs were returned. An attempt to obtain "control" data from cabin crew who flew on other 747s led to a response rate of only 15.3% (38 of 248). A third survey had a response rate of 7.6% (65 of 850).

TABLE 6-2 Sources of Data on Health Outcomes Attributed to Exposure to Aircraft Cabin Air

Planned Studies, Systematic Reviews of Records, and Presentations at Conferences

Respiratory Symptoms of Flight Attendants During High-Altitude Flight, Tashkin et al. (1983)
— Analysis of survey conducted by flight attendant union in response to concerns about O_3 exposure.
— Response frequencies to surveys ranged from 7.6% to 55%.
— No direct measurements of exposure.
Flight Attendant Health Survey, Cone and Cameron (1984)
— Flights from San Francisco to Honolulu.
— Monitored ozone, nitrogen oxides, sulfur dioxide, phosphoric acid esters.
— Surveyed crew for symptoms.
— 95% response frequency (683/720), but total eligible not given.
Air Quality on Commercial Aircraft, Pierce et al. (1999)
— Cabin air monitoring of 8 Boeing 777s; comfort survey included.
— No sampling plan (930 (43%) of passengers; 27 (26%) of flight attendants).
American Society of Heating, Refrigerating and Air-Conditioning Engineers (ASHRAE) Project 957-RP, ASHRAE/CSS (1999)
— Systematic monitoring of environment of Boeing 777 with recirculation and O_3 converter.
— Passenger and crew comfort questionnaire; 930 (43%) passengers and 27 (26%) of cabin crew completed survey.
— Appendix contains literature review with references to small planned studies related to physiology.
O_3 and Relative Humidity on Airline Cabins on Polar Routes, De Ree et al. (2000)
— Comparison of ozone-related symptoms in crew of planes with and without O_3 catalytic converters.
— O_3 and relative humidity measurements.
— Variable response rates from pilots (79-94%) and cabin crew (66-71%).
— Nonsystematic survey of passengers.
— Weak statistical analysis.
Questionnaire Survey to Evaluate the Health and Comfort of Cabin Crew, Lee et al. (2000)
— Nonsystematic sampling of cabin crew on Cathay Pacific flights.
— Number eligible and response rate not reported.
Passenger Comfort and the Effect of Air Quality, Rankin et al. (2000)
— Self-administered survey of passengers on six types of aircraft with air recirculation of 0 to 50%.
— Only 43% response rate (3,630/8,517).

Smoke/fumes in the Cockpit, Rayman and McNaughton (1983)
— Review of 89 incidents of smoke and fumes in cockpit in U.S. Air Force aircraft.
— Apparent complete ascertainment.

Air Quality and Health Effects Associated with the Operation of BAe 146-200 Aircraft, van Netten (1998)
— Review of clinical assessments and accident reports over 4 mo filled out by crew on two BAe 146-200 that had experienced oil-seal failures.
— Completeness of data unknown.

Aerotoxic Syndrome, Winder and Balouet (2000a)
— Published presentation at Proceedings of the International Congress on Occupational Health, Brisbane, 2000.
— Summary of "aerotoxic syndrome."
— No primary data or references.

In-Cabin Trace Chemicals and Crew Health Issues, Balouet (1998); Airborne Chemicals in Aircraft Cabins, Balouet and Winder (2000a)
— Data based on 350 selected reports of symptoms supposedly related to documented leak events.
— No details on methods for selection of reports or documentation of leaks and exposure.
— Cites common words to describe symptoms from diverse and unrelated incidence episodes in support of constellation of symptoms.
— Citation of impending investigations as implicit evidence of problem.
— "Aerotoxic syndrome."

Outbreak of Influenza Aboard a Commercial Airliner, Moser et al. (1979)
— 72% of passengers at risk became ill.
— Flight was grounded for over 3 h and air-conditioning system was inoperative.

Outbreak of Influenza A/Taiwan/1/86 Infections at a Naval Base and Association with Airplane Travel, Klontz et al. (1989)
— Influenza among 114-member squadron within 72 h after flights from Puerto Rico to Key West, Florida.
— Evidence of transmission of influenza was occurring before flights.

Imported Measles in the United States, Amler et al. (1982)
— Three cases of measles in children on flight from Venezuela to Miami, Florida (one child had prodromal symptoms during the flight).
— No information on relationships between children, their proximity, or contact before, during, or after flight; two secondary cases could have been infected before flight.

Surveillance Report of Measles Transmission in an Airport, CDC (1983)
— One person appeared to infect one passenger on same flight and five others within airport.

(Continued)

Measles Outbreak on Flight from New York to Tel Aviv, Slater et al. (1995)
— 8 of 306 passengers developed measles.
— No information on location of passengers in aircraft.
— Waiting area was congested, and loaded plane was grounded for 2 h for repairs.

Tuberculosis Transmission on a Flight from London to Minneapolis, Minnesota, McFarland et al. (1993)
— Contact investigation for passenger with pulmonary tuberculosis.
— Alternative explanations could be found for all cases with positive skin tests.

Tuberculosis Transmission among Flight Crew, Driver et al. (1994)
— Flight attendant flew for 6 mo while symptomatic.
— Crew members on same flights as infected attendant were compared with control group of crew on other flights.
— Greater prevalence of positive skin tests among crew that had contact with infected attendant.

Tuberculosis Transmission on Airline Flights, Kenyon et al. (1996)
— Symptomatic passenger flew on four flights; other passengers and crew were investigated.
— Greatest number of positive skin tests were found on fourth flight.
— Six passengers that had no other risk factors were seated in same section as infected person.

Tuberculosis Transmission on Airline Flights, Miller et al. (1996)
— Contact investigation of passengers and crew from two flights with passenger who had pulmonary tuberculosis.
— Data not available on 35% of potential contacts.
— Two people with positive skin test (and no other risk factors) did not sit near or have contact with infected passenger.

Tuberculosis Transmission on Airline Flights, Moore et al. (1996)
— Infected person flew on two short flights (about 1 h).
— Contact investigation of passengers and crew.
— Data available on only 53% of potential contacts.

Tuberculosis Transmission on a Flight, Wang (2000)
— Contact investigation of passengers and crew on flight with infected passenger.
— Three subjects with skin-test conversions were not seated in same section as infected passenger.

In-flight Arterial Saturation in Pilots, Cottrell et al. (1995)
— Oxygen saturation was measured with continuous-reading pulse oximeters during flights of about 4 h.

In-flight Arterial Saturation in Subjects with Chronic Obstructive Lung Disease, Schwartz et al. (1984)

— Measurements taken in unpressurized aircraft cabin before and during flight.
Arterial Saturation in Subjects with Chronic Obstructive Lung Disease, Dillard et al.
(1989), Naughton et al. (1995)
— Simulation performed in hypobaric chamber.
Air Travel in Patients with Chronic Obstructive Lung Disease, Dillard et al. (1991)
— 44 of 100 subjects reported traveling by air within previous 2 yr.
— 8 subjects reported symptoms.
Oxygen Saturation in Cabin Crew, Ross (2001)
— "Spot" measurements were taken periodically with pulse oximeters.
— No data on saturations related to cabin altitude and work activities of crew.

Testimony to Various Committees and Unpublished Summaries

Commonwealth of Australia Proof Committee Hansard Senate–Air Safety—BAe 146
Cabin Air Quality, Senate Rural and Regional Affairs and Transport References
Committee (1999)
— Case-report testimony based on exposures to smoke, mist, fumes.
— "Old socks" odors.
— Association between odors and short-term irritation of mucous membranes,
  nausea, shortness of breath.
Case Study of "Cabin Crew Syndrome" Presented to ASHRAE Aviation
Subcommittee, Wright and Clarke (1999)
— Suggestion, but no documentation, of an incident (sweet smell).
— Single case with multiple signs and symptoms over many months.
— Inference based on toxicology of organophosphates.
Association of Flight Attendants (AFA) Reports of Health Effects Related to
Exposure to Insecticide Spraying, Witkowski (1999); Cone and Das (2001)
— Sample of cases submitted by AFA.
— Lists of disability claims.
Unreferenced Reports on Symptoms and Symptom Complexes by Balouet and
Colleagues
— Description of "aerotoxic syndrome" (Balouet and Winder 1999).
  – Anecdotal presentation of symptom complex.
  – No exposure data.
  – No information on how subject data were obtained.
  – Presentation of case studies.
— Report on symptoms associated with "exposure" to chemicals in aircraft
  (Balouet and Winder 2000b).
  – Apparent tabulation of data from several databases to which reports are
    filed.
  – No information on techniques for selection of data.
  – No information on quality of data reports.

---

2. No O_3 concentrations were measured on any flight for which survey data were available.

3. The association with possible O_3 exposure during flight was based on the subjective assessment of three experts who decided which reported symptoms were most likely due to O_3 exposure. The only objective measurements were of pulmonary function in 21 attendants (no criteria for selection given). All results were normal.

The possibility of selection bias in the study means that the results cannot be interpreted as supporting or refuting the relation of symptoms to O_3 exposure on high-altitude flights. Moreover, many of the symptoms considered "definitely" or "probably" related to O_3 exposure are nonspecific and prevent use of the study as an assessment tool for estimating effects of O_3 exposure.

The Occupational Health Clinic of San Francisco General Hospital was commissioned by the AFA and American Airlines to conduct a study of the cause of symptoms reported by flight crews on flights between San Francisco and Honolulu in association with an odor described as that of "dirty socks" (Cone and Cameron 1984). Three types of aircraft—Boeing 747, DC-10-10, and DC10-30—were evaluated. Questionnaires were distributed to crew members. A wide array of symptoms were reported more frequently on the B747 and the DC-10-10 (eye, nose, throat, and sinus irritation; dry or watery eyes; shortness of breath; dizziness; and lightheadedness). Symptoms of eye, nose, and sinus irritation; headache; and chest symptoms (burning, difficulty in breathing, and cough) were reported more frequently when odors were noted. Of the agents monitored—O_3, nitrogen oxides, sulfur dioxide, and phosphoric acid esters—only episodic increases in nitrogen oxides were observed, and association with odor was reported only on one occasion. No other exposure data were related to symptoms. The authors concluded that there was evidence of exposure to a "powerful mucous membrane irritant" whose etiology could not be determined from the study. They speculated that vaporization and pyrolysis products of aircraft fluids were a possible cause. Although a high response rate was reported (95%, 683 of 720), the total number of eligible crew members was not provided, nor were any data provided on the relation of chronic respiratory symptoms to smoking (32% were current smokers) or allergy (36% had unspecified allergy). The statistical analysis did not take into account crew differences in the factors that could have contributed to the results.

The ASHRAE commissioned a study to monitor the cabin air of Boeing 777s (Pierce et al. 1999). A cabin-comfort survey administered to passengers and cabin crew had a number of components, one of which related specifically to symptoms. For virtually every symptom, cabin crews were more likely to report the presence of a symptom and greater severity than passengers (Table 6-3). The study has several serious problems that make it difficult to interpret the data. The response rates of passengers and crew were very low (Table 6-2). The percentages for the reported symptoms do not add to 100% (Table 6-3), so it is difficult to know what is being reported. Finally, no formal analysis links the environmental measurements with the symptom reports.

In response to complaints from cabin and cockpit crews of a variety of symptoms thought to be related to O_3 exposure, DeRee and colleagues (2000) carried out a monitoring and symptom study in two European airlines. The planes of airline A were fitted with O_3 converters but lacked humidifiers; airline B had humidifiers but no O_3 converters. O_3 and relative-humidity measurements were made on the flight deck. Symptom questionnaires were completed by flight and cabin crew at the beginning and end of 24 polar flights (12 flights per airline) monitored in February-May 1998. Cabin crews were "encouraged" to record symptoms reported by passengers, but no systematic surveillance was conducted. Background information on smoking and upper respiratory symptoms was obtained. Participation rates of cabin crews were 71% (187) on airline A and 66% (222) on airline B. Mean O_3 concentrations during cruise on airlines A and B were 2-40 and 43-177 ppb (not stated if sea-

TABLE 6-3 Report of Selected Symptoms by Passengers and Cabin Crew

Symptoms	Passengers (n = 930)[a]			Cabin Crew (n = 27)[a]		
	Great Extent	Some-what	Not At All	Great Extent	Some-what	Not At All
Dry, itchy, or irritated eyes	7.3	17.3	33.0	22.2	37.0	3.7
Dry, stuffy nose	9.7	19.9	30.5	29.6	37.0	7.4
Sore, dry throat	4.8	12.8	37.0	11.1	33.3	22.2
Shortness of breath	0.9	3.8	45.3	3.7	22.2	37.0
Sinus pain	2.9	7.3	41.9	18.5	7.5	44.4
Skin dryness or irritation	4.4	10.5	38.4	37.0	33.3	0.0

[a] Percentages do not add to 100% for each symptom and no explanation given in Pierce et al. (1999) or ASHRAE/CSS (1999).

Source: Adapted from Pierce et al. (1999).

level equivalent), respectively. Mean relative humidity was 8-12% and 5-28% on airlines A and B, respectively. Of the 16 symptoms included in the questionnaire, four were considered to be O_3-related (coughing, tightness of the chest, shortness of breath, and "breathing hurts"). Symptoms were reported equally by cabin crew of the two airlines before flight (68%) and similarly after flight (A, 95%; B, 91%). Symptom reporting was also similar in percentages of cabin-crew members who reported worsening (about 30%) and improvement in preflight O_3-related symptoms. Similar findings were observed for aggregate nonspecific symptoms (e.g., headache and watery and stinging eyes); however, no data are provided on symptoms specifically related to mucosa irritation. No correlation (Pearson product-moment) was observed for changes in O_3-related or nonspecific symptoms and measured O_3 concentrations (it was not stated which O_3 metric was used—mean or maximum) on airline A; weak, statistically nonsignificant correlations were observed for O_3-related (0.18) and nonspecific (0.21) symptoms on airline B. The authors concluded that there was "not a straightforward relationship between O_3 levels and relative humidity on the one hand and reported symptoms on the other." Although the study had a reasonable design, the presentation of the data is problematic. The assumption that the four symptoms were O_3-related was not unreasonable, given findings from controlled human exposure in environmental chambers (Folinsbee et al. 1988, 1994). However, those symptoms are not specific for O_3 exposure. No data are provided on the specific symptoms, and the basic data analysis is weak. The use of Pearson correlations probably is not justified, given the distributional characteristics of the symptom data (binomial). Other analytic strategies, such as logistic regression, would have been more appropriate and could have been used to evaluate symptom response in relation to O_3, relative humidity, smoking, and preflight medical conditions. Moreover, that there were so few measurements of independent O_3 and relative humidity gave the study relatively little precision in estimating health effects related to O_3.

A survey of symptoms among 185 cabin-crew members of an Asian airline was carried out from September 1996 to March 1997 (Lee et al. 2000). The number of crew members eligible to participate and the method of flight selection are not provided, although the authors state that the survey was "compulsory." Various mucosal, respiratory and nonspecific symptoms (e.g., headache and faintness) were evaluated. Air quality (carbon dioxide, relative humidity, temperature, noise level, PM_{10}, and carbon monoxide [CO]) was monitored on all flights on which questionnaires were distributed. The monitoring instruments were placed in the center of the economy class, but the details of the monitoring protocol are not provided. Flights were 1.5-18 h long, and smoking was permitted. The results are presented in terms of acceptability ratings of various components of air quality and health symptoms; no formal analysis plan is provided. Table 6-4 summarizes the reported symptoms. Over 50% of the crew members reported at least one symptom of skin or mucosal irritation during the flight. No data are reported on any of the following: the relationship between monitoring data and reports of symptoms or ratings of air-quality acceptability, the presence of and relationship between preflight symptoms and symptoms experienced during flight, the relationship between flight duration and symptoms, and the relationship between the amount of smoking on each flight and the reports of symptoms.

A survey of passenger comfort and cabin air quality on standard and wide-body aircraft with and without recirculated air was carried out by an aircraft manufacturer (Rankin et al. 2000). A series of health-related questions were included in the survey. Self-administered questionnaires were completed by 3,630 passengers (43% of 8,517 distributed) on 71 flights in March, April, and June 1997. Of the questionnaires completed, 57% were from flights of 2-3 h, 32% from flights of 6-7 h, and 11% from flights of 10-12 h. The contribution of the flight to the presence of symptoms was rated on a scale of 1 (great extent) to 7 (not at all). Figure 6-1 summarizes the data. Overall health before the flight (no data given on responses and on the format of the variable) was most closely correlated with overall health during the flight ($R = 0.80$; type of correlation coefficient not specified). A regression analysis showed that only

TABLE 6-4 Frequency of Symptoms Reported by Cabin Crew

Symptoms	None to Mild, %	Moderate to Severe, %
Dry, stuffy nose	26	74
Irritation, dryness, itchiness in eyes	36	64
Dry, sore throat	41	59
Dry, irritated skin	43	57
Stomach discomfort (indigestion, gas)	57	43
Ear problems	66	34
Dizziness, faintness, lightheadedness	67	33
Headache	68	32
Nausea, motion sickness	71	29
Shortness of breath	73	27

Source: Adapted from Lee et al. (2000).

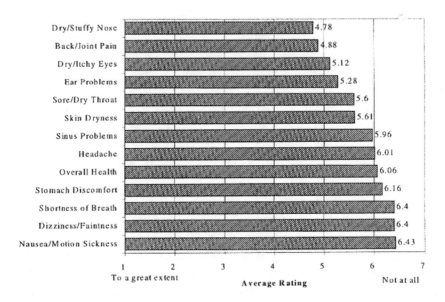

FIGURE 6-1 Average ratings of health-related symptoms. Source: Rankin et al. 2000. Reprinted with permission by American Society for Testing and Materials, copyright 2000.

five factors (health before trip; eye irritation and dryness; back, muscle, and joint pain; headache; and skin dryness and irritation) were related to "overall health," most of the variance being associated with health status before the trip. One of the standard-body planes evaluated used 100% bleed air for ventilation, and two used 50% and 30% recirculated air. The results are summarized in Table 6-5. For all symptoms but one, the ratings for symptoms were slightly more favorable (fewer cases) for the plane with 100% bleed air (no recirculation). However, the differences might not be meaningful and might be explained by the slightly better rating for "health before flight" given by passengers on the planes with no recirculated air. Overall health ratings tended to decrease with increasing flight duration. For all health variables, ratings were lower for flights of 6-7 h than those of 2-3 h. For most symptoms, there was a further decrement for flights of 10-12 h. The lowest ratings occurred for mucosal and skin symptoms. Although this study is large and relatively comprehensive in its presentation of the data, there is a possibility of substantial bias due to a participation frequency of less than 50%. Moreover, the validity of some of the analytic techniques (e.g., the regression analysis

TABLE 6-5 Average Passenger Ratings of Selected Symptoms in Relation to Percentage of Recirculated Air on Standard-Body Aircraft

	Percentage Recirculation[a]		
	0% (n = 235)	30% (n = 515)	50% (n = 423)
Dry, irritated, itchy eyes	5.49	5.16	5.18
Dry, stuffy nose	5.19	4.85	4.85
Sore, dry throat	5.84	5.64	5.67
Dry, irritated skin	5.99	5.83	5.71
Shortness of breath	6.56	6.38	6.42
Dizziness, faintness	6.54	6.39	6.43
Sinus pain	6.13	6.05	6.03
Health before flight	6.31	6.15	6.27
Health during flight	6.21	6.08	6.12

[a] Scores based on the extent to which flight contributed to symptoms: 1 = "to a great extent," 7 = "not at all."
Source: Adapted from Rankin et al. (2000).

cited previously) is questionable, given the type of data reported. It would have been useful to provide the percentages of passengers who reported no individual symptoms and no symptoms. Nonetheless, the study results show the need to consider the pre-existing health of passengers in any survey of health-related symptoms during flight.

Two reports specifically tried to relate symptoms to smoke or fume exposures in the aircraft. Rayman and McNaughton (1983) evaluated 89 reported incidents of smoke and fumes in the cockpits of military aircraft in 1970-1980. Multiple causes of the fumes or smoke were identified. Most symptoms were related to the central nervous system, the most common being dizziness (42 of 89)[1]; irritated eyes and mucous membranes (31 of 89); nausea and vomiting (31 of 89); confusion, disorientation, and performance decrement (23 of 89); headache (22 of 89); and decreased visual acuity (10 of 89). Chest pain, respiratory distress, and cough were reported in six or fewer instances. Although this report provides data that are complete with respect to frequency of occurrence of incidents, no information is provided on the extent to which the reports of symptoms were obtained by standardized methods or on the duration of the exposures; nor were objective exposure measures mentioned.

---

[1] Percentages are not given because it is not clear whether the denominator refers to events or persons.

In 1998, van Netten summarized accident reports and clinical assessments of complaints of ill health from flight crews who flew on the British Airways BAe 146 aircraft. The reports were collected over 4 mo. This aircraft had experienced episodes of oil-seal failures. Exposures were measured during flight and under test conditions on aircraft that had experienced such leaks. Of the 200 crew members at risk, 112 (56%) from five aircraft (35 flights each) reported symptoms. The most frequently reported symptoms were headache (25.9%), burning eyes (24.1%), burning throat (42.9%), and disorientation (14.3%). Lower respiratory complaints were reported by less than 3% of crew members. Paresthesias were reported in five instances. Carboxyhemoglobin measurements obtained 4 h after an incident in four subjects (criteria for selection not given) were 2% or less (the highest value was observed in a smoker). The author reported that all symptoms abated in 24 h, and no chronic symptoms were reported. However, the duration of followup and the occurrence of a systematic survey of chronic symptoms were not reported. This study has the clear advantage of having measured exposure on aircraft implicated in the occurrence of symptoms in crew members. Although symptom reports coincided with oil leaks, the specific agents responsible, the relationship between duration of exposure and symptoms, and the relationship between specific symptoms and locations and activities on the aircraft cannot be ascertained from the data. Test flights and sampling with one aircraft on the tarmac with engines running (both aircraft had been implicated in incidents with odors and symptoms in crew) identified a variety of volatile organic compounds, a "distinct oil odor," and CO concentrations of 1-2 ppm (a single reading was 3 ppm). No crew members were present during the testing, and specific symptoms previously reported by the crew could not be related to compounds identified in the analyses.

The "Aerotoxic Syndrome"

Balouet and Winder have argued in a series of documents for the existence of a stereotypical symptom complex, "aerotoxic syndrome," which attends exposure of cabin crew to hydraulic fluids, engine oil, and their pyrolysis products (Balouet 1998; Balouet and Winder 1999, 2000a,b; Winder and Balouet 2000b). Their papers repeat many data, so the committee's evaluation of them focuses on the one document (Balouet and Winder 2000b) that has the clearest presentation of the authors' contention that such a syndrome exists. The authors argue that in-cabin leaks, smoke, and fume events could expose up to 40,000 passengers and crews worldwide each year, although the committee was unable to verify the source for this assertion.

Support for the existence of the syndrome is derived from the work of Rayman and McNaughton (1983), Tashkin et al. (1983), and Van Netten (1998), which is evaluated above. Balouet and Winder (2000b) state that there "are common themes in symptom clusters in these studies." However, that claim does not appear to be supported by the data presented. For example, of the three most common symptoms (eye irritation, pain on deep breathing, and shortness of breath) in Tashkin et al. (1983) (the largest of the three studies), at least two are not reported in either of the other two studies. In fact, only three symptoms (headache, sinus congestion, and nausea) are reported in all three studies, and there is rather little agreement on their prevalence. Six case studies are also cited; however, the committee found it difficult to interpret them, given the lack of selection criteria, the sources of the material used in the case summaries, and the incomplete and qualitative nature of the summaries.

Thus, the committee concludes that evidence does not warrant the designation of a specific syndrome related to exposure to various physical agents (e.g., mists and smoke) and decomposition products derived from leaks of engine oil and hydraulic fluids. The committee recommends that until such information is available, the designation "aerotoxic syndrome" not be used for symptoms reported in coincidence with cabin air contamination.

Disinsection Studies

The details of current disinsection practices and the attendant toxicology are discussed in Chapters 3 and 5, respectively. The committee evaluated a report prepared by the California Department of Health Services Occupational Health Branch,[2] which reviewed reports from over 100 flights to Australia from January 1990 to September 2000. The symptoms reported by flight attendants were headache, sore throat, skin rash, nausea, runny nose, eye burning, difficulty in breathing, cough, dizziness, shortness of breath, and sinus problems. The total pool of at-risk flight attendants who could have made a report was not provided. The report noted an increase in the report of symptoms over the 10-yr period and particularly in 2000. The authors considered

---

[2]Presented to the committee as part of a public-access document submitted by AFA. The report was prepared by James Cone and Ruppa Das in response to a July 2000 request by AFA. The report was submitted in March 2001.

the increase in reporting to be due to increased awareness rather than increased risk. In particular, the July 2000 release of a specific health-problem reporting form by AFA was noted. Although the report acknowledged that disinsection was not a desirable practice, no data were provided on the risks posed by various types and frequencies of exposure.

Transmission of Infectious Agents

There is evidence that the environment of an aircraft cabin does not contain higher concentrations of microorganisms than do comparable public environments (see Chapter 4, Table 4-2). However, concerns persist that transmission of infectious agents is a potential hazard associated with travel in commercial aircraft, especially in aircraft that recirculate a portion of the cabin air (see Chapter 4 for more complete discussion). Reports of transmission of influenza and measles viruses and *Mycobacterium tuberculosis* during flights of commercial airlines have fueled the concern. Careful evaluation of the published studies suggests that even in those cases the perception of risk may have been overemphasized.

Influenza Virus

Two reports serve as the primary sources of data on transmission of influenza virus. Moser et al. (1979) reported transmission of influenza virus during a 1977 flight from Homer to Anchorage, Alaska. The circumstances surrounding the flight were unusual: passengers were on the grounded plane for over 3 h, during which the air-conditioning system was inoperative because of an engine failure. Of the 53 passengers at risk, 38 (72%) became ill, and the risk of illness increased with time spent on the grounded aircraft. The risk of transmission probably was not different from that occurring in other relatively confined spaces with poor ventilation and contact between people. This study is not useful for risk assessment associated with normal flight circumstances. Although the study points out the need to maintain adequate ventilation during conditions of prolonged passenger time on grounded aircraft, it also indicates that contact between passengers, if they are moving around in the grounded plane as in this case, makes viral transmission quite likely.

A study (Klontz et al. 1989) of a squadron of 114 U.S. naval personnel who developed influenza within 72 h of a 1986 flight from Puerto Rico to Key

West, Florida, also suggests transmission of influenza virus in aircraft. Careful evaluation of the data presented indicates that extensive transmission of influenza was taking place in the squadron before the time of the flight. Because many cases were identified in the first 24 h after the flight, transmission was as likely to be a coincidental result of exposure before boarding as a result of exposure during the flight. Consequently, the results of the study are of little use for risk assessment.

Those two studies do not constitute an adequate database from which to determine whether the risk of transmission of influenza virus is heightened in the aircraft cabin environment relative to other closed environmental spaces under conditions of normal or abnormal operation of the environmental control systems (ECSs).

Measles Virus

The evidence is weak that the commercial aircraft cabin constitutes an environment for enhanced transmission of measles virus. A 1982 review by the Centers for Disease Control and Prevention (CDC) of cases of measles imported into the United States (Amler et al. 1982) identified a single instance in which three children who traveled from Venezuela to Miami, Florida, had measles. One had measles prodromal symptoms on the plane, and the other two developed measles rash within 14 d of the flight. No data were provided on the relationships of the children, their proximity, or contact before, during, or after the flight. Because the measles rash usually follows the clinical onset of disease by several days (Slater et al. 1995), the two children with "secondary" cases could have been infected before the time of the flight.

A 1983 surveillance report by CDC (1983) identified a U.S. naval officer who infected nine people with measles. One of the secondary cases was in a passenger on the same flight from Seattle to San Diego as the officer. No details were given about the proximity of that passenger to the officer on the plane or while passengers were in the airport waiting for the departure of the flight. Five of the other patients with secondary cases who were not on the same flight as the officer had "visited at least one of the three departure gates visited by the officer that day." Thus, it is possible that the infected fellow passenger was also infected in the airport.

Slater et al. (1995) described an outbreak of eight cases of measles among 360 passengers on a 1994 flight from New York to Tel Aviv, Israel. The source case was not identified. The authors noted that the waiting area for the flight was so crowded that airport personnel permitted access to the area only to passengers. In addition, the loaded plane was delayed on the ground for 2 h for repairs, during which time the air-conditioning was not working. Thus, some of the secondary infections could have taken place before people boarded the aircraft. Additional transmission could have been facilitated by the conditions of poor ventilation on the aircraft. No data were provided on the proximity of the seating of passengers or on their contact with each other while the aircraft was grounded for repairs.

Those data are of little value in the assessment of the risk of measles transmission on a commercial aircraft, and they do not provide definitive evidence of transmission on the aircraft.

Mycobacterium tuberculosis

A number of published studies have evaluated the risk of transmission of *Mycobacterium tuberculosis* on commercial flights. Several of those were summarized in a 1995 CDC report (CDC 1995). This report will not be discussed; rather, the individual investigations will be presented chronologically.

The Minnesota Department of Health reported (in a letter) a contact investigation of a person who had smear-positive cavitary-pulmonary tuberculosis and who traveled on a 1992 flight from London to Minneapolis, Minnesota (McFarland et al. 1993). Of the 343 people on the flight, contact was successful for only 59 (61%) of the 97 U.S. citizens and 55 (22%) of the 246 noncitizens. The authors could find alternative explanations for all the persons reported to have positive skin tests for *M. tuberculosis* and concluded that there was no evidence of transmission on the flight. With so many data missing, it is difficult to accept this conclusion without reservation.

In 1994, CDC reported the results of an investigation of the work contacts of a flight attendant who had a diagnosis of pulmonary tuberculosis in November 1992 and who had flown in May-October 1992 while she was symptomatic (Driver et al. 1994). All crew members (274) who flew with the attendant while she was symptomatic were contacted and evaluated by skin test and questionnaire. A group of flight-crew members (355) who had not flown with the attendant was used to compare the prevalence of positive tuberculin skin tests with tests of the exposed contacts. There was a significantly greater prevalence of positive skin tests in the contacts of the index case, especially in crew members who flew with the attendant during August-October 1992. The prevalence of apparent exposure increased with the number of hours of flight time that a crew member shared with the attendant. During that later period, contacts had about 2.3 h of flight time for each hour of ground time with the attendant. Although it is clear that the contacts of the attendant had an increased risk of infection with *M. tuberculosis*, the data do not provide any useful information for risk assessment of cabin air. *M. tuberculosis* is known to be spread by prolonged personal contact between infected and susceptible people. The fact that such contact took place to a great extent in an aircraft cabin does not permit any conclusion about the aircraft environment to transmission risk. Moreover, crew members often share rooms during layovers and that also provides close contact. No data were provided on the relationship between room-sharing and the prevalence of positive skin tests. Therefore, the data in this paper are of little use for risk assessment.

One of the most frequently cited studies with regard to transmission of *M. tuberculosis* on commercial aircraft is that of Kenyon and colleagues (Kenyon et al. 1996). Their study reports an investigation of passenger and crew contacts of a Korean woman who flew on four flights while symptomatic and who died with extensive pulmonary tuberculosis shortly after the last flight (a Boeing 747-100 with 50% recirculated air). The investigation was limited to U.S. and Canadian citizens. Skin-test data were available on 87% (802 of 925) of the contacts, and the final analysis was limited to 82% (760) of the contacts. Of the 760, six (0.8%), all from the last of the four flights (2.4% of the contacts on the fourth flight), had skin-test conversions—two were attributed to the booster phenomenon. Two of the four subjects with true conversions sat within two seats of the woman with tuberculosis, and all the passengers who had positive skin tests but no other risk factors or conversion sat within two rows of the woman or visited with friends who were sitting near the woman. The one crew member with a positive skin test and no risk factors was stationed in the rear galley near the woman's seat. This study provides credible evidence that transmission took place on the fourth flight. However, given the proximity of the passengers to the woman (either by seat assignment or by time spent during the flight), it cannot be determined whether transmission was due to close personal contact or environmental conditions specific to aircraft passenger cabins.

A study reported by Miller et al. (1996) on transmission risk related to a passenger with active pulmonary tuberculosis (who flew from Moscow, Russia, through Frankfurt, Germany and New York, to Cleveland, Ohio) is not informative. Data on 35% of the potential contacts were missing. Neither of the two contacts who had positive tuberculin skin tests and no other risk factor for tuberculosis sat near or had contact with the index passenger. Thus, although transmission of *M. tuberculosis* to the two passengers could have occurred, there is no direct evidence that it did.

A study of contacts of a patient who had highly infectious laryngeal tuberculosis and who took two flights of about 1 h each failed to find evidence of transmission (Moore et al. 1996). However, data on only 53% of potential contacts were available.

A study of transmission on a Boeing 747-400 flight, which originated in Taiwan, was able to evaluate 73% (225 of 308) of the Taiwanese passenger and crew contacts of a passenger with tuberculosis (Wang 2000). Three subjects, none of whom sat in the same section as the passenger with tuberculosis, were considered to have true skin-test conversions on the basis of a rigorous protocol that eliminated the possibility of false positives due to the booster effect. Although the conversions could have been due to exposure during the flight, it should be pointed out that the prevalence of positive tuberculin skin tests among the Taiwanese passengers on the flight was 77%. Given that high level of infection, it is difficult to eliminate the possibility that the conversions were due to exposure unrelated to the flight.

Of all the studies, the one by Kenyon and colleagues (Kenyon et al. 1996) provides the most credible evidence that *M. tuberculosis* can be transmitted during long commercial flights. However, the study provided no data that permit an evaluation of the extent to which current designs for the environmental control systems (ECS) on aircraft contribute to this risk. In fact, the data are most consistent with the idea that the ECS added little or nothing to transmission risk.

Although it is reasonable to assume that infectious agents are transmitted during commercial airline flights, it has not been possible to determine conclusively whether transmission is related to close personal contact or environmental conditions specific to passenger cabins (see Chapters 2 and 4 for more details). The available data are of little use for the evaluation of the performance of the ECS with regard to the specific infections discussed above or for transmission of infectious agents in general. To improve future investigations of possible exposure to nationally notifiable diseases (CDC 2001) during air travel, physicians should notify local health authorities of patients who recently traveled while infectious, and airlines should collect sufficient contact information on all passengers to allow them to be notified of possible exposure to an infectious person during air travel and the potential need for medical evaluation.

Cabin Air Pressure and Health Risks

Current standards permit pressurization of the cabins of commercial aircraft up to an equivalent altitude of 2,450 m (about 8,000 ft) under normal operating conditions (FAR 25.841). The health concerns associated with current cabin pressurization are reviewed in Chapter 5. The 1986 NRC report summarized the effects of altitude on partial pressure of oxygen (PO_2) and recommended that passengers with heart, lung, and middle ear diseases be educated about the potential risks of flight. However, that 1986 committee had few direct data on the oxygen (O_2) saturation that might be expected in passengers and cabin crew under normal conditions of modern commercial airline flight.

Cottrell and colleagues (1995) used continuous-reading pulse oximeters to measure O_2 saturation in 38 pilots on 21 flights about 4 h long. Maximal and minimal O_2 saturations were 95-99% (mean, 97%) and 80-93% (mean, 88.6% ± 2.9%), respectively. Of the subjects, 53% developed an O_2 saturation of less than 90% at some time during the flight (duration below 90% was not given). Schwartz et al. (1984) studied subjects with severe chronic obstructive pulmonary disease (COPD) (average resting arterial PO_2, 68.0 ± 7.3 mm Hg) during flights at altitudes of 1,650-2,250 m (about 5,400 to 7,400 ft) in unpressurized aircraft cabins. PO_2 decreased to an average of 51.0 ± 9.1 mm Hg at an altitude of 1,650 m and little further change at 2,250 m. Several simulation studies have been carried out in hypobaric chambers with COPD patients and healthy subjects (Dillard et al. 1989; Naughton et al. 1995). Declines in PO_2 were observed in all patients at rest; it fell to below 50 mm Hg in many subjects and was made worse by light exercise. Of 100 COPD patients, 44 reported traveling by air in the 2 yr before the interview (Dillard et al. 1991). Eight of the patients reported increased symptoms during flight (no direct physiological measurements were available); five of these patients experienced shortness of breath when walking in the cabin, and two requested supplemental O_2 for their symptoms.

As part of a study for British Airways, Building Research Establishment Ltd. examined the effects of cabin air pressure on O_2 saturation in cabin crew (Ross 2001). Saturation in cabin crew of Boeing 777s and 747s was measured with a pulse oximeter. Cabin pressure was measured hourly during flights, and oximetry readings in cabin crews were "instantaneous spot measurements." On each of 16 flights, 10-15 crew members were studied. Cabin altitudes were 1,585-2,286 m (5,200-7,500 ft). Symptom questionnaires were distributed. Ross (2001) reported that O_2 saturations of 90% or less occurred in 16 of the subjects (maximal number of crew was 240, but the exact number sampled not given) and often were followed or preceded by readings above 90%. This study, as presented, is of little value for several reasons:

- The collection method for the oximetry data is not adequate. Spot measurements are not useful unless related specifically to the cabin altitude in the aircraft and the activities of the crew when the measurement is made.
- No quality-control criteria are given to ensure that the recorded readings represent a stable measurement over some prespecified number of heartbeats. In fact, the data on subjects who had at least one saturation value of 90% or less suggest that no such criteria were applied; that is, the readings were preceded or followed by much higher readings.
- The data are presented as average 1-h values, apparently representing the averaging of data on several people over a given period in that only spot measurements were made for each crew member.
- No data are presented on the relationship between saturation and cabin altitude during various flight segments or during various work activities of the crew. Those data gaps preclude an assessment of the relationship between cabin altitude and O_2 saturation in crew members during the course of the flights.

Under ordinary conditions of commercial flight, it is clear that reductions in PO_2 to the point of hypoxia can take place at rest or in situations of minimal exertion. As discussed in Chapter 5, PO_2 levels are such that reduced arterial O_2 content could pose a definite health risk for persons with underlying pulmonary or cardiac disease or untreated or partially treated anemia.

OTHER SOURCES OF HEALTH EFFECTS INFORMATION

A considerable amount of material was submitted to the committee in support of health effects related to incidents of smoke, fumes, and unusual odors in the cabins of commercial aircraft both in association with and independent of known or suspected episodes of leaks of hydraulic fluid, engine oil, or other sources of contaminants in the cabin. Those materials consisted primarily of testimony and summaries of unpublished data. Three examples are presented here as typical of the data presented.

The Australian Senate Rural and Regional Affairs and Transport Committee (1999, 2000) conducted an inquiry into reported cabin air quality and health-related consequences related to incidents of engine oil leaks that involved British Airways BAe 146 aircraft. The Senate committee summarized testimony on symptoms related to oil fumes in aircraft cabin air, the "aerotoxic syndrome." The report (Australian Senate 2000) cited testimony that questioned the specificity of this symptom complex and the fact that such nonspecific symptoms "are present at any one time in 10 percent of the population." The report also highlighted other testimony that questioned the validity of claims of chronic symptoms related to acute exposures to products of engine oil leaks and fumes. In relation to health effects, the Australian Senate (2000) concluded the following:

> It appears to the Committee that contamination of cabin aircraft air on the BAe 146 aircraft has led to short-term and medium-term health problems. . . . Some scientists link these health problems to contaminants, although the link has not yet been definitively established. . . . This remains a question to be further investigated and assessed.
> The Committee is also convinced that there is sufficient evidence before this inquiry to justify further examination of the following factors:
> • the effects on human health of the introduction into the aircraft cabin . . . of engine oil, by-products of engine oil combustion and other compounds as a result of leaking seals and bearings; and
> • the cumulative physical effect of exposure to these substances which can affect particular individuals.

However, the Australian Senate went on to recommend that "aerotoxic syndrome" be included "in appropriate codes as a matter of reference for future Workers Compensation and other insurance cases," although this appeared to contradict its summary of the evidence.

Two sets of unpublished data illustrate the difficulty with the data used to link specific symptoms to putative incident exposures in aircraft cabin air. The first set of unpublished data provided a review of the evidence of health effects related to cabin air incidents (C. Witkowski, Association of Flight Attendants, unpublished data, January 24, 1999), including all incidents reported in relation to a particular airline with a focus on the MD-80 aircraft. Flight reports, insurance claims, OSHA reports, medical records, and mechanical reports were evaluated. The criteria for the selection of specific reports were not provided. An air-quality incident was defined as "a specific mechanical problem resulting in smoke, mist, odors, fumes or smells, or those incidents where there was a complaint made regarding symptoms experienced by flight attendants or passengers." The results of the review are presented in Table 6-6. Although an association between reports of various symptoms and incidents on flights is inferred, few data are provided in terms of the persons at risk, the percentage who reported symptoms, or the temporal pattern of onset and recovery. Moreover, no data are provided on similar symptoms reported on flights during which no odors were detected and no observable conditions occurred or on the health status of those affected. It is not possible to distinguish symptoms reported by passengers and those reported by flight attendants.

The second set of unpublished data centered on a case study of a flight attendant with acute and chronic symptoms that occurred in relation to a presumed leak of hydraulic fluid ("hydraulic pressure light illuminated in flight") (Wright and Clarke 1999). Details of a variety of neurological symptoms over a 6-mo period are reported and related, by analogy, to those which might be expected in connection with exposure to toxic concentrations of organophosphates—specifically organophosphate-induced delayed neuropathy (see Chapter 5 for more information on this toxic effect). No specific exposure data are provided. The criteria for selection and the representativeness of this single anecdote are impossible to determine. Allegedly, the "entire crew" were ill, and passengers reported nausea and vomiting (the number at risk and the percentage with symptoms were not provided).

CURRENT HEALTH-RELATED DATA
COLLECTION SYSTEMS

The federal government maintains several databases that are potentially relevant to the identification of health-related problems associated with cabin air; a database summary is available at the Federal Aviation Administration (FAA) web site (http://intraweb.nasdac.faa.gov/learn_about/dat_learn.asp). Several databases of particular relevance are reviewed below.

- FAA Accident/Incident Data System (AIDS): Collects descriptive data on incidents (e.g., fumes, smoke, and odors) that do not meet aircraft-damage or personal-injury thresholds for the National Transportation Safety Board definition of an accident. Includes steps taken for remediation. Data are available on 1978 to the present. No health-related data are provided.

TABLE 6-6 Summary of 760[a] Incidents Reported by the Association of Flight Attendants

Source of Incident Reports	
Passenger report of symptoms	(9.9%)
Flight-attendant report of symptoms	(64.7%)
Reports of observable conditions,	
such as smoke and mist fumes in cabin	(25.4%)

Reported Incidence of Episodes for Airline
    7.6 incidents per 10,000 flights for airline under consideration from July 1989 to August 1998
Character of Odor Reported in Incidents with Observed Conditions
(data from 83% of incidents):
    Unusual (29%),[b] burning (18.6%), sweet (12.9%), nail polish (9.3%), toxic (7.7%), dirty socks (5.1%), 17.6% of incidents identified as hydraulic fluid or oil
Symptom Reports (based on 925 flight attendants involved in 760 incidents)[c]
    97.2% of incidents involved reports of disorientation, confusion, "spacey," euphoria for incidents with smoke or haze
        Landing: difficulty in breathing, coughing, irritation of eyes, throat, and lungs
        Mid-flight: headaches, disorientation, giddiness, difficulty in concentrating, nausea, vomiting
        Ground and takeoff: same as for mid-flight

[a] See text for definition of "incident."
[b] Number of fight attendants and/or passengers involved not given. Not clear whether percentage refers to proportion of people or events.
[c] Reports prepared by 492 people. Specific percentages and number of persons involved for each symptom for various flight segments not provided.
Source: C. Witkowski, Association of Flight Attendants, unpublished data, January 24, 1999.

- FAA Service Difficulty Reporting System (SDRS): Repository of Service Difficulty Reports, Malfunction and Defect Reports, and Maintenance Difficulty Reports. Data are available on 1986 to the present. No health-related data are provided.
- Aviation Safety Reporting System (ASRS): Voluntary confidential incident reporting system administered by National Aeronautics and Space Administration (NASA). Has narrative descriptions of the conditions of an incident and any health-related information that the people who filed the reports chose to provide. The data summaries do not indicate any systematic data recording structure. ASRS data are anonymous and confidential and

therefore are not useful for retrospective assessments of health effects of reported incidents. Moreover, because the data are reported as unstructured narratives, they are not particularly useful even when linked to more specific information about incidents in the AIDS and SDRS databases. Finally, given that reporting to ASRS is voluntary, the database is not useful as a numerator to estimate what fraction of incidents reported to the other databases are associated with reports of adverse health outcomes. Data are available on 1988 to the present.

No information was available to the committee on the details of operation of incident reporting systems or maintenance record reporting systems for any commercial airline.

AFA has distributed a standard event reporting form to its members as part of its Safety and Health Database Initiative (AFA 2001). The form is intended to be completed primarily by cabin crew, but may also be completed by passengers, pilots, or mechanics in response to a perceived air-quality incident, including required pesticide spraying on international flights. Form submission to the database (which will eventually be available on line) is voluntary. It is not designed to obtain nonincident data ("baseline" symptom frequency under ordinary operating conditions of flight) against which incident-related symptoms could be evaluated, but rather will function to provide AFA with a tabulation of problems reported by its members. The listing of health-related symptoms sometimes requires that the person completing the form make what amounts to a diagnosis (e.g., "allergic reaction," "ear inflammation . . . /damage," "infectious agent"), which is a less than optimal way to collect such data. On the basis of the little information available to the committee, this data system is not intended to and will not remedy the need for more systematic data on health-related symptoms under nonincident conditions. Moreover, the system does not have a built-in systematic sampling strategy to ensure comprehensive assessment of health effects on crew and passengers under incident-related conditions (J. Murawski, AFA, personal communication, 2001).

CONCLUSIONS

- None of the current systems for reporting signs and symptoms that could reflect health-related responses to aircraft cabin air quality are standardized with regard to how potentially affected individuals are surveyed. There is also a lack of standardization with respect to how specific data are obtained.

Of the relevant databases maintained by the federal government, only the NASA ASRS has health-related data as a focus.

• Although acute health effects potentially stem from incidents involving exposure to engines oil, hydraulic fluids, and their pyrolysis products, the existing database is inadequate to ascertain adequately any of the following in relation to documented incidents: the frequency of individual symptoms or constellations of symptoms, the characteristics of the cabin crew and passengers who might be at greatest risk from these exposures, and the long-term sequelae of single high-exposure incidents or recurrent low-level exposure.

• With respect to repeated exposure of cabin crew to cabin air during routine conditions of flight (nonincidents) and in the absence of smoking, the situation is even more difficult to assess, inasmuch as there are very few valid systematic surveys. The data that are available are insufficient to quantify the risk posed by cabin air-quality conditions with regard to acute symptoms or chronic sequelae.

• Data on exposure of passengers and related health effects are so sparse that further comment is not warranted.

• Data that have become available since the 1986 NRC report raise questions about the safety of current regulations related to cabin pressure, especially in light of the increased number of older and younger people flying, including children, infants, and adults with cardiovascular or pulmonary disease.

RECOMMENDATIONS

• Because the committee concludes that there is insufficient evidence to warrant designation of a specific syndrome related to exposure to leaks of engine oil or hydraulic fluids, the committee recommends that the term *aerotoxic syndrome* not be used for symptoms associated with incidents of cabin air contamination.

• Current regulations for cabin pressure should be reviewed to determine whether they are adequate for protecting people who might be unusually susceptible to changes in air pressure, such as the elderly, infants, children, and people with cardiovascular or pulmonary disease. Consideration should be given to whether air pressure regulations can be adjusted and to what management options are available to deal with potential problems (e.g., equipping some seats with supplemental oxygen). FAA and the airlines should work with medical organizations, such as the American Medical Association and the Aerospace Medical Association, to improve health professionals' awareness of the need to advise patients of the risks that might be posed by flying.

• Current systems for the collection of health data in relation to cabin air quality are woefully inadequate and do not permit any quantitative assessment of the relationship between cabin exposure and potential health effects on cabin crew or passengers. A program for the systematic collection, analysis, and reporting of health data in relation to cabin air quality needs to be implemented to resolve many of the issues raised in this report.

REFERENCES

AFA (Association of Flight Attendants). 2001. AFA Aircraft Air Quality Reporting Form. Association of Flight Attendants Safety and Health Initiative.

Amler, R.W., A.B. Bloch, W.A. Orenstein, K.J. Bart, P.M. Turner Jr, and A.R. Hinman. 1982. Imported measles in the United States. JAMA 248(17):2129-2133.

ASHRAE/CSS (American Society of Heating Refrigerating and Air-conditioning Engineers and /Consolidated Safety Services). 1999. Relate Air Quality and Other Factors to Symptoms Reported by Passengers and Crew on Commercial Transport Category Aircraft. Final Report. ASHRAE Research Project 957-RP. Results of Cooperative Research Between the American Society of Heating, Refrigerating and Air-Conditioning Engineers, Inc., and Consolidated Services, Inc. February 1999.

Australian Senate. 1999. Air Safety-BAe 146 Air Quality. The Proof and Official Hansard transcripts of Senate Rural and Regional Affairs and Transport References Committee, Commonwealth of Australia Senate, Canberra, Australia. November 1, 1999.

Australian Senate. 2000. Air Safety and Cabin Air Quality in the BAe 146 Aircraft. Report by the Senate Rural and Regional Affairs and Transport References Committee, Parliament House, Parliament of the Commonwealth of Australia, Canberra. October 2000.

Balouet, J.C. 1998. In-Cabin Trace Chemicals and Crew Health Issues. Presentation at Annual Meeting of the Aerospace Medical Association, Seattle, WA, May 20, 1998.

Balouet, J.C., and C. Winder. 1999. Aerotoxic syndrome in air crew as a result of exposure to airborne contaminants in aircraft. American Society of Testing and Materials Symposium on Air Quality and Comfort in Airliner Cabins, New Orleans, October 27-28, 1999.

Balouet, J.C., and C. Winder. 2000a. Airborne Chemicals in Aircraft Cabins. Presentation to Aerospace Medical Association. Air Transport Medicine Symposium, Houston, May 7, 2000.

Balouet, J.C., and C. Winder. 2000b. Symptoms of Irritation and Toxicity in Air Crew as a Result of Exposure to Airborne Chemicals in aircraft. [Online]. Available: http://www.aopis.org/balouetSymptomsofIrritationandToxicityinAirCrew.html [October 16, 2001].

CDC (Centers for Disease Control and Prevention). 1983. Epidemiological notes and reports: Interstate importation of measles following transmission in an airport—California, Washington, 1982. MMWR 32(16):210, 215-216.

CDC (Centers for Disease Control and Prevention). 1995. Exposure of passengers and flight crew to Mycobacterium tuberculosis on commercial aircraft, 1992-1995. MMWR 44(08):137-140.

CDC (Centers for Disease Control and Prevention). 2001. Summary of Notifiable Diseases, United States 1999. MMWR 48(53):1-104. [Online}. Available: http://www.cdc.gov/mmwr/PDF/wk/mm4853.pdf [November 19, 2001].

Cone, J.E., and B. Cameron. 1984. Flight Attendant Health Survey. APFA. Occupational Health Clinic, San Francisco General Hospital. Interim Report No. 2. June 1984.

Cottrell, J.J., B.L. Lebovitz, R.G. Fennell, and G.M. Kohn. 1995. Inflight arterial saturation: continuous monitoring by pulse oximetry. Aviat. Space Environ. Med. 66(2):126-130.

Daughtrey, W., R. Biles, B. Jortner, and M. Ehrich. 1996. Subchronic delayed neurotoxicity evaluation of jet engine lubricants containing phosphorus additives. Fundam. Appl. Toxicol. 32(2):244-249.

De Ree, H., M. Bagshaw, R. Simons, and R.A. Brown. 2000. Ozone and relative humidity in airline cabins on polar routes: measurements and physical symptoms. Pp. 243-258 in Air Quality and Comfort in Airliner Cabins, N. Nagda, ed. West Conshohocken, PA: American Society for Testing and Materials.

Dillard, T.A., W.A. Beninati, and B.W. Berg. 1991. Air travel in patients with chronic obstructive pulmonary disease. Arch. Intern. Med. 151(9):1793-1795.

Dillard, T.A., B.W. Berg, K.R. Rajagopal, J.W. Dooley, and W.J. Mehm. 1989. Hypoxemia during air travel in patients with chronic obstructive pulmonary disease. Ann. Inter. Med. 111(5):362-367.

Driver, C.R., S.E. Valway, W.M. Morgan, I.M. Onorato, and K.G. Castro. 1994. Transmission of Mycobacterium tuberculosis associated with air travel. JAMA. 272(13):1031-1035.

Folinsbee, L.J., D.H. Horstman, H.R. Kehrl, S. Harder, S. Abdul-Salaam, and P.J. Ives. 1994. Respiratory responses to repeated prolonged exposure to 0.12 ppm ozone. Am. J. Respir. Crit. Care Med. 149(1):98-105.

Folinsbee, L.J., W.F. McDonnell, and D.H. Horstman. 1988. Pulmonary function and symptom responses after 6.6-hour exposure to 0.12 ppm ozone with moderate exercise. JAPCA 38(1): 28-35.

Kenyon, T.A., S.E. Valway, W.W. Ihle, I.M. Onorato, and K.G. Castro. 1996. Transmission of multidrug-resistant Mycobacterium tuberculosis during a long airplane flight. N. Engl. J. Med. 334(15):933-938.

Klontz, K.C., N.A. Hynes, R.A. Gunn, M.H. Wilder, M.W. Harmon, and A.P. Kendal. 1989. An outbreak of influenza A/Taiwan/1/86 (H1N1) infections at a naval base and association with airplane travel. Am. J. Epidemiol. 129(2):341-348.

Lee, S.C., C.S. Poon, X.D. Li, F. Luk, M. Chang, and S. Lam. 2000. Questionnaire survey to evaluate the health and comfort of cabin crew. Pp. 259-268 in Air Quality and Comfort in Airliner Cabins, N.L. Nagda, ed. West Conshohocken, PA: American Society for Testing and Materials.

McFarland, J.W., C. Hickman, M. Osterholm, and K.L. MacDonald. 1993. Exposure to Mycobacterium tuberculosis during air travel. Lancet. 342(8863):112-113.

Miller, M.A., S. Valway, and I.M. Onorato. 1996. Tuberculosis risk after exposure on airplanes. Tuber. Lung Dis. 77:414-419.

Moore, M., K.S. Fleming, and L. Sands. 1996. A passenger with pulmonary/laryngeal tuberculosis: no evidence of transmission on two short flights. Aviat. Space Environ. Med. 67(11):1097-1100.

Moser, R.M., T.R. Bender, H.S. Margolis, G.R. Noble, A.P. Kendal, and D.G. Ritter. 1979. An outbreak of influenza aboard a commercial airliner. Am. J. Epidemiol. 110(1):1-6.

Naughton, M.T., P.D. Rochford, J.J. Pretto, R.J. Pierce, N.F. Cain, and L.B. Irving. 1995. Is normobaric simulation of hypobaric hypoxia accurate in chronic airflow limitation? Am. J. Respir. Crit. Care Med. 152(6 Pt 1):1956-1960.

NRC (National Research Council). 1986. The Airliner Cabin Environment: Air Quality and Safety. Washington, DC: National Academy Press.

Pierce, W., J. Janczewski, B. Roethlisberger, and M. Janczewski. 1999. Air quality on commercial aircraft. ASHRAE J. (Sept.):26-34.

Rankin, W.L., D. R.Space, and N.L. Nagda. 2000. Passenger comfort and the effect of air quality. Pp. 269-290 in Air Quality and Comfort in Airline Cabins, N.L. Nagda, ed. West Conshohocken: American Society for Testing and Materials.

Rayman, R.B., and G.B. McNaughton. 1983. Smoke/fumes in the cockpit. Aviat. Space Environ. Med. 54(8):738-740.

Ross, D. 2001. Air Quality in Passenger Cabins of Commercial Aircraft. BRE Pub. Draft No. 204743. Building Research Establishment, Ltd, Watford, UK. May 23, 2001.

Schwartz, J.S., H.Z. Bencowitz, and K.M. Moser. 1984. Air travel hypoxemia with chronic obstructive pulmonary disease. Ann. Intern. Med. 100(4):473-477.

Slater, P.E., E. Anis, and A. Bashary. 1995. An outbreak of measles associated with a New York/Tel Aviv flight. Travel Med. Int. 13(3):92-95.

Tashkin, D.P., A.H. Coulson, M.S. Simmons, and G.H. Spivey. 1983. Respiratory symptoms of flight attendants during high-altitude flight: possible relation to cabin ozone exposure. Int. Arch. Occup. Environ. Health. 52(2):117-137.

van Netten, C. 1998. Air quality and health effects associated with the operation of BAe 146-200 aircraft. Appl. Occup. Environ. Hyg. 13(10):733-739.

Wang, P.D. 2000. Two-step Tuberculin testing of passengers and crew on a commercial airplane. Am. J. Infect. Control 28(3):233-238.

Wick, R.L., and L.A. Irvine. 1995. The microbiological composition of airliner cabin air. Aviat. Space Environ. Med. 66(3):220-224.

Winder, C., and J.C. Balouet. 2000a. Aerotoxic syndrome: adverse health effects following exposure to jet oil mist during commercial flights. Pp. 196-199 in Towards a Safe and Civil Society, Proceedings of International Congress on Occupational Health Conference, Brisbane, Australia, Sept. 4-6, 2000, I. Eddington, ed. Brisbane: ICOH.

Winder, C. and J.C. Balouet. 2000b. Aerotoxic Syndrome: A New Occupational Disease. Draft Report. School of Safety Science, University of New South Wales, Sydney, Australia and Environment International, Nogent Sur Marne, France. June 2000.

7

Air-Quality Measurement
Techniques and Applications

As discussed in Chapters 3 and 6, many studies have attempted to show a link between aircraft cabin air quality and health effects. Such an association has been difficult to demonstrate, in part because air quality has been measured in only a small portion of aircraft flights and in part because studies have varied considerably in sampling strategy, environmental conditions measured, and measurement methods used. At present, only air temperature and barometric pressure are routinely measured in commercial aircraft cabins, and only the pressure measurements are recorded as part of the flight data. Furthermore, because most flight data recorders retain only data from the most recent 30 min of operation, current recording practices do not permit assessing variations in cabin pressure throughout a flight or, therefore, identifying periods during which partial pressure of oxygen (PO_2) is low.

In this report, *measurement* refers to the quantitative determination of airborne concentration of a contaminant or of air temperature, relative humidity, or pressure with a suitable instrument. *Monitoring* refers to measurement over a relevant period (e.g., the duration of a flight segment) coupled with the creation of a durable record of the data thus obtained. For example, measurement of cabin air pressure and its indication on a display in the cockpit would not be considered monitoring, but recording the measurement as a function of time in a form that can be reviewed and analyzed later would constitute monitoring. *Continuous monitoring* refers to measurement without interruption during the period of interest and the display or recording of results nearly instantaneously. Although practical instruments do not provide truly instanta-

254

neous data, they can generate measurements that are averaged over periods of a few seconds to a few minutes and produce a record of consecutive short-term averages that span the entire period of interest. *Integrated monitoring* relies on instruments that collect a sample of air or of a contaminant in air over a longer period (several minutes to more than an hour, depending on instrument design); the resulting concentration is the integrated, or time-weighted average, concentration over the period of sample collection.

Air temperature is measured and controlled in all commercial aircraft for the comfort of passengers and crew and to help provide cooling capacity to maintain appropriate operating temperatures for electronic and mechanical equipment. Because thermal loads are not the same in all parts of the aircraft, control zones are used. Each zone has an independent temperature sensor and adjustable supply of conditioned air. For example, thermal conditioning in the cockpit is controlled separately from that in the passenger cabin, which may be divided into two or more control zones. The latter subdivision helps to minimize longitudinal movement of air in the cabin (Lorengo and Porter 1986; Stevenson 1994; Hunt et al. 1995).

The location of air temperature sensors varies, but the temperature generally appears to be measured in the supply air as it enters a zone. In some instances, the temperature is measured in the cabin or cockpit air after the supplied air mixes with the resident air. Although air temperature is automatically controlled, the set point can be changed by cockpit crew in response to reports of thermal discomfort from cabin occupants. (See Chapter 2 for additional details on temperature control in aircraft).

Barometric pressure in the pressure hull of the fuselage is measured continuously and is under precise control of an automatic system. The supply of compressed air from the environmental control system (ECS) and release of air through an exhaust valve are balanced automatically to maintain cabin pressure. The system is designed to operate so that the pressure difference across the pressure hull does not exceed a specified limit and to ensure that, at least under routine conditions, the barometric pressure in the pressure hull does not fall below a cabin pressure altitude of 2,440 m (8,000 ft) (ASHRAE 1999a). Those two requirements limit the maximal altitude of the aircraft (Stevenson 1994; Hunt et al. 1995; ASHRAE 1999a).

The PO_2 in the cabin is not measured routinely. However, because the aircraft ECS does not alter the fraction of oxygen (O_2) in outside air, and human occupants in a plane do not reduce the O_2 concentration by an amount that is physiologically important, the PO_2 in the cabin will be a fixed fraction of the total pressure (Arnold et al. 2000). (See Chapter 2 for additional details on cabin air pressure.)

The lack of data on cabin air quality other than temperature and pressure during routine and nonroutine operations of aircraft imposes severe limitations on the ability of the Federal Aviation Administration (FAA), the airlines, and their staff to determine the causes of and measures needed to reduce incidents of health effects and complaints from passengers and cabin crew. To investigate the presumed association between cabin air quality and health effects, quantitative assessment of exposures is needed, and this requires systematic and extensive collection and recording of data on numerous characteristics of the aircraft environment. Without such detailed air-quality information derived from flights in which the ECS operates as designed and those in which mechanical problems (e.g., fluid-seal failures) occur, critical evaluation of the link between air quality and complaints or health problems in crew or passengers will be impossible. Accordingly, aircraft measurements should be expanded by incorporating several simple, reliable instrument systems for monitoring relevant characteristics of cabin air quality.

Depending on the program objectives (see Chapter 8), several air-quality characteristics may need to be monitored continuously. The characteristics may include air temperature, pressure, ozone (O_3), carbon monoxide (CO), carbon dioxide (CO_2), relative humidity, and fine particulate matter (PM). Integrated samples of PM may also need to be collected and analyzed to determine the concentration of toxic components of airborne PM.

Techniques used to monitor the characteristics that are not currently monitored on aircraft are discussed in the following sections. The final sections of this chapter address the location of sampling ports, data processing, and the committee's conclusions and recommendations.

OZONE

To meet the FAA O_3 limits, commercial aircraft that fly "high-O_3" routes often use devices to remove O_3 from the cabin supply air. Such devices are usually catalytic converters, but charcoal adsorption has also been used (SAE 1965; Boeing 2000). On planes that use O_3 converters, current practice requires replacing the catalyst infrequently (e.g., once every 2-6 yr depending on model for Boeing aircraft [D. Space, Boeing, personal communication, February 12, 2001] or after about 12,000 flight hours for Airbus aircraft [M. Dechow, Airbus, personal communication, February 9, 2001]). Yet numerous contaminants have the potential to come into contact with and poison the catalyst during daily operations. Real-time O_3 monitoring will indicate whether the catalyst is functioning as intended and will alert crew when maintenance or replacement is required. In fact, the 1986 National Research Council (NRC) committee stated in its report on aircraft air quality: "Because catalytic converters are subject to contamination and loss of efficiency, it is suggested that FAA establish policies for periodic removal and testing, so that the effective life of these units can be established. A program of monitoring is needed, to establish compliance with the existing standard and to determine whether the catalytic converters are operating normally and effectively. These data should be maintained in such a manner that they can be used for reference on passenger and crew exposures to O_3 and to document the concentrations of O_3" (NRC 1986).

Continuous O_3 monitoring would also test the assertion that O_3-removing devices are not necessary on some routes. The available information on O_3 concentrations during different flight segments is far from complete. The present committee understands that O_3-destroying catalysts are most often used on high-altitude polar flights. However, data collected by the National Aeronautics and Space Administration (NASA) in the late 1970s demonstrate that high O_3 can also be encountered on flights at lower latitudes, especially during late winter and early spring (Holdeman et al. 1984). Table 7-1 presents data from nine flights in 1978 chosen to illustrate that fact. The data are based on simultaneous measurements of O_3 in the ambient air and supply air for a United Airlines 747SP, and the reported values are means of all O_3 measurements taken on each flight. For each flight, mean O_3 concentration in the cabin exceeded 0.20 ppm, well in excess of the 1-h national ambient air-quality standard (NAAQS) set by the Environmental Protection Agency (EPA). Furthermore, because the data are means, they do not reveal the magnitude of short-term peak concentrations during the sampled flights. (See Chapter 3 for a detailed discussion of O_3 in aircraft and methods of controlling O_3 in aircraft.)

Available Technology

The feasibility of monitoring O_3 in commercial aircraft has been demonstrated in a sampling program launched in 1983 by European scientists called Measurement of Ozone and Water Vapor by Airbus In-Service Aircraft (MOZAIC). Its aim is to measure O_3 and water vapor in the atmosphere by

TABLE 7-1 Simultaneous Ozone Measurements in Supply Air and Outdoor Air on Selected 747 Flights, January-March 1978

		Ozone Concentration, ppm	
		---	---
Date	Route[a]	Supply Air	Outside Air
1/25/78	JFK-SFO	0.215	0.471
1/29/78	CHI-JFK	0.380	0.774
2/22/78	CHI-DEN	0.223	0.538
2/25/78	JFK-CHI	0.233	0.643
3/8/78	DEN-CHI	0.280	0.653
3/8/78	CHI-DEN	0.231	0.440
3/12/78	SFO-JFK	0.233	0.528
3/13/78	JFK-CHI	0.236	0.473
3/15/78	CHI-LAX	0.227	0.570

[a] JFK, New York; SFO, San Francisco; CHI, Chicago; DEN, Denver; LAX, Los Angeles.

using commercial long-range aircraft. MOZAIC uses fully automatic instruments installed on five long-range Airbus 340 aircraft that are in normal service. The participating carriers are Air France, Sabena, Lufthansa, and Austrian Airlines. By the end of December 1997, 7,500 flights using the instruments had been completed, and 54,000 flight hours of observation had been automatically recorded (Marenco et al. 1998). Although MOZAIC is focused on O_3 outside the aircraft, the same general approach could be used to monitor O_3 and other characteristics inside the aircraft.

The MOZAIC program uses a dual beam ultraviolet-photometric instrument to measure O_3. The instrument monitors the absorbance of O_3 at 254 nm. Similar instrumentation could be used for routine monitoring of O_3 in an aircraft. It is rugged, reliable, accurate, and sufficiently sensitive (about 1 ppb with a 10-cm pathlength). It uses no consumables and holds calibration well. Similar instrumentation was also used in the earlier NASA studies (Holdeman et al. 1984). Given the constraints on instrumentation aboard an aircraft, ultraviolet-photometric monitoring is more readily adapted to this environment than alternative approaches, such as chemiluminescence or electrochemical methods.

Application to Aircraft Cabin

The primary sampling point for measuring O_3 should be in the air supplied to the passenger cabin.[1] Sampling at that location will provide the best indication of the efficacy of the O_3 catalyst and of the worst case for occupant exposure, in that supply air is typically diluted with recirculated air. Furthermore, exposure of the average occupant will be lower because of O_3 decomposition in the cabin. The reading could also be used to alert pilots when the aircraft is passing through air masses that have high O_3 concentrations. Although air masses with high O_3 are probably large, the pilots may be able to seek cleaner air or may temporarily decrease the amount of outdoor air brought into the aircraft. Such readings, archived for a suitable period, would also be useful in evaluating incident complaints from passengers or crew.

A secondary sampling point for supplemental monitoring of O_3 might be in the exhaust air. The aircraft selected for supplemental monitoring would be ones that fly routes with potentially higher O_3. When values measured in the exhaust air are compared with values measured in the supply air, the difference will provide information on the rate of surface removal in the aircraft. Such information is helpful in specifying design requirements for O_3-removing devices. (See Chapter 3 for a discussion of retention ratios in Box 3-2.) Improved knowledge of the magnitude of the surface removal rate would also facilitate calculation of O_3 concentration in the aircraft cabin, given O_3 concentration in the supply air and a known air-exchange rate.

CARBON MONOXIDE

Data from limited research investigations suggest that CO concentrations in aircraft cabin air are generally well below those associated with health effects (Nagda et al. 2000). However, a few reports suggest that operation of aircraft under nonroutine conditions (e.g., when an engine-seal leak permits

---

[1]CFR Section 121.578 "Cabin ozone concentration" states that "no certificate holder may operate an airplane above the following flight levels unless it is successfully demonstrated to the Administrator that the concentration of O_3 inside the cabin will not exceed. . . ." Note that the section does not specify where in the cabin the reading should be taken.

engine oil or hydraulic fluid to enter bleed air) may lead to the production of CO and contamination of cabin air to concentrations that are associated with health risks (Rayman and McNaughton 1983; van Netten 1998; Pierce et al. 1999; Balouet 2000; van Netten and Leung 2000, 2001).

The acceptable limits for CO in air, as summarized in Table 7-2, vary widely, depending on the organization setting the limit and the population to be protected. The limits set by the National Institute for Occupational Safety and Health (NIOSH), the Occupational Safety and Health Administration (OSHA), and the American Conference of Governmental Industrial Hygienists (ACGIH) are intended for workplace exposures of healthy adults; they are not intended to apply in situations where infants, children, the elderly, or those with pre-existing cardiovascular or pulmonary disease might be exposed. The latter subpopulations are addressed by the EPA NAAQSs.

Available Technology

Portable instruments for continuous monitoring of CO have been in use for several years in buildings and in occupational settings. They use electrochemical sensors that have sufficient accuracy in the range of concentrations of interest (1-100 ppm), and they have been used in a few research investigations

TABLE 7-2 Recommended Limits for Carbon Monoxide in the United States

Standard	Concentration, ppm	Duration, h	Reference
EPA NAAQS	35	1	EPA (1985)
EPA NAAQS	9	8	EPA (1985)
OSHA PEL	50	8	OSHA (1989)
NIOSH REL	35	8	NIOSH (1972)
NIOSH REL	200	—[a]	NIOSH (1972)
ACGIH TLV	25	8	ACGIH (1999)
FAA AWS	50	Not specified	FAA (1996)

[a] Not to be exceeded at any time.

Abbreviations: PEL, permissible exposure limit; REL, recommended exposure limit; TLV, Threshold Limit Value; AWS, air worthiness standard.

aboard aircraft (Nagda et al. 2001). The available CO instruments provide analogue voltage output signals suitable for recording or digital logging by computer. An example is the model 190 CO monitor manufactured by the Draeger Corporation (Woebkenberg and McCammon 1995).

More sophisticated instruments based on nondispersive infrared absorption have also been developed. These instruments have substantially better accuracy and precision than electrochemical methods, but they are much larger, are more expensive, and require larger power supplies. Their superior accuracy and precision might not be warranted for aircraft cabin air monitoring (Parish et al. 1994).

Application to Aircraft Cabin

Because events leading to production of CO in bleed air are likely to be rare, their evaluation would require continuous monitoring and recording of concentrations on a large number of commercial flights. However, a continuous monitor could provide a warning of hazardous conditions in time for appropriate measures to reduce or prevent excessive exposures.

CO monitors should be placed in air supply ducts leading to each of the cabin air ventilation zones. Electrochemical devices can be operated in the active mode, in which a small air pump draws a continuous sample from the air supply ducts to the sensor via flexible tubing, or in the passive mode, in which the sensor is directly in the air stream to be sampled and the CO enters the sensor by diffusion through a semipermeable membrane (Woebkenberg and McCammon 1995).

CARBON DIOXIDE

Except in the case of fire, CO_2 does not pose a health hazard at the concentrations likely to be encountered in commercial aircraft; however, CO_2 is a useful surrogate indicator of substandard ventilation of a space with outside air, as has been shown in numerous studies of building-related symptoms (Cain et al. 1983; Berg-Munch et al. 1986). (See Appendix B for definition and discussion of building-related symptoms.) When fresh-air ventilation is substandard, trace contaminants (e.g., bioeffluents) generated in an occupied space accumulate to concentrations that might trigger complaints from the occupants. Because human occupants are the major source of indoor CO_2, this gas is a useful indicator of the buildup of the trace contaminants generated by humans and human activity. The recommended upper limit for CO_2 concentration on the basis of health effects is 5,000 ppm as established by OSHA, NIOSH, ACGIH, and FAA. However, much lower limits are required when CO_2 is used as a surrogate for other bioeffluents. In that context, building ventilation guidelines are set so that the indoor CO_2 concentration does not exceed 700 ppm above the concentration in outside air (ASHRAE 1999b), or about 1,100 ppm. Recent evidence suggests that building-related symptoms decrease with decreasing CO_2 even when the CO_2 is below 800 ppm (Seppänen et al. 1999). Accordingly, maintenance of an upper limit of 800-1,000 ppm appears to be necessary to minimize the frequency of complaints of poor air quality from the occupants of a building (Seppänen et al. 1999). However, because the aircraft cabin environment differs substantially from that of buildings (e.g., the occupant density and air exchange rates are much higher in aircraft), the use of indoor air-quality guidelines for CO_2 in air might not be appropriate for commercial aircraft.

Available Technology

Instruments of a size suitable for use in continuous monitoring on aircraft have been developed and used in research investigations (Nagda et al. 2000). They are nondispersive infrared photometers that use light-emitting diodes as the infrared sources. Such instruments have acceptable accuracy for CO_2 concentrations of 100-50,000 ppm (0.01-5% by volume). An example is the Telaire model 7001 instrument manufactured by Englehard Corporation (Woebkenberg and McCammon 1995).

Application to Aircraft Cabin

One or more CO_2 sensors should be placed in the exhaust ducts from the ventilation zones of the aircraft to monitor the effectiveness of fresh-air ventilation to each zone. Effective ventilation depends on the volume of fresh air delivered to the space and the extent to which the incoming air is mixed with the resident air by turbulence and natural convection. Because outside air contains CO_2 at about 370 ppm, any increase in cabin air concentration above this reflects the contribution of interior sources.

The use of dry ice (frozen CO_2) for keeping foods and beverages cool on board aircraft may occasionally lead to difficulty in interpreting continuous-monitoring records of CO_2 concentration. To avoid misinterpretation of data because of interference from sublimation of dry ice, it must first be determined whether air exhausted from the galley is directed overboard and not recirculated.

RELATIVE HUMIDITY

As described in Chapter 2, moist outside air is dehumidified before it is supplied to the cabin to prevent excessive humidity in the cabin. However, when the outside air contains little moisture, as is the case at typical cruise altitudes, air supplied to the passenger cabin is not humidified. The main source of humidity is the occupants (moisture from exhaled air and evaporation of perspiration from skin). Evaporation from food and drinks may also contribute a modest amount of moisture. As explained in Chapter 2, relative humidity is inversely related to the outside-air ventilation rate in this situation.

Relative humidity is not routinely monitored in commercial aircraft, but some measurements have been reported as part of several research investigations involving small numbers of flights. At cruise altitudes, the results of those measurements are consistent with expected humidity and indicate the absence of other major sources of moisture in flight (Arnold et al. 2000; De Ree et al. 2000; Lee et al. 2000; Nagda et al. 2001).

Although low relative humidity has not been associated with increased susceptibility to infection or other health effects, it appears to be related to complaints of irritation of eyes and mucous membranes among passengers and crew (Lindgren et al. 2000). (See Chapter 5 for additional discussion on health effects of low humidity.) Therefore, relative humidity should be monitored during flights so that the relationship between all air-quality characteristics and the health and comfort of passengers and crew can be fully evaluated.

Available Technology

Portable instruments for monitoring relative humidity or dewpoint temperature have been evaluated by several investigators and have been shown to have sufficient accuracy and precision for monitoring relative humidity between 2.5% and 80% (Freitag et al. 1994; Lafarie 1985). The most commonly used methods incorporate a thin hygroscopic polymer film whose electrical capacitance varies with relative humidity or an electrolyte solution whose electrical impedance varies with relative humidity. These instruments have an accuracy of approximately ±2.5% at relative humidity below 80% if calibrated periodically. Newer devices use a surface acoustic wave sensor and have better accuracy and precision (Hoummady et al. 1995), but their added cost may outweigh their greater accuracy.

Application to Aircraft Cabin

As with many of the instruments described above, relative-humidity monitors should be placed so as to sample the air leaving the cabin or cockpit. Because the outside air at cruise altitude contains essentially no moisture, the moisture content of cabin air will provide an indication of the balance between the indoor sources of moisture (e.g., occupants, food, and beverages) and the outside air delivered by the ECS. Air-temperature sensors should be co-located with the relative-humidity sensors so that relative-humidity measurements can be accurately converted to absolute-humidity values needed to complete this evaluation.

PARTICULATE MATTER

Continuous Monitoring for Fine Particles

Fine particles (particles with diameters of approximately 0.2-2.0 μm) are generated by combustion and can indicate nonroutine events, such as the pyrolysis of hydraulic fluids and engine oil that have accidentally entered bleed air. However, the ambient air that enters the cabin can also be a source of fine particles, particularly when the aircraft is on the ground or during takeoff and landing, as can effluents from the galley during meal service. Therefore, correct interpretation of fine-particle measurements will require that they be compared with data on the phase of the flight and the timing of galley activity.

Available Technology

A nephelometer (a continuous monitor of light scattered by suspended fine particles) can be used to monitor air that leaves the passenger cabin and is directed to the recirculation or exhaust system (ACGIH 2001). It would provide a continuous and on-line indication and recording of the mass concentration of fine particles and thus the potential exposure of passengers and cabin attendants to combustion byproducts. Although coarse particles (particles with diameters greater than 2 μm) from resuspended dust on carpets, seats, luggage, and occupants' clothing may also be present in the cabin air, they are less efficient in scattering light and will contribute less than the fine particles, per unit mass, to the measured light scattering.

Several companies manufacture and distribute portable nephelometers that could be used to monitor fine-particle concentrations in aircraft cabins. The most rugged and reliable use light-emitting diodes as light sources and solid-state photodetectors to collect the scattered light from particles passing through the sensing zone (Jensen and O'Brien 1993; Watson and Chow 1993). Some of them also have built-in data loggers that can accumulate and average concentration readings over preselected intervals ranging from seconds to many minutes and can retain several thousand individual measurements for analysis of temporal patterns of concentration. One of the most compact is the MIE Corp. Personal DataRam; another suitable instrument is the Dustrak (TSI Corp.).

Application to Aircraft Cabin

If the sensing zone of a nephelometer is located in the air leaving the passenger cabin, the monitoring system can determine the highest particle concentration in the cabin. Although particles are lost to surfaces in the cabin, the exhausted air contains particles originating from sources that influence supply air and sources in the cabin.

The output signals from the nephelometers should be connected to the aircraft's air-quality data recorder for analysis and correlation with reported health problems. Although the recordings will not provide information on particle composition, the temporal record of particle concentrations throughout the flight—in conjunction with other available data on aircraft system operations, observations in the cabin during the flight, and later physical investigation—may alert crew to nonroutine events and help to identify the nature of a problem.

Integrated Monitoring of Particulate Matter

Although a nephelometer can provide real-time data on particle concentrations throughout a flight segment, it cannot provide information on the chemical or biological composition of the particles. When such information is needed for forensic evaluation of air-quality problems that may have contributed to passenger or crew illnesses or incapacitation, it can be obtained only from PM samples that are collected on filters during the flight and then analyzed with sensitive laboratory techniques.

The collection of such filter samples from the air exhausted from the passenger cabin is relatively simple and inexpensive. However, the analysis of such samples is usually expensive. Such analyses would normally be performed only in the rare cases when poor air quality is considered a likely cause of passenger or crew illness or incapacitation. The occurrence of nonroutine conditions during a flight is an important example in which a record of airborne PM might be valuable. The exposure information could be linked to health data to evaluate the impact of nonroutine operation of the ECS.

Available Technology

Sequential air samplers that collect particles on segments of a continuous filter-tape reel are commercially available. The tape can be advanced at programmed intervals to present a fresh surface for each phase of sampling. The previous samples accumulate on the takeup reel for laboratory analysis as needed. A relatively compact and rugged unit of this type is marketed by MDA, Inc. (ACGIH 2001).

Application to Aircraft Cabin

A modified version of the MDA sequential sampler might be used that advances the filter tape in response to altimeter readings, with operational temporal segments collected for the following flight segments:

- The ground-based phase (boarding of flight crew through aircraft takeoff).

- The interval from takeoff through reaching cruise altitude.
- The interval during flight at cruise altitude.
- The interval covering descent, landing, and taxi to gate.

The filter tapes containing appropriate identification codes for locating specific flight-segment samples should be archived for some period in case a forensic or research investigation is needed. If no air-quality problems are identified and no analyses are needed, the filter tapes could be transferred to long-term storage or discarded.

PESTICIDES

Measurement of exposure to pesticides during and after aircraft disinsection may require air monitoring, as well as other analytic techniques because exposures may occur as a result of surface contamination. For the airborne route, analysis of integrated PM samples for pesticides would provide a measure of exposure to airborne particles resulting from direct spraying or resuspension of settled material. However, the assessment of dermal or oral exposure resulting from contaminated surfaces in the cabin poses considerable difficulty.

The methods available to assess the noninhalation routes include analyzing samples removed from aircraft cabin surfaces and from skin and sampling of body fluids or tissues. Surface measuring techniques have recently been reviewed (Schneider et al. 2000) and consist of hand-washing, surface suction, or surface-wipe sampling followed by analysis of the removed material for the chemical of concern. The removal efficiency of these sampling methods is highly variable, depending on surface characteristics, time since surface deposition, and other factors. Therefore, any application of the methods to aircraft cabin surfaces or to the skin of passengers and crew requires careful prior determination of removal efficiency. None of the few reports of surface sampling of aircraft for pesticides, such as that by Murawski (2001), has included measurements of removal efficiency. It appears that a reliable method for surface monitoring of pesticide exposure is not yet available.

Biological monitoring for pesticide exposure of cabin occupants (e.g., collection and analysis of voided urine after a flight) may be useful. Metabolites of several pyrethroid insecticides can be found in the urine of exposed persons within 24 h of exposure (Lauwerys and Hoet 1993). A more detailed discussion of biological monitoring techniques and their advantages and disadvantages for aircraft occupants is provided in Appendix D.

OTHER MONITORING METHODS

Many other continuous and integrated monitoring techniques have been used to investigate air quality other than that on commercial aircraft. The techniques include analysis for sulfur, phosphorus, volatile organic compounds (VOCs), and semivolatile organic compounds and analysis of human body fluids or tissues to assess systemic exposure. Although their implementation in aircraft has not been demonstrated, and their widespread application might not be technically feasible, their use in selected research applications could be valuable. A more detailed discussion of several techniques is presented in Appendix D.

SAMPLING LOCATIONS

Except for O_3 and CO, the most appropriate location for instrument sampling points is in the cabin or cockpit air outlet ducts. Air at these locations best reflects the mixed concentration in the ventilated space and therefore yields the best indication of the average exposure of passengers and crew. Although drawing samples from the inlet ducts would give a better measure of the quality of the supply air, it does not reflect what the passengers and crew are directly exposed to because supply air is mixed to a variable extent with the resident air in the cabin or cockpit. The committee notes that sampling from points in the air outlet ducts might not provide useful information when aircraft doors are open during loading and unloading because the cabin ventilation system might not be operating.

In some cases, sampling multiple locations in the aircraft may be necessary. For example, in aircraft with separate ventilation zones for cockpit, first-class, and economy sections, multiple sampling points would be needed to give a complete picture of air quality in the occupied spaces. It may also be useful in some cases to sample bleed air or recirculated air after filtration or passage through O_3 or VOC scrubber devices to evaluate their performance.

DATA PROCESSING

Cockpit Indicators

Analogue or digital signals from all or some of the continuous instruments could be displayed in the cockpit. Such displays would be useful in evaluating the performance of the ECS during various flight phases and could also yield early indications of seal failures, fluid leaks, or other problems that could require changes in the flight path or maintenance after landing. Cockpit crew members would need to be trained in evaluation of such signals.

Logging and Storage of Data

To meet the purposes of air-quality monitoring, data from each instrument must be recorded and stored in a form that can be retrieved and examined. Each of the continuous instruments described above provides electronic signals suitable for recording and storage as time-resolved data. Those signals could be added to the flight-data recording system if it has sufficient capacity, or they could be logged with a separate dedicated recorder for analysis during investigations of equipment failure or health complaints from passengers and crew. The following are several examples of potential application of air-quality monitoring:

- Cockpit indicators could alert a pilot to high O_3 concentrations (greater than 0.25 ppm, according to FAR). Logged O_3 data could provide information for maintenance crews on the efficiency of O_3 converters and indicate when such devices should be refurbished or replaced.
- The CO_2 indicator could alert a pilot to conditions under which the air-exchange rate should be increased.
- The CO and nephelometer (PM) data could alert maintenance personnel to leaking fluid seals, especially at cruise, when no other sources of fine particles are expected.

Available Technology

With the exception of integrated PM sampling, each instrument type described above has the ability to sample and analyze air for the target contaminant or property on a time scale of seconds to minutes, so their output can be regarded as "continuous." In the cockpit, these instruments can provide analogue or sometimes digital outputs that are appropriate (Ness 1991). Their signals provide up-to-the-minute data on absolute concentrations, and instrument response is fast enough to permit accurate determination of the rate of change in concentration.

For data recording and storage, relatively few continuous monitors for gaseous chemicals with data-logging capabilities are available, and an external storage device may therefore be needed (Gressel et al. 1988). Such devices are modified computers that collect, store, and deliver data to other computers or display devices. They are available with a wide variety of data capacities and are capable of collecting signals from multiple instruments nearly simultaneously (Ness 1991).

Data stored during a flight should include flight number, date, and continuous data on cabin air characteristics with elapsed time indicated. They could be transferred to an archive file on a larger computer and stored for a selected period.

Figure 7-1 provides an example of continuous monitoring that was conducted on a Boeing 767 aircraft with a 98% load factor and the type of data that can be obtained from sampling instruments (Spengler et al. 1997; Dumyahn et al. 2000). Measurements that were taken throughout the flight from boarding through deplaning included barometric pressure, temperature, relative humidity, and CO_2. The vertical dashed lines indicate the various states of flight: boarding, takeoff, cruise, landing, and deplaning. The horizontal axis indicates the time when concentrations were observed.

CONCLUSIONS

- Current air-quality measurement practices on commercial aircraft include only indicators of temperature and pressure. These practices are insufficient to determine all cases when the ECS is not working properly or when air-quality incidents occur, and they do not allow evaluation of the possible link between exposures and health effects.

- Although continuous air monitoring has not been implemented, it is technically feasible for a number of air-quality characteristics on commercial aircraft, including temperature, barometric pressure, O_3, CO, CO_2, relative humidity, and fine PM. Collecting filter samples of suspended PM that could be archived for analysis is also feasible.

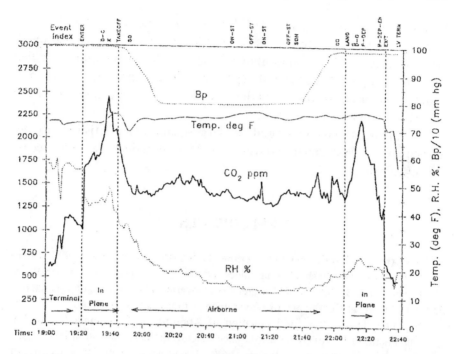

FIGURE 7-1 Environmental conditions on a Boeing 767 with 98% load factor. Event index: D-C (door close), x (taxi), SO (seatbelt off), ON-ST (on seat), OFF-ST (off seat), SON (seatbelt on). Abbreviations: Bp, barometric pressure; RH, relative humidity.

• Although air-quality monitoring techniques for additional agents, such as pesticides, are available, their applicability to aircraft may require further research and development.

RECOMMENDATIONS

• Instruments for monitoring O_3, CO, CO_2, temperature, cabin pressure, relative humidity, and PM should be used in the surveillance or research investigations aboard commercial aircraft as described in Chapter 8.

• Because of the committee's concern that O_3 can exceed health standards on routine flights that are not expected to encounter high O_3 concentrations, it recommends that FAA take effective measures to ensure that the current FAR for O_3 (i.e., average concentrations not to exceed 0.1 ppm above 27,000 ft, and peak concentrations not to exceed 0.25 ppm above 32,000 ft) is met on all flights, regardless of altitude. These measures should include a requirement that O_3 converters be installed, used, and maintained on all aircraft capable of flying at or above those altitudes, or a requirement that strict operating limits be set with regard to altitudes and routes for aircraft without converters to ensure that the O_3 concentrations are not exceeded in reasonable worst-case scenarios.

- Methods for monitoring additional air-quality characteristics and other measures of cabin-occupant exposure should be investigated as indicated by the data needs.

REFERENCES

ACGIH (American Conference of Governmental Industrial Hygienists). 1999. Documentation of the Threshold Limit Values and Biological Exposure Indices, Sixth Ed. Cincinnati, OH: American Conference of Governmental Industrial Hygienists.

ACGIH (American Conference of Governmental Industrial Hygienists). 2001. Air Sampling Instruments for Evaluation of Atmospheric Contaminants, 9th Ed. Cincinnati, OH: American Conference of Governmental Industrial Hygienists.

Arnold, K., S. Crha, and L. Patten. 2000. The effect of recirculation on aircraft cabin air quality—In-flight tests and simulation study. Pp. 26-44 in Air Quality and Comfort in Airliner Cabins, N. Nagda, ed. West Conshohocken, PA: American Society for Testing and Materials.

ASHRAE (American Society of Heating Refrigerating and Air-Conditioning Engineers). 1999a. Chapter 9 in Aircraft in 1999 ASHRAE Handbook: Heating, Ventilating, and Air- Conditioning Applications. American Society of Heating, Refrigerating, and Air-Conditioning Engineers, Atlanta, GA.

ASHRAE (American Society of Heating Refrigerating and Air-Conditioning Engineers). 1999b. ASHRAE Standard—Ventilation for Acceptable Indoor Air Quality. ANSI/ASHRAE 62-1999. American Society of Heating Refrigerating and Air-Conditioning Engineers, Atlanta, GA.

Balouet, J. 2000. Letter from J.C. Balouet, Environment International, Nogent/Marne, France to the National Research Council. In-Cabin Airborne Chemicals: A Short List. July 26, 2000.

Berg-Munch, B., G. Clausen, and P. Fanger. 1986. Ventilation requirements for the control of body odor in spaces occupied by women. Environ. Int. 12(1-4):195-200.

Boeing. 2000. Airplanes 201. The Airplane Cabin Environment. Boeing Commercial Airplanes Group, Seattle, WA.

Cain, W., B.P. Leaderer, R. Isseroff, L.G. Berglund, R.J. Huey, E.D. Lipsitt, and D. Perlman. 1983. Ventilation requirements in buildings—I. Control of occupancy odor and tobacco smoke odor. Atmos. Environ. 17(6):1183-1197.

De Ree, H., M. Bagshaw, R. Simons, and R.A. Brown. 2000. Ozone and relative humidity in airline cabins on polar routes: Measurements and physical symptoms. Pp. 243-258 in Air Quality and Comfort in Airliner Cabins, N. Nagda, ed. West Conshohocken, PA: American Society for Testing and Materials.

Dumyahn, T.S., J.D. Spengler, H.A. Burge, and M. Muilenburg. 2000. Comparison of the environments of transportation vehicles: results of two surveys. Pp. 13-25 in Air Quality and Comfort in Airliner Cabins, N.L. Nagda, ed. West Conshohocken, PA: American Society for Testing and Materials.

EPA (U.S. Environmental Protection Agency). 1985. National Primary Ambient Air Quality Standards for Carbon Monoxide. 40 CRF 50.8. Fed. Regist. 50(Sept. 13):37501.

FAA (Federal Aviation Regulation). 1996. Part 25. Airworthiness Standards: Transport Category Airplanes. Subpart D- Design and Construction, Ventilation and Heating. Sec. 25.831. Ventilation. Fed. Regist. 61(Dec. 2):63956.

Freitag, H., Y. Feng, L. Mangum, M. McPhaden, J. Neander, and L. Stratton. 1994. Pp. 1-18 in Calibration Procedures and Instrumental Accuracy Estimates of Tropical Atmosphere Ocean Temperature, Relative Humidity and Radiation Measurements. NOAA Technical Memorandum ERl PMEL 104. Seattle, WA: U.S. Dept. of Commerce, National Atmospheric and Oceanic Administration, Environmental Research Laboratory, Pacific Marine Environmental Laboratory.

Gressel, M.G., W.A. Heitbrink, J.B. McGlothlin, and T.J. Fischbach. 1988. Advantages of real-time data acquisition for exposure assessment. Appl. Ind. Hyg. 3(11):316-320.

Holdeman, J.D., L.C. Papathakos, G.J. Higgins, and G.D. Nastrom. 1984. Simultaneous Cabin and Ambient Ozone Measurements on two Boeing 747 Airplanes. Vol. II—January to October, 1978. NASA Technical Memorandum 81733 FAA-EE-83-7. NASA, U.S. Department of Transportation. March 1984.

Hoummady, M., C. Bonjour, J. Collin, F. Lardet-Vieudrin, and G. Martin. 1995. Surface acousting wave (SAW) dew point sensor: Application to dew point hydrometry. Sensors and Actuators B 27(1):315-317.

Hunt, E.H., D.H. Reid, D.R. Space, and F.E. Tilton. 1995. Commercial Airliner Environmental Control System, Engineering Aspects of Cabin Air Quality. Presented at the Aerospace Medical Association Annual Meeting, Anaheim, CA.

Jensen, P., and D. O'Brien. 1993. Industrial hygiene. Pp. 537-559 in Aerosol Measurement: Principles, Techniques, and Applications, K. Willeke, and P. Baron, eds. New York: Wiley.

Lafarie, J.P. 1985. Relative humidity measurement: A review of two state of the art sensors. Pp. 875-889 in Moisture and Humidity 1985: Measurement and Control in Science and Industry. Research Triangle Park, NC: Instrument Society of America.

Lauwerys, R., and P. Hoet. 1993. Industrial Chemical Exposure: Guidelines for Biological Monitoring, 2nd. Ed. Boca Raton, FL: Lewis.

Lee, S.C., C.S. Poon, X.D. Li, F. Luk, M. Chang, and S. Lam. 2000. Air quality measurements on sixteen commercial aircraft. Pp. 45-60 in Air Quality and Comfort in

Airliner Cabins, N. Nagda, ed. West Conshohocken, PA: American Society for Testing and Materials.

Lindgren, T., D. Norback, K. Andersson, and B.G. Dammstrom. 2000. Cabin environment and perception of cabin air quality among commercial aircrew. Aviat. Space Environ. Med. 71(8):774-782.

Lorengo, D., and A. Porter. 1986. Aircraft Ventilation Systems Study. Final Report. DTFA-03-84-C-0084. DOT/FAA/CT-TN86/41-I. Federal Aviation Administration, U.S. Department of Transportation. September 1986.

Marenco, A., V. Thouret, P. Nedelec, H. Smit, M. Helten, D. Kley, F. Karcher, P. Simon, K. Law, J. Pyle, G. Poschmann, R. Von Wrede, C. Hume, and T. Cook. 1998. Measurement of ozone and water vapor by Airbus in-service aircraft: The MOZAIC airborne program, an overview. J. Geophys. Res. 103(19):25631-25642.

Murawski, J. 2001. Letter from Judith Murawski, Association of Flight Attendants to Eileen Abt, National Research Council, regarding wipe sample data (permethrin) collected on a recently-sprayed aircraft. March 5, 2001.

Nagda, N.L., H.E. Rector, Z. Li, and E.H. Hunt. 2001. Determine Aircraft Supply Air Contaminants in the Engine Bleed Air Supply System on Commercial Aircraft. ENERGEN Report AS20151. Prepared for American Society of Heating, Refrigerating, and Air-Conditioning Engineers, Atlanta, GA, by ENERGEN Consulting, Inc., Germantown, MD. March 2001.

Nagda, N., H. Rector, Z. Li, and D. Space. 2000. Aircraft cabin air quality: A critical review of past monitoring studies. Pp. 215-239 in Air Quality and Comfort in Airliner Cabins, N. Nagda, ed. West Conshohocken, PA: American Society for Testing and Materials.

Ness, S.A. 1991. Pp. 145-295 in Air Monitoring for Toxic Exposures: An Integrated Approach. New York: Van Nostrand Reinhold.

NIOSH (National Institute for Occupational Safety and Health). 1972. Criteria for a Recommended Standard: Occupational Exposure to Carbon Monoxide. DHEW (HSM) 73-11000. Rockville, MD: U.S. Dept. of Health, Education, and Welfare, Health Services and Mental Health Administration, National Institute for Occupational Safety and Health.

NRC (National Research Council). 1986. The Airliner Cabin Environment: Air Quality and Safety. Washington, DC: National Academy Press.

OSHA(Occupational Safety and Health Administration). 1989. OSHA Regulations, Standards-29 CFR 19100.1000 Air Contaminants, Toxic and Hazardous Substances Final Rule, Table Z-1. Fed. Regist. 54:(Sept. 5)36767. [Online]. Available: http://www.osha-slc.gov/FedReg_osha_data/FED19890905.html [October 18, 2001].

Parish, D.D., J.S. Holloway, and F.C. Fehsenfeld. 1994. Routine continuous measurement of carbon monoxide with parts per billion precision. Environ. Sci. Tech. 28(9):1615-1618.

Pierce, W., J. Janczewski, B. Roethlisberger, and M. Janczewski. 1999. Air quality on commercial aircraft. ASHRAE J. (Sept.):26-34.

Rayman, R.B, and G.B. McNaughton. 1983. Smoke/fumes in the cockpit. Aviat. Space Environ. Med. 54(8):738-740.

SAE International. (Society of Automotive Engineers International.). 1965. Ozone in High Altitude Aircraft. SAE AIR910. Society of Automotive Engineers International.

Seppanen, O.A., W.J. Fisk, and M.J. Mendell. 1999. Association of ventilation rates and CO2 concentrations with health and other responses in commercial and institutional buildings. Indoor Air 9(4):226-252.

Schneider, T., J.W. Cherrie, R. Vermeulen, and H. Kromhout. 2000. Dermal exposure assessment. Ann. Occup. Hyg. 44(7):493-499.

Spengler, J., H. Burge, T. Dumyahn, M. Muilenberg, and D. Forester. 1997. Environmental Survey on Aircraft and Ground-Based Commercial Transportation Vehicles. Prepared by Department of Environmental Health, Harvard University School of Public Health, Boston, MA, for Commercial Airplane Group, The Boeing Company, Seattle, WA. May 31, 1997.

Stevenson, G. 1994. Environmental control for aircraft. Pp. 97-101 in World Aerospace Technology '94. London: Sterling.

van Netten, C. 1998. Air quality and health effects associated with the operation of BAe 146-200 aircraft. Appl. Occup. Environ. Hyg. 13(10):733-739.

van Netten, C., and V. Leung. 2000. Comparison of the constituents of two jet engine lubricatin goils and their volatile pyrolytic degradation products. Appl. Occup. Environ. Hyg. 15(3):277-283.

van Netten, C., and V. Leung. 2001. Hydraulic fluids and jet engine oils: Pyrolysis and aircraft air quality. Arch. Environ. Health 56(2):181-186.

Watson, J., and J. Chow. 1993. Ambient air sampling. Pp. 622-639 in Aerosol Measurement. Principles, Techniques, and Applications, K. Willeke, and P. Baron eds. New York: Van Nostrand Reinhold.

Woebkenberg, M.L, and C.S. McCammon. 1995. Direct-reading gas and vapor instruments. Pp. 439-510 in Air Sampling Instruments for Evaluation of Amosperic Contaminants, 8th Ed., B. Cohen, and S. Hering, eds. Cincinnati, OH: American Conference of Governmental Industrial Hygienists.

8

Surveillance and Research Programs on Cabin Air Quality

The analyses presented in previous chapters have repeatedly led to the conclusion that available air-quality data are not adequate to address specific questions on aircraft cabin air quality and its possible effects on cabin occupant health. On the basis of its review, the committee has identified the following critical questions:

1. Do current aircraft, as operated, comply with the Federal Aviation Administration (FAA) design and operational limits for specific chemical contaminants—ozone (O_3), carbon monoxide (CO), and carbon dioxide (CO_2)—and for ventilation rate? Are the existing federal aviation regulations (FARs) for air quality adequate to protect health and ensure the comfort of passengers and cabin crew?
2. What is the association, if any, between exposure to cabin air contaminants and reports or observations of adverse health effects in cabin crew and passengers?
3. What are the frequency and severity of air-quality incidents (nonroutine conditions) that might lead to deterioration of cabin air quality by introduction of air contaminants, such as pyrolysis products of engine oil?

Studies have been published that report detailed monitoring of several aspects of aircraft air quality. However, the studies have involved a very small fraction of the total commercial aircraft flights and therefore cannot be portrayed as representing the full range of conditions during routine and nonroutine operations. A summary and critical review of nine reports published during the last 13 years (Nagda et al. 2000) showed that the studies varied widely in sample selection, pollutants monitored, measurement methods, and quality control. The number of flights sampled ranged from less than 10 to 158; only three studies sampled at least 50 flights.[1] Furthermore, only three studies claimed random selection of flights, and only one provided supporting details. Study duration ranged from 1 month to 1 year. None of the studies included more than four aircraft types. Three studies included measurements of only one contaminant; the others included measurements of six to nine contaminants or contaminant groups. For logistical reasons, state-of-the-art methods for air-quality monitoring were not used in all studies, and only one study provided a description of calibration procedures. Two additional studies (Haghighat et al. 1999; Lee et al. 2000), which were not reviewed by Nagda et al. (2000), involved 16 and 43 flights. Both appear to have limitations in sampling methods similar to those described above. Subject to those limitations, the published data lead to the tentative conclusion that the concentrations of CO and CO_2 under routine operations most likely do not exceed the FAA guidelines, but O_3 concentrations might exceed the guidelines on some flights. (See Chapter 3 for a more detailed critique of the O_3 data.)

No published reports include measurements of air quality during flights involving nonroutine events, such as leaks of hydraulic fluid or engine oil into bleed air. Because some of the constituents or pyrolysis products of those fluids have high toxicity (Wyman et al. 1993; Wright 1996; van Netten and Leung 2000), obtaining exposure data during air-quality incidents is critical. Little information is available that would permit an estimate of the frequency of such events. (See Chapter 3 for further discussion of air-quality incidents and their possible frequency.)

To address the important unresolved questions regarding aircraft cabin air quality and its possible effects on occupant health, the committee recommends two complementary approaches: a surveillance program and a research program. The primary goals of the surveillance program are to determine aircraft compliance with FAA cabin air-quality regulations, to characterize air-quality characteristics and establish temporal trends in them, and to estimate the frequency of nonroutine incidents in which air quality is seriously degraded. This

---

[1]To put the numbers of flights sampled in perspective, there were about 8 million departures of passenger jets equipped with at least 30 seats operated by U.S. domestic aircraft companies in 1999 (DOT 2001).

program also includes a health surveillance component to determine the incidence of health effects in cabin crew and passengers and to identify possible associations between air quality and health effects. The research program is designed to focus more narrowly on specific unresolved issues of cabin air quality. The two approaches are described in more detail on the following pages.

AIR-QUALITY SURVEILLANCE

A program of systematic surveillance of air quality is needed to determine compliance with FARs for air quality and to establish temporal trends in air-quality measures. Accurate characterization of the variation in air-quality characteristics during routine operations, coupled with health-surveillance information, would provide insight into the possible association between air quality and reported health effects in cabin occupants.

At a minimum, the surveillance program should include continuous monitoring and recording of O_3, CO, CO_2, fine particulate matter (PM), cabin pressure, temperature, and relative humidity. An adequate surveillance program must be extensive and sample a large number of flight segments over a relatively short period. Estimates based on the studies reviewed to date (Nagda et al. 2001) reveal that, depending on the specific design of the surveillance program, samples from at least 100 flight segments over 1-2 years might be needed. Steps should be taken to ensure that a variety of aircraft types, aircraft companies, and routes are represented in the flights monitored. A standard instrument package should be developed that incorporates continuous monitors and data-recording hardware suitable for installation on commercial aircraft. (See Chapter 7 for a discussion of current technology that is adequate for accomplishing this goal.) Because aircraft equipment design and operating procedures evolve continuously, a continuing program will be required to characterize air quality in contemporary aircraft and to evaluate changes in air quality as aircraft equipment ages or is upgraded. The air-quality surveillance program must be coupled with a health surveillance program.

HEALTH SURVEILLANCE

As described in Chapter 6, the current systems for reporting signs and symptoms, which could reflect health-related responses to the cabin environment or the cabin air, are not standardized with respect to the methods of surveying potentially affected individuals or how specific data are obtained. Among the relevant databases maintained by the federal government, only the National Aeronautics and Space Administration (NASA) Aviation Safety Reporting System focuses on health-related data. However, as noted in Chapter 6, the data are collected as unstandardized narratives that do not appear to be abstracted in a standardized manner and stored in a format that would facilitate analysis. The committee also notes that the survey instrument released recently by the Association of Flight Attendants has a number of design flaws that could impede its use (see Chapter 6).

Although implementing all the details of an ideal system might not be practical, defining the minimal characteristics of a rigorous system for collecting, storing, analyzing, and disseminating health outcomes related to routine or nonroutine conditions is essential (see Table 8-1). Such definition should provide the guidance to improve the completeness and validity of current data on cabin air quality and health.

On the basis of exposure and self-interest, the cabin crew appears to be the logical vehicle through which a routine health surveillance system should operate. Self-interest should work to minimize nonparticipation and to mitigate problems of selection bias that severely compromise current systems for data collection. Securing unbiased estimates of passenger symptoms on any continuing basis seems less practical, although a serious marketing effort to educate passengers on the need for this information could make sampling passengers possible. Ideally, passengers would be surveyed on flights on which the crew is scheduled to be surveyed.

Central to any valid surveillance system are a plan for systematic sampling across all possible exposures under routine flight conditions and a set of standardized procedures for reporting all suspected incidents. Because many different aircraft are used on many flight routes, some form of multistage sampling would be required for monitoring health-related symptoms under routine flight conditions. However sophisticated the sampling scheme, the validity of the data will ultimately depend on the thoroughness with which the final-stage sampling units complete the requisite forms (cabin crew for routine conditions, and cabin crew, cockpit crew, and passengers for incident conditions). All data forms must be relatively short, have few or no text fields, and permit direct entry through scanning. The use of personal digital assistants was considered, but it probably is not practical because of cost, limitations of programming, and the need to have a network that ensures easy access.

Some combination of options would most likely be necessary to generate

TABLE 8-1 Desirable Properties of a Health Surveillance System for Commercial Aircraft

Under Routine Flight Conditions	
Systematic sampling of cabin crew	Multistage cluster sampling could be conducted with aircraft type as primary unit and flight type as secondary unit. Sampling could also be conducted with airline as primary unit, aircraft type as secondary unit, and flight type as tertiary unit. All cabin crew on sampled flights must complete preflight and postflight surveys.
Collection of data	Standardized pencil-and-paper forms: Structured to be completed within 5 min. Formatted to be scanned into database. Formatted for self-return to centralized data processing center.
Management of database and reporting	Central data center receives, edits, and processes all data; prepares reports at regular intervals by type of aircraft and flight; and distributes reports to airlines, FAA, manufacturers, flight attendant unions, and pilot union. Mechanisms to release data to public and organization to release data to be determined.
Actions based on surveillance data	To be decided by interested regulatory agencies, airlines, and crew.
Under Incident Conditions	
Ad hoc sampling survey for suspected incident	Would supplement routine health surveillance monitoring.
Collection of data	All cabin *and* cockpit crew complete surveys. Standardized pencil-and-paper form for incident description (same properties as above). Standardized health form to supplement incident form. Standardized supplemental form to be distributed to all passengers to report incident-related symptoms. Standardized form for health followup of crew. Standardized maintenance evaluation form to be linked to database.

Management of database and reporting	Same procedure as above. Central data manager sends followup health form at 6 and 12 mo. Findings are distributed to groups noted above plus responsible maintenance monitoring databases (FAA Accident/Incident Data System and Service Difficulty Reporting System).
Actions based on surveillance data	To be decided by interested regulatory agencies, airlines, and crew.

an effective solution to the problem of passenger and crew complaints of cabin air quality and its possible association with health effects. As noted above, air-quality monitoring remains an essential approach and must be coordinated with the health surveillance system to address the questions regarding possible links between air quality and health effects.

AIR-QUALITY RESEARCH

Apart from the surveillance program, the committee recommends a series of research investigations, each aimed at a specific aspect of cabin air quality. The investigations would have a more limited scope than the surveillance program, but could involve more intensive air sampling for one or two contaminants in selected aircraft or on selected flights. The research program must be coordinated with the surveillance program discussed above. When it is appropriate, data collected in the surveillance program can be used to formulate research questions. The following pages outline seven research topics on which further information is critical for assessing cabin air quality. Suggestions for collecting data on exposure to biological agents and related health effects were provided in Chapter 4.

Ozone

The committee identified several questions relevant to O_3 in aircraft cabins that should be addressed. How is the concentration of O_3 in aircraft cabins affected by factors such as ambient O_3 concentration, deposition on surfaces in the aircraft, the presence and effectiveness of the catalytic converter, the maintenance and replacement schedule for the converter, and chemical reactions of O_3 with component surfaces in the cabin? What is the strength of the relationship between cabin O_3 concentration, both short-term and averaged over the duration of a flight, and health effects in the occupants? (See Chapter 3 for discussion of factors affecting O_3 concentration and Chapter 5 for discussion of health effects of O_3.) O_3 concentrations for this study should be monitored in both supply and exhaust air.

Cabin Pressure and Oxygen Partial Pressure

The committee identified several critical unanswered questions regarding cabin pressure: Is a maximal cabin pressure altitude of 2,440 m (8,000 ft) appropriate for avoiding hypoxia? Are passengers with pulmonary or cardiovascular disease exposed to unacceptable risk at that pressure? How does cabin pressure vary with factors such as flight duration, altitude, and aircraft type?

Because partial pressure of oxygen (PO_2) in air is proportional to air pressure, continuous monitoring and recording of cabin pressure altitude is sufficient to determine the temporal variation in inhaled PO_2. However, the relationship between inhaled and arterial PO_2, and therefore hemoglobin saturation, are influenced by the functional status of the cardiovascular and respiratory systems (Slonim and Hamilton 1971; Murray 1976; Robson et al. 2000). Therefore, use of pulse oximetry to assess hemoglobin saturation in crew and passengers should be considered. (See Chapters 5 and 6 for further discussion of cabin pressure and health risk.)

Outside-Air Ventilation

The committee was unable to resolve several questions related to ventilation on aircraft: Is the current FAA design requirement for outside-air ventilation (FAR 25.831) adequate to minimize complaints? Are environmental control systems (ECSs) in aircraft operated so that they meet the outside air-ventilation rate in the FAA design requirements at all times during normal operation? To what extent are complaints from passenger and crew associated with the outside-air ventilation rate? How do factors such as passenger loading and recirculation of cabin air affect the amount of outside-air ventilation needed? Can the conclusion drawn in Chapter 4 that infectious-disease agents are transmitted primarily between people in close proximity be verified? What effects, if any, do outside-air ventilation rate, total airflow (outside air plus recirculation air), and airflow patterns in the cabin have on disease transmission? Is American Society of Heating, Refrigerating and Air-Conditioning Engineers Standard 62 (ASHRAE 1999) appropriate for aircraft, and is it adequate to avoid air-quality complaints on aircraft?

The CO_2 concentration in an aircraft cabin depends primarily on the outside-air ventilation rate per occupant and, in the absence of confounding factors (e.g., dry ice in galleys), may be used as a surrogate for outside-air ventilation rate. Outside-air ventilation flow rates may be measured directly, but this would be extremely difficult on an aircraft. Monitoring should be conducted continuously from "crew on" to "crew off" and in conjunction with corresponding evaluations of cabin occupant comfort and health variables, as noted above.

Air-Quality Incidents

Questions identified by the committee regarding contaminants entering bleed air because of oil leaks or equipment malfunctions include the following: How does the frequency vary with the type of engine or bleed-air system? What is the toxicity of the constituents and pyrolyzed products of the materials? What is their relationship to reported health effects? How are the oil, fluids, and pyrolyzed products distributed from the engines, into the ECS, and throughout the cabin environment?

A complex approach to monitoring may be required to evaluate these questions and may include careful evaluation of maintenance records to identify aircraft in which fluid is lost or aircraft components that require frequent service of fluid seals and cabin air monitoring to detect airborne products of leakage, such as fine PM and CO. Fine PM could indicate a leak in which aerosols are produced, and CO could indicate a leak in which incomplete combustion of fluids occurs at high temperatures. Integrated filter sampling of airborne particles during flight would be necessary for determination of the chemical composition of the materials released by oil leaks or equipment malfunctions.[2] The samples could be stored and analyzed later. Because nonroutine conditions appear to be rare, sampling a large number of flight segments may be required to accumulate sufficient data to characterize the role of fluid leaks in affecting cabin air quality. For example, if the frequency of the events is no more than 1 per 1,000 flights, acquiring data from 100 events would involve sampling a minimum of 100,000 flight segments. Sampling such a large number of flights is not feasible, and more appropriate approaches are discussed below.

- *Focus on "problem" aircraft.* Use surveillance data to identify aircraft types in which problems are especially frequent, and conduct intensive air monitoring on them. Only a few aircraft or engine types may be responsible for most serious health complaints reported to date in nonroutine incidents. Intensive sampling with all the techniques outlined in Chapter 7 may be possible for investigating the aircraft and thus identifying the causes and remedies of problems.

- *Review maintenance and repair records.* Identify aircraft that have recurrent problems (e.g., excessive loss of engine oil or hydraulic fluids and failed fluid seals) that might be associated with deterioration in cabin air quality. Linking the results of such an investigation to the health surveillance program might make it possible to evaluate the reported association between cabin air quality and health problems.

- *Investigate the need for additional air-contaminant control devices.* The goal is to capture contaminants that might enter through the ECS before they can enter the aircraft cabin. Potential approaches could include additional air-cleaning devices in the air supply lines upstream of the points where bleed air is mixed with recirculated air or upstream of the points where air is directed to the cabin in aircraft without recirculation. When the air contaminant control devices are being investigated, some objective measures of their performance must be defined to validate expected improvements in cabin air quality.

- *Investigate only aircraft or flights on which health complaints have been reported.* This approach could be based on the health surveillance

---

[2]The committee has not suggested sampling for volatile organic compounds or semivolatile organic compounds because no available sampling techniques are practical or feasible on aircraft (see Appendix D). However, if new techniques become available, their application to this research should be considered.

system described above. Although air-quality measurements might not be available for the particular flights in question, careful review of the aircraft itinerary, operations log, maintenance history, and other characteristics might reveal information that would suggest a cause of the complaints and a suitable remedy.

Although the implementation of a health surveillance system for routine conditions of flight is relatively straightforward, some unique complexities exist for the evaluation of incidents. Given the concerns of cabin crew, the most important need would be to develop a standard procedure for followup of cabin crew who have worked on a flight on which an incident has been reported. The intervals for data collection, the mechanisms to maximize compliance and to protect privacy, and the total duration of followup after an incident all require consideration. The need to link data collected in an incident surveillance system with existing data systems (FAA's Accident/Incident Data System, NASA's Aviation Safety Reporting System, FAA's Service Difficulty Reporting System) and the need to link maintenance findings and corrective actions to the system also require consideration. A simple sampling protocol for incident conditions and the responsibility for the maintenance and updating of a central data repository would be required. The repository would be responsible for the maintenance of data accuracy and privacy.

Pesticide Exposure

The committee identified several questions related to pesticide exposure: What exposure concentrations and chemical constituents are observed in commercial aircraft as a result of disinsection? How do the exposures depend on the pesticide application method (blocks away, top of descent, on arrival, residual treatment, and preembarkation; see Chapter 3)? What health risks, if any, are associated with such exposures?

Evaluation of pesticide exposure may require air monitoring and other analytic techniques. For the airborne route, analysis of integrated PM samples would indicate exposure to airborne particles resulting from direct spraying or resuspension of settled material. Methods available to assess the noninhalation routes include analysis of samples removed from aircraft surfaces and from skin and sampling of body fluids or tissues. (See Chapter 7 and Appendix D for further discussion of the sampling techniques.)

Relative Humidity

The committee identified several questions regarding the issue of relative humidity: What is the contribution of low relative humidity to the perception of dryness? Do other factors cause or contribute to the irritation attributed to the dry cabin environment during flight?

As noted previously in this report, low relative humidity occurs on nearly all flights during cruise. Therefore, the focus of this research effort is to determine the relationship between relative humidity and the complaints of irritation of eyes and mucous membranes among passengers and crew, not to determine the cabin relative humidity itself.

Fine Particulate Matter

Although fine PM is not a major focus of the research program, the committee identified several questions related to fine PM: What is the role of fine PM in inducing complaints and health effects in cabin occupants? What are the probable sources of the fine PM? What are the important chemical or biological components of fine PM in the cabin? To what extent do filters or other air-cleaning devices reduce PM concentrations in aircraft cabins? What factors influence the effectiveness of those devices?

Careful study of this topic requires reliable monitoring of fine PM that is both continuous and time-integrated (averaged over the period of a flight). (See Chapter 7 for a discussion of techniques for monitoring fine PM.)

STAGING

Feasibility Demonstration of Air-Quality Monitoring Program

Air-quality monitoring will be required for both surveillance and research. One possible approach to implementing the necessary monitoring is a feasibility study. A prototype instrument package could be assembled by using commercial instruments with modifications as necessary to meet power, space, and safety requirements. The package should include direct-reading monitors for temperature, cabin pressure, relative humidity, O_3, CO, CO_2, and fine PM. It could be installed in nonrevenue-generating space in selected commercial aircraft before their initial delivery or during a periodic overhaul. The presence of passengers and crew in the aircraft during flight will be necessary to evaluate the performance of the monitors in the presence of human sources of several important contaminants. Hardware and software for data collection and storage should be included in the demonstration system. It may ultimately be feasible to incorporate the data generated by the monitoring systems into the flight data recording system already in place on all commercial aircraft.

It will be especially important to demonstrate that the air-quality monitoring package, including all pumps and air-sampling tubing, can be installed and operated without disrupting normal aircraft operation. The committee recognizes that the visibility of the monitoring equipment in the passenger cabin must be minimized to avoid raising health and safety concerns among passengers.

Development of Standard Monitoring Package

When the prototype package has been developed and tested on selected flights, a standard package can be designed for installation on any commercial aircraft. The committee emphasizes that the collection of monitoring data from a large number of flights for the air-quality characteristics described will have great value for compliance and enforcement of existing air-quality regulations, as well as for research on the safety and comfort goals of aircraft ECSs. The ability to evaluate correlations between objective measurements of exposure to contaminants or their surrogates and reports or measures of health and comfort problems in aircraft cabin occupants would facilitate a considerable advance over current knowledge. This information would answer many of the questions that have arisen about the causes and frequencies of relatively rare but possibly severe health events in commercial aviation.

Another advantage of the committee's recommended surveillance program is that continuous monitoring data could provide justification for cockpit and cabin crew to try to minimize excessive exposures in flight. For example, if CO_2 were found to be high in the cabin, the pilot could increase the ventilation rate. However, when in-flight corrections cannot be made, off-specification or unexpected performance of the ECS can be documented, and appropriate maintenance can be promptly scheduled. That practice would minimize exposures that would otherwise continue or worsen during later flights. The committee, however, notes that adding instrumental indicators in the cockpit that might draw cockpit-crew attention may place an unreasonable additional burden on those responsible for safe operation during the flight. An appropriate balance must be sought between automated monitoring and recording without crew attention and the need for human observation and evaluation of the data as they emerge.

Finally, the capacity for collection and storage of extensive air-quality monitoring data would prove useful in the investigation of events and complaints. Airline companies and FAA could use the data to verify performance of aircraft ECSs or to develop appropriate solutions to environmental problems that might be revealed only through routine continuous monitoring. However, access to data and the duration of its storage are important issues, and a clear policy on these matters should be developed before an air-quality monitoring program begins.

CONCLUSIONS

• Existing air-quality data are inadequate for evaluating the possible association between air contaminants in routine operations and health problems or complaints from crew and passengers. Measurements have been made on a very small fraction of the total flight segments with methods that often lack acceptable accuracy and precision. Although CO and CO_2 concentrations do not appear to exceed FAA guidelines under routine operating conditions, O_3 concentrations probably do exceed the guidelines on some flights.

• Virtually no air-quality measurements are available for assessing the nature, severity, or frequency of nonroutine incidents aboard aircraft. Although complaints of foul odor and a variety of more serious health conditions have been reported, any relationship to deterioration in cabin air quality cannot be determined. There is some information on various possible toxic components produced by pyrolysis of aircraft fluids, but their presence in cabin air has not been documented.

• Equipment for control of airborne contaminants is available. Examples include high-efficiency particle filters and charcoal adsorbers. Systematic collection of continuous monitoring data on selected contaminants would reveal cases in which proper selection of additional control equipment can be made.

• In addition to the lack of exposure information, a major difficulty in the evaluation of the potential effects of cabin air quality on the health of passengers and cabin crew is the lack of standardized health surveillance systems for obtaining health-related data during normal and nonroutine operating conditions.

• Although airlines require frequent medical evaluations of pilots, regular health evaluations of cabin crew are not required. Lack of such data precludes any systematic assessment of the extent to which occupational exposures of cabin crew are associated with chronic health conditions that follow acute exposures during incident conditions.

RECOMMENDATIONS

• Air quality in commercial aircraft should be monitored with a dual approach that includes a routine surveillance program and a more focused research program.

• Routine surveillance of a number of air-quality characteristics (O_3, CO, CO_2, fine PM, cabin pressure, relative humidity, and temperature) should be implemented in a continuing program to characterize the range of air quality found in aircraft.

• A detailed research program is needed to investigate specific questions about the possible association between air contaminants and observed or reported health effects. Relevant subjects include factors that affect O_3 concentrations in cabin air, the need to lower cabin pressure altitude to prevent hypoxia in susceptible cabin occupants, the adequacy of outside-air ventilation flow rates, the severity of events in which contaminants enter bleed air from oil-seal leaks or other equipment malfunctions, the potential for pesticide exposure due to current disinsection practices, the contribution of low relative humidity to the perception of dryness, and the role of fine PM in generating health complaints.

• Health surveillance should be integrated into the air monitoring programs. Health surveillance is needed for the systematic collection, analysis, and reporting of health outcomes related to routine and nonroutine conditions in commercial aircraft. On the basis of self-interest and exposure, the cabin crew should be the vehicle through which the surveillance system would operate.

• Congress should designate a lead federal agency and provide sufficient funds to conduct or direct the research program that is aimed at filling major knowledge gaps identified in this report. An independent advisory committee with appropriate scientific, medical, and engineering expertise should be constituted to oversee the research program to ensure that its objectives are met.

REFERENCES

ASHRAE (American Society of Heating Refrigerating and Air-Conditioning Engineers). 1999. Chapter 9 in Aircraft in 1999 ASHRAE Handbook: Heating, Ventilating, and Air-Conditioning Applications. American Society of Heating, Refrigerating and Air-Conditioning Engineers, Atlanta, GA.

DOT (U.S. Department of Transportation). 2001. Airport Activity Statistics of Certificated Air Carriers, Summary Tables, Twelve Months Ending December 31, 1999. BTS01-03. Office of Airline Information, Bureau of Transportation Statistics, U.S. Department of Transportation, Washington, DC. [Online]. Available: http://www.bts.gov/publications/airactstats/index.html [October 18, 2001].

Haghighat, F., F. Allard, and R. Shimotakahara. 1999. Measurement of thermal comfort and indoor air quality aboard 43 flights on commercial airlines. Indoor Built Environ. 8(1):58-66.

Lee, S., C. Poon, X. Li, F. Luk, M. Chang, and S. Lam. 2000. Air quality measurements on sixteen commercial aircraft. Pp. 45-60 in Air Quality and Comfort in Airliner Cabins, N. Nagda, ed. West Conshohocken, PA: American Society for Testing and Materials.

Murray, J.F. 1976. Pp. 223-276 in The Normal Lung: The Basis for Diagnosis and Treatment of Pulmonary Disease. Philadelphia: Saunders.

Nagda, N.L., H.E. Rector, Z. Li, and E.H. Hunt. 2001. Determine Aircraft Supply Air Contaminants in the Engine Bleed Air Supply System on Commercial Aircraft. ENERGEN Report AS20151. Prepared for American Society of Heating, Refrigerating, and Air-Conditioning Engineers, Atlanta, GA, by ENERGEN Consulting, Inc., Germantown, MD. March 2001.

Nagda, N., H. Rector, Z. Li, and D. Space. 2000. Aircraft cabin air quality: a critical review of past monitoring studies. Pp. 215-239 in Air Quality and Comfort in Airliner Cabins, N. Nagda, ed. West Conshohocken, PA: American Society for Testing and Materials.

Robson, A.G., T.K. Hartung, and J.A. Innes. 2000. Laboratory assessment of fitness to fly in patients with lung disease: a practical approach. Eur. Respir. J. 16(2):214-219.

Slonim, N., and L. Hamilton. 1971. Pp. 150-174 in Respiratory Physiology, 2nd Ed. St. Louis: Mosby.

Van Netten, C., and V. Leung. 2000. Comparison of the constituents of two jet engine lubricating oils and their volatile pyrolytic degradation products. Appl. Occup. Environ. Hyg. 15(3):277-283.

Wright, R. 1996. Formation of the neurotoxin TMPP from TMPE-phosphate formation. Tribology Transactions 39:827-834.

Wyman, J., E. Pitzer, F. Williams, J. Rivera, A. Durkin, J. Gehringer, P. Serve, D. von Minden, and D. Macys. 1993. Evaluation of shipboard formation of a neurotoxicant (trimethyolpropane phosphate) from thermal decomposition of synthetic aircraft engine lubricant. Am. Ind. Hyg. Assoc. J. 54(10):584-592.

Appendixes

Appendix A

Biographical Information on the Committee on Air Quality in Passenger Cabins of Commercial Aircraft

MORTON LIPPMANN *(Chair)* is professor of environmental medicine and director of the Center for Particulate Matter Health Effects Research and of the Human Exposure and Health Effects Research Program at New York University School of Medicine. He earned a bachelors degree in chemical engineering at the Cooper Union, a masters degree in industrial hygiene from Harvard University, and a Ph.D. in environmental health science from New York University. Dr. Lippman is a member of the Executive Committee of the Science Advisory Board of the Environmental Protection Agency and chairs several nongovernmental scientific advisory committees, including those of the National Environmental Respiratory Center and the University of Southern California Medical School's Study of the Health Effects of Air Pollution in Children. He has also chaired or served on several NRC committees.

HARRIET A. BURGE is associate professor of environmental microbiology at the Harvard School of Public Health. She earned a Ph.D. in botany from the University of Michigan and continued her postdoctoral training there in aeroallergens. Among several major areas of research, Dr. Burge's current focus is on the role of environmental exposures in the development of asthma and evaluating exposure to fungi, dust mite, cockroach, and cat allergens in three separate epidemiology studies assessing risk factors for the development of asthma. Dr. Burge has served on a number of NRC committees, including a study on airliner cabin air quality and a recent IOM study on asthma and indoor air quality.

BYRON JONES is associate dean for Research and Graduate Programs and director of the Engineering Experiment Station at the College of Engineering, Kansas State University. He earned his Ph.D. in mechanical engineering from the Oklahoma State University. Dr. Jones' research interests are in heat and mass transfer, human thermal systems simulation, and thermal measurements and instrumentation. He was recently appointed chairman of an American Society of Heating, Refrigerating and Air-Conditioning Engineers (ASHRAE) standards committee that is developing standards for the aircraft cabin environment.

JANET M. MACHER is an air pollution research specialist with the Division of Environmental and Occupational Disease Control of the California Department of Health Services. She holds a masters degree in industrial hygiene and microbiology from the University of California, Berkeley, and an Sc.D. in environmental health science from the Harvard School of Public Health. Her research has focused on the evaluation of methods to collect and identify airborne biological material and on engineering measures to control airborne infectious and hypersensitivity diseases.

MICHAEL S. MORGAN is a professor in the Department of Environmental Health, Industrial Hygiene and Safety Program of the University of Washington and serves as director of the Northwest Center for Occupational Health and Safety (a NIOSH-funded education and research center). Dr. Morgan holds a Sc.D. in chemical engineering from the Massachusetts Institute of Technology. His research is focused on human response to inhalation of air contaminants, including the products of combustion and volatile solvents, and has encompassed both ambient air contaminants and occupational environmental health hazards.

WILLIAM W. NAZAROFF is professor of environmental engineering in the Department of Civil and Environmental Engineering of the University of California, Berkeley. He received his Ph.D. in environmental engineering science from the California Institute of Technology. His main research interest is indoor air quality, with emphasis on pollutant-surface interactions, transport/mixing phenomena, aerosols, environmental tobacco smoke, source characterization, exposure assessment, and control techniques. Dr. Nazaroff has served as associate editor of *Health Physics* and currently serves in a similar capacity for the *Journal of the Air & Waste Management Association*. He is on the editorial board of *Indoor Air.*

RUSSELL B. RAYMAN, currently executive director of the Aerospace Medical Association in Alexandria, VA, retired from the U.S. Air Force in 1989 with the rank of colonel after a military medical career. Dr. Rayman earned his M.D. from the University of Michigan. As a member of the Air Force, he served in many medical positions both abroad and in the United States and held a number of academic appointments at universities in Texas and Ohio.

JOHN D. SPENGLER is the Akira Yamaguchi Professor of Environmental Health and Human Habitation and director of the Environmental Science and Engineering Program at the Harvard School of Public Health. He received his Ph.D. from the State University of New York in Albany. Dr. Spengler's research is focused on assessment of population exposures to environmental contaminants (air, water, food, and soil) that occur in homes, offices, schools, and during transit, as well as in the outdoor environment. He has served on several NRC committees including Airliner Cabin Air Quality, Passive Smoking, and Risk Assessment of Hazardous Air Pollutants.

IRA B. TAGER is professor of epidemiology in the Division of Public Health, Biology, and Epidemiology at the University of California, Berkeley, and is co-director and principal investigator for the Center for Family and Community Health. He holds an M.D. from the University of Rochester School of Medicine and an M.P.H from the Harvard School of Public Health. Dr. Tager's research interests include, among others, the development of exposure assessment instruments for studies of health effects of chronic ambient ozone exposure in childhood and adolescence, effects of ozone exposure on pulmonary function, and the effects of oxidant and particulate air pollution on cardio-respiratory morbidity and mortality and morbidity from asthma in children.

CHRISTIAAN VAN NETTEN is an associate professor in the Department of Health Care and Epidemiology at the University of British Columbia and head of the Division of Occupational and Environmental Health. He received his Ph.D. in biochemistry/electrophysiology from Simon Fraser University. Dr.

van Netten's research interests include environmental toxicology, the use of electrodiagnostics to monitor worker exposure to agents that affect the peripheral nervous system, and identification of indoor air pollution associated with health problems. In addition, he has conducted research on air quality in commercial aircraft.

BERNARD WEISS is professor of environmental medicine and pediatrics at the University of Rochester School of Medicine and Dentistry. He received his Ph.D. in psychology from the University of Rochester. His special interests and publications lie primarily in areas that involve chemical influences on behavior. These include the neurobehavioral toxicology of metals such as lead, mercury, and manganese, solvents such as toluene and methanol, drugs such as cocaine, endocrine disruptors such as dioxin, and air pollutants such as ozone.

CHARLES J. WESCHLER served for more than 25 years as a research scientist at Bell Labs and Bellcore (now Telcordia Technologies), including as program manager for the Advanced Environmental Strategies Group. He recently became an adjunct professor in the Department of Environmental and Community Medicine at the University of Medicine and Dentistry of New Jersey, Robert Wood Johnson Medical School/Rutgers. Dr. Weschler earned a Ph.D. in chemistry from the University of Chicago. His research interests include chemical interactions among indoor pollutants, the chemistry of the outdoor environment as it impacts the indoor environment, indoor-outdoor relationships for vapor and condensed phase species, indoor airborne particles and their inorganic and organic constituents, indoor chemistry as a potential source of particles, understanding the factors that influence the concentrations, transport, and surface accumulations of indoor pollutants, and impacts of indoor pollutants.

HANSPETER WITSCHI is professor of toxicology and associate director of the Institute for Toxicology and Environmental Health at the University of California, Davis. He earned his M.D. from the University of Berne, Switzerland. He is a diplomate of the American Board of Toxicology and the Academy of Toxicological Sciences. Dr. Witschi's research interests include experimental toxicology, biochemical pathology, interaction of drugs and toxic agents with organ function at the cellular level, pulmonary carcinogenesis, and air pollutants and lung disease.

Appendix B

Building-Related Symptoms

People spend far more time indoors than outdoors—on average, more than 90% (Nelson et al. 1994). However, regulatory standards for environmental health hazards, other than those in the workplace,[1] focus on the outdoor environment. The unregulated indoor environment includes homes, schools, commercial spaces (e.g., shopping malls), recreation areas, and highly specialized settings such as aircraft cabins.

Aircraft cabins have environmental problems similar to those of modern offices. Both aircraft and office buildings attempt to balance energy efficiency with other needs such as adequate ventilation, clean air, and acceptable temperature and humidity levels. Office buildings and aircraft may achieve the goal of energy efficiency by decreasing the amount of outside air drawn into the ventilation system. If the ventilation rate for outside air is reduced, and the outside air is mixed with recirculated air, this can result in elevated concentrations of carbon dioxide (CO_2), volatile organic compounds (VOCs), and odors from internal sources such as cabin or building occupants. Chemical contaminants in outdoor air (e.g., motor vehicle exhaust) may enter a building. Contaminants may also originate from indoor sources, such as carpeting, adhesives, upholstery, wood paneling, office machines, cleaning agents, combustion prod-

---

[1]With recognition of BRS as an occupational health concern, the Occupational Safety and Health Administration proposed a set of rules for workplace environments on the basis that ". . . air contamination and other air-quality factors can act to present a significant risk of material impairment to employees working in indoor environments" (Fed. Regist. 59:15969).

ucts, and biological contaminants. VOCs may be emitted from construction materials such as wall panels, furniture, and office equipment (e.g., computers, printers, copiers, and fax machines). In the aircraft cabin, VOCs may come from internal sources (e.g., passengers, their belongings, aircraft component materials, cleaning materials) or enter the cabin in the bleed air (see Chapter 3). In addition to VOCs, contamination by bacteria, viruses, and fungi is another persistent challenge to indoor air quality. Office buildings provide many opportunities for microbial growth such as faulty or inadequately maintained air-circulation systems, and bioaerosols emitted by occupants. Chapter 4 discusses microbial contamination in aircraft.

Although office buildings and passenger cabins have very different external environments, the building environment may be a valuable research model for studying cabin air quality. In the sections below, some characteristics of building-related symptoms (BRS) that may provide information on aircraft cabin air quality are described.

CHARACTERIZATION OF BUILDING-RELATED SYMPTOMS

In buildings, the combination of reduced ventilation and contaminant emissions can result in serious costly, unexpected, and often unexplained health complaints by building occupants. Those complaints may arise with the installation of new office equipment or when people move into a space. BRS is a term applied to a group of complaints from a substantial number of employees or residents. This term replaces the widely used term sick-building syndrome. BRS is used to describe nonspecific symptoms (e.g., eye, nose, or throat irritation, headache, fatigue, or other discomfort) that cannot be associated with a well-defined cause but that appear to be linked to time spent in particular buildings. Although poorly defined, BRS is distinct from building-related illnesses, which are diagnosable diseases that can be directly attributed to specific indoor exposures (Menzies et al. 1995; Hedge 1995; Hodgson 1995; Menzies and Bourbeau 1997). BRS can be uncomfortable, even disabling, but permanent sequelae are rare (Redlich et al. 1997). Although there are several physiological markers for eye and mucosal effects, objective physiological abnormalities generally are not found.

BRS is characterized by the following attributes:

• Most complaints can be categorized as neurobehavioral disruption (e.g., impaired memory), sensory irritation (especially eye, nose, and throat), skin irritation, unspecific hypersensitivity reactions, and aberrant odor and taste sensations.

- Lower airway or internal organ symptoms are infrequent.
- More symptoms are reported in one building or a part of one building than elsewhere.
- Symptoms resolve shortly after leaving the building.

INVESTIGATIONS OF BRS

Investigations of BRS have focused on describing or solving a particular situation, although some have more general applicability. The California Healthy Building Study examined relationships between employee health complaints and several building, workspace, job, and personal factors (Mendell et al. 1996). Most complaints (40.3%) were of eye, nose, or throat irritation. Other complaints included fatigue and sleepiness (33.2%), headache (19.8%), dry and itchy skin (10.8%), chest tightness (7.5%), and chills and fever (4.5%). Among the 12 buildings surveyed, the prevalence of complaint differed. Occupants of the mechanically ventilated and air-conditioned buildings reported more symptoms than those of naturally ventilated buildings. Carpeting, carbonless-copy paper (which emits VOCs), and photocopiers seemed to be substantial contributors. Nonchemical factors, such as space sharing and distance from a window, also seemed to increase complaint prevalence.

Elevated CO_2 derived from occupant respiration and bioeffluents appears to be an important contributor to complaints of BRS. A review of BRS studies encompassing about 30,000 subjects (Seppänen et al. 1999) indicated a progressive reduction in symptoms as CO_2 concentrations were reduced to below 800 ppm. Apte et al. (2000) analyzed the correlation between indoor CO_2 concentrations and BRS symptoms in 41 office buildings. They relied on two measures: the daily average and the peak (1-h) differences between indoor and outdoor CO_2 concentrations. Statistical analyses were conducted by categorizing BRS complaints into mucous membrane or chest and breathing difficulties. These analyses demonstrated a significant dose-response relationship between indoor CO_2 and the incidence of BRS symptoms involving mucous membrane irritation, chest tightness, and wheezing. As CO_2 concentrations decreased below 800 ppm, complaints also decreased. The authors emphasized that CO_2 is an indicator of the building ventilation rate and not necessarily a direct cause of BRS.

Although many reports of BRS and remediation measures are available, almost no experimental data exist on the relationship between contaminant types and occupant complaints. In one experiment by U.S. Environmental Protection Agency (Otto et al. 1990), investigators synthesized a mixture of 22 VOCs and exposed volunteers at 25 mg/m³ of the mixture or to clean air. During a 240-minute exposure (longer than most domestic flights but shorter than most overseas flights), the volunteers reported increasing discomfort (irritation of eyes, nose, and throat) whether exposed to air or VOCs; however, the discomfort ratings were greater with exposure to VOCs at all times. Although a brief conventional, neuropsychological test battery revealed no significant performance deficits, a prolonged performance assay might offer a more suitable criterion for assessing such potential deficits. The sleep deprivation literature indicates that tests of longer duration are more sensitive to environmental disruption than those of shorter duration.

Such an approach was used by Wargocki and coworkers (2000), who recruited 30 females (five groups of six each) to perform simulated office work for 4.6 hours under outdoor flow rates of 3, 10, or 30 L/s per person. The performance measures included text typing, addition, proof-reading, and writing down alternative uses for common objects, which was designed to measure creative thinking. Six subjective measures were also assessed: air quality, odor intensity, eye-nose-throat irritation, environmental conditions (e.g., humidity), recognized BRS symptoms, and effort required to perform tasks. Simulated work performance, BRS complaints, and rated air quality all improved significantly with increased ventilation.

ANALOGIES WITH PASSENGER AIR CABINS

The committee recognizes that there are differences between the physical structures and operations of buildings and aircraft; however, they are both enclosed spaces occupied by people. Therefore, the committee decided that it was appropriate to discuss below some of the studies of responses of cabin crews and passengers to cabin air quality and compares them to studies on building occupants. The committee also compares contaminant concentrations in aircraft and buildings.

Cabin Crew

Unlike passengers, cabin crew are often engaged in high-level activity; they board the aircraft when exposure to aircraft engine emissions may be high, and they deal chronically with stressful conditions such as disrupted circadian rhythms. A compilation of cabin crew and passengers' complaints is presented in Table B-1 (Pelletier 1998). These complaints are similar to those reported by occupants of offices associated with BRS, however, they have not received the same scrutiny, because systematic surveys of cabin crew, comparable to those in the BRS literature, are relatively rare (see Chapter 6). Lee et al. (2000) designed a questionnaire to evaluate cabin crew responses to a variety of conditions including health complaints. Health complaints included eye, nose, and throat irritation; indices of dizziness; breathing difficulties; skin dryness; gastrointestinal problems; and nausea. The three symptoms reported most often by the crew were dry, itchy, or irritated eyes; dry or stuffy nose; and skin dryness or irritation. One-third or more of the crew rated irritation and dryness as the most severe symptoms.

For a larger Swedish study on air quality (Lindgren et al. 2000), 1,513 aircraft crew members (i.e., pilots and cabin crew) and 168 office workers were recruited. Aircrew had a higher incidence of complaints about poor air quality on aircraft than office workers did about building air quality. A major portion of such aircrew complaints is likely attributable to the extremely low relative humidity in the cabin (5%) during the international flights, compared with humidity levels of 20% or higher in buildings. For reference, the lower bound of the American Society of Heating, Refrigerating and Air-conditioning Engineers (ASHRAE) standard for relative humidity ranges from 20% to 30% depending on temperature (ASHRAE 1992). Lindgren et al. note that general complaints about the work environment, more common among flight crew members than office workers, may be attributable to other factors such as work stress.

Passengers

A survey of passengers who completed self-administered questionnaires on a variety of aircraft and flight lengths, used a 7-point scale from "poor" to "excellent" to obtain comfort ratings for a variety of characteristics including air quality (Rankin et al. 2000). Dryness and irritation, along with back or joint pain, received the lowest comfort ratings, but the mean ratings were average (4.0) or better; the variability for these ratings were not given. Although no major problems with air quality were reported by passengers, the authors noted the desirability of obtaining objective measures of air quality that can be correlated with passenger comfort ratings. Addressing this data gap is critical

TABLE B-1 Complaints of Flight Attendants and Passengers, 1993-1997

Major Symptoms	Flight Attendants		Passengers		Total	
	%	No. of complaints	%	No. of complaints	%	No. of complaints
Headache (severe)	15	279	12	67	15	346
Difficulty breathing (great)[a]	15	266	13	72	14	338
Nausea[b]	12	225	15	86	13	311
Dizziness	14	260	7	38	13	298
Fatigue (sudden)[c]	13	227	11	63	12	290
Throat problem[d]	7	132	5	30	7	162
Stuffiness/excessive temperature[e]	5	86	12	68	6	154
Lightheadness	8	136	0	1	6	137
Chemical odor problem[f]	4	78	5	29	4	107
General air-quality complaint	2	45	11	61	4	106
Eye problem[g]	2	45	3	20	3	65
Fainted	0.3	6	5	31	2	375
Heart palpitations	1	17	1	6	1	23
Others[h]	0.4	8	0	0	0.3	8
Total	100	1,810	100	572	100	2,382

[a] Difficulty breathing (great) includes lack of air or oxygen, shortness of breath, catching breath, gasping for air, chest pains, and pressure on chest.
[b] Nausea includes stomach pain, vomiting, and malaise.
[c] Fatigue (sudden) includes tiredness, weakness, faintness, and exhaustion.
[d] Throat problem includes sore throat, nose and sinus problems, ear ringing, and congestion.
[e] Stuffiness and excessive temperature includes sweating.
[f] Chemical odor problem includes strong odor, fumes, and foul smell.
[g] Eye problem includes swelling, dryness, soreness, itchiness, and blurred vision.
[h] Others includes paleness.
Source: Adapted from Pelletier (1998).

because such information would help determine the validity of the rating scale.

One common complaint of crew and passengers is eye irritation, which is also a frequent complaint in BRS situations (e.g., Wargocki et al. 2000). In response to such complaints, Backman and Haghighat (2000) surveyed air quality on 15 different aircraft at different times and altitudes. The authors periodically measured CO_2, temperature, and humidity. The data convinced them that poor air quality can cause contact lenses intolerance and eye irritation. The authors urged ventilation for aircraft be improved to reduce such symptoms.

Comparisons of Contaminant Levels in Aircraft and Buildings

Only a few studies have measured speciated organics within aircraft cabins. As indicated above, similar contaminants have been identified in both the aircraft and building environments. Table B-2 compares selected VOCs concentrations measured in aircraft cabins with concentrations measured in public, commercial, and residential buildings. The "Aircraft (1994)" column is based on measurements on 22 flights with nine different types of aircraft, and the "Aircraft (1996)" column stems from measurements on 27 flights with a single aircraft type (Boeing 777). The last four columns in the table present typical concentrations of organic compounds found in different buildings. The concentrations in aircraft do not appear to differ significantly from those found in buildings. The one exception is ethanol, which is found at significantly higher concentrations in aircraft. The ethanol concentrations probably reflect volatilization from alcoholic beverages served on the aircraft and exhalation by passengers who have consumed those beverages. However, the elevated ethanol concentrations are not expected to be a health or comfort concern as the threshold-limit value for ethanol is 1,000 ppm.

Table B-3 compares the concentrations of formaldehyde, some inorganic gases, particles, bacteria, and fungi measured in aircraft cabins with concentrations measured in public, commercial, and residential buildings. The air-quality measurements are a subset of the more complete listing provided in Chapter 1, Table 1-2. The data represent four recently published studies and, therefore, are measurements made well after smoking was banned on domestic flights. The building data are measurements from public buildings and exclude situations with either smoking or cooking. The nitrogen oxide concentrations reported in the National Institution for Occupational Safety and Health aircraft study are high relative to concentrations typically measured within buildings.

TABLE B-2 Concentrations (mg/m^3) of Selected VOCs in Aircraft Compared with Public, Commercial, and Residential Buildings

VOC	Aircraft (1994), Range[a]	Aircraft (1996), Range[a]	Building Mix, Weighted Average[b]	Office Buildings, Geo. Mean[c]	Office Buildings, Median[d]	Office Buildings, Range, Median[e]
Ethanol	280-4,300	290-2,600	50-100	36	—	7.1-220, 29
Acetone	74-150	52-140	20-50	10.2	—	—
2-Propanol	—	12-43	—	5.6	—	—
Toluene	0-29	9-19	20-50	9.8	6	1.6-360, 9
1,1,1-Trichloroethane	0-3	0-5	20-50	24.3	—	0.6-450, 3.6
meta- & para-Xylene	0-8	2-4	10-20	9.1	5	0.8-96, 5.2
Ethyl acetate	0-4	0-26	5-10	1.1	—	0.2-65
n-Decane	0-6	2-5	5-10	2.9	6	0.3-50
n-Undecane	0-20	4-20	1-5	7	9	0.6-58, 3.7
1,2,4-Trimethylbenzene	0-4	0-2	5-10	3.9	5	0.3-25
2-Butanone	3-16	4-8	1-5	—	—	0.7-18
Benzene	1-6	—	5-10	3.2	—	0.6-17, 3.7
Tetrachloroethylene	0-16	5-28	5-10	2.7	4	0.3-50
ortho-Xylene	0-3	0-2	5-10	3	2	0.3-38
n-Hexane	—	0-20	1-5	1.8	—	0.6-21, 2.9
d-Limonene	12-24	2-45	20-50	6.7	6	0.3-140, 7.1

[a] Source: Dumyahn et al. (2000).

[b] Based on a summary of 50 studies conducted in more than 1,200 buildings between 1978 and 1990. The indoor geometric mean concentrations from multiple studies are summarized as a "weighted average" of geometric means. Source: Brown et al. (1994).

[c] Geometric means of the most frequently identified VOCs in 12 California office buildings, selected without regard to worker complaints. Included are naturally and mechanically ventilated buildings. Source: Daisey et al. (1994).

[d] Selected VOCs identified in the administrative facilities (11 of the 70 buildings) and their median concentrations are reported. Source: Shields et al. (1996).

[e] Range of indoor concentrations for selected VOCs identified in the EPA BASE (Building Assessment Survey Evaluation) study. For several compounds, median concentrations are also reported. The cited database contains measurements from 56 buildings. Source: Girman et al. (1999).

TABLE B-3 Concentrations of Formaldehyde, Selected Inorganic Gases, Particles, Bacteria, and Fungi in Aircraft Compared with Public, Commercial and Residential Buildings

Contaminant	Aircraft, Range[a]	Aircraft, Range[b]	Aircraft, Range[c]	Aircraft, Range[d]	Office Buildings, Median or Range
Formaldehyde	—	—	<1-70 ppb	0-<0.07 ppb	11.4 ppb[e]
Nitric oxide (NO)	—	0-81 ppb	—	(NO & NO$_2$)	0-100 ppb[f]
Nitrogen dioxide (NO$_2$)	23-60 ppb	4-32 ppb	—	<200-3,100 ppb	30 ppb[f]
Sulfur dioxide	—	1-3 ppb	—	—	1 ppb[g]
Carbon monoxide	0.8-1.3 ppm	1.9-2.4 ppm	<0.1-7 ppm	<0.2-9.4 ppm	<1-6 ppm[h]
Carbon dioxide	1,200-1,800 ppm	418-4,752 ppm	330-3,157 ppm	310-1,600 ppm	400-1,400 ppm[i]
Ozone	2-10 ppb	0-90 ppb	2-122 ppb	<50-1,000 ppb	0-120 ppb[j]
Particles	3-10 mg/m^3	7.6 mg/m^3	25-200 mg/m^3	30-380 mg/m^3	25 mg/m^3 [h]
Bacteria	—	44-93 CFU/m^3	0-1,763 CFU/m^3	—	<7-1,000 (20) CFU/m^3 [e]
Fungi	—	17-107 CFU/m^3	0-450 CFU/m^3	—	<7-5,000 (35) CFU/m^3 [e]

[a] Source: Spenger et al. (1997).
[b] Source: Lee et al. (1999).
[c] Source: Nagda et al. (2001).
[d] Source: Waters et al. (2001).
[e] Source: Girman et al. (1999); for bacteria and fungi median values are in parentheses; CFU, colony forming units.
[f] Source: Wilson et al. (1993).
[g] Source: Brauer et al. (1991).
[h] In the absence of indoor combustion appliances, indoor carbon monoxide concentrations will be comparable to outdoor levels. In the absence of smoking, cooking or other strong particle sources, indoor particle concentrations are comparable to or lower than outdoor concentrations. In 1999, the U.S. average concentration for particulate matter with an aerodynamic diameter of 10 μm was 25 mg/m^3.
Source: Adapted from EPA (2000).
[i] Source: Nabinger et al. (1994).
[j] Source: Weschler (2000).

Other aircraft studies have not reported similarly high nitrogen oxide concentrations. The high end of CO_2 measurements in aircraft cabins appears to be above that typically measured within buildings. In aircraft, O_3 concentrations are also higher than those typically found in buildings. The results of Nagda et al. (2001) and Waters et al. (2001) indicate that the particle concentrations in aircraft may occasionally be higher than those encountered in buildings in the absence of smoking and cooking. This finding may reflect particles generated during inflight meal preparation or those brought into the cabin when the aircraft is on the ground, especially on the runway waiting for takeoff.

CONCLUSIONS

Although this appendix outlines similarities between building and aircraft environments, a comparison is limited by the lack of data on exposures in the aircraft environment. More data on exposures to contaminants on aircraft are needed. Unlike the volume of BRS research in buildings, research on the association of cabin air quality with health complaints of passengers and crew is sparse. The bulk of information about symptoms or complaints (e.g., irritation) comes from reports filed by air crews. These reports are not solicited by a regulatory or health agency or gathered in a systematic manner, and are primarily a response to air-quality incidents. As emphasized in Chapter 6, systematic information about symptoms must be acquired using appropriate methods and tools for measuring subjective health and comfort variables such as irritation and fatigue.

REFERENCES

ASHRAE (American Society of Heating Refrigerating and Air-Conditioning Engineers). 1992. Thermal Environmental Conditions for Human Occupancy. ANSI/ASHRAE 55-1992. American Society of Heating, Refrigerating, and Air-Conditioning Engineers, Atlanta, GA.

Apte, M.G., W.J. Fisk, and J.M. Daisey. 2000. Associations between indoor CO2 concentrations and sick building syndrome symptoms in U.S. office buildings: an analysis of the 1994-1996 BASE study data. Indoor Air 10(4):246-257.

Backman H., and F. Haghighat. 2000. Air quality and ocular discomfort aboard commercial aircraft. Optometry 71(10):653-656.

Brauer, M., P. Koutrakis, G. J. Keeler, and J.D. Spengler. 1991. Indoor and outdoor concentration of inorganic acidic aerosols and gases. J. Air Waste Manage. Assoc. 41(2):171-181.

Brown, S.K., M.R. Sim, M.J. Abramson, and C.N. Gray. 1994. Concentrations of volatile organic compounds in indoor air - a review. Indoor Air. 4:123-134.

Daisey, J.M., A.T. Hodgson, W.J. Fisk, M.J. Mendell, and J. Ten Brinke. 1994. Volatile organic compounds in twelve California office buildings: classes, concentrations and sources. Atmos. Environ. 28(22):3557-3562.

Dumyahn, T.S., J.D. Spengler, H.A. Burge, and M. Muilenburg. 2000. Comparison of the environments of transportation vehicles: results of two surveys. Pp. 3-25 in Air Quality and Comfort in Airliner Cabins, N. Nagda, ed. West Conshohocken, PA: American Society for Testing and Materials.

EPA (U.S. Environmental Protection Agency). 2000. Latest Findings on National Air Quality: 1999 Status and Trends. EPA-454/F-00-002. Office of Air Quality, U.S. Environmental Protection Agency. August 2000.

Girman, J.R., G.E. Hadwen, L.E. Burton, S.E. Womble and J.F. McCarthy. 1999. Individual volatile organic compound prevalence and concentrations in 56 buildings of the Building Assessment and Evaluation (BASE) study. Pp. 460-465 in Indoor Air 99, Vol. 2., G. Raw, C. Aizlewood, and P. Warren, eds. London: Construction Research Communications.

Hedge, A. 1995. In defence of "the sick building syndrome." Indoor Environ. 4(5):251-253.

Hodgson, M. 1995. The sick-building syndrom. Occup. Med. 10(1):167-175.

Lee, S.-C., C.-S. Poon, X.-D. Li, and F. Luk. 1999. Indoor air quality investigation on commercial aircraft. Indoor Air 9(3):180-187.

Lee, S.C., C.S. Poon, X.D. Li, F. Luk, M. Chang, and S. Lam. 2000. Questionnaire survey to evaluate the health and comfort of cabin crew. Pp. 259-268 in Air Quality and Comfort in Airliner Cabins, N.L. Nagda, ed. West Conshohocken, PA: American Society for Testing and Materials.

Lindgren, T., D. Norback, K. Andersson, and B.G. Dammstrom. 2000. Cabin environment and perception of cabin air quality among commercial aircrew. Aviat. Space Environ. Med. 71(8):774-782.

Mendell, M.J., W.J. Fisk, J.A. Deddens, W.G. Seavey, A.H. Smith, D.F. Smith, A.T. Hodgson, J.M. Daisey, and L.R. Goldman. 1996. Elevated symptom prevalence associated with ventilation type in office buildings. Epidemiology 7(6):583-589.

Menzies, D., and J. Bourbeau. 1997. Building-related illnesses. N. Engl. J. Med. 337(21):1524-1531.

Menzies, R., J. Pasztor, J. Leduc, and F. Nunes. 1995. The "sick building" -a misleading term that should be abandoned. Pp. 47-58 in IAQ '94: Engineering Indoor Environments, E.L. Besch, ed. Atlanta: American Society of Heating, Refrigerating and Air-Conditioning Engineers, Inc.

Nabinger, S.J., A.K. Persily, and W.S. Dols. 1994. Study of ventilation and carbon dioxide in an office building. ASHRAE Trans. 100(2).

Nagda, N.L., H.E. Rector, Z. Li, and E.H. Hunt. 2001. Determine Aircraft Supply Air

Contaminants in the Engine Bleed Air Supply System on Commercial Aircraft. ENERGEN Report AS20151. Prepared for American Society of Heating, Refrigerating, and Air-Conditioning Engineers, Atlanta, GA, by ENERGEN Consulting, Inc., Germantown, MD. March 2001.

Nelson, W.C., W.R. Ott, and J.P. Robinson. 1994. The National Human Activity Pattern Survey (NHAPS): Use of Nationwide Activity Data for Human Exposure Assessment. Atmospheric Research and Exposure Assessment Lab. Prepared in cooperation with Environmental Protection Agency, Research Triangle Park, NC. Presented at the Air and Waste Management Association 87th Annual Meeting, Cincinnati, OH. June 1994. NTIS PB94-197464. 17pp.

Otto, D., L. Molhave, G. Rose, H.K. Hudnell, and D. House. 1990. Neurobehavioral and sensory irritant effects of controlled exposure to a complex mixture of volatile organic compounds. Neurotoxicol. Teratol. 12(6):649-652.

Pelletier, F. 1998. The Perspective of Canadian Flight Attendants on Cabin Air Issues. A Presentation to the ASHRAE Aviation Subcommittee of TC 9.3 Transportation, June 23, 1998. Health and Safety Chairperson, Airline Division of the Canadian Union of Public Employees. [Online]. Available: http://www.airdiv-cupe.org/health.phtml [Oct. 4, 2001].

Rankin, W.L., D.R. Space, and N.L. Nagda. 2000. Passenger comfort and effect of air quality. Pp. 269-290 in Air Quality and Comfort in Airliner Cabins, N.L. Nagda, ed. West Conshohocken, PA: American Society for Testing and Materials.

Redlich, C.A., J. Sparer, and M.R. Cullen. 1997. Occupational medicine: sick-building syndrom. Lancet. 349(9057):1013-1016.

Seppanen, O.A., W.J. Fisk, and M.J. Mendell. 1999. Association of ventilation rates and CO2 concentrations with health and other responses in commercial and institutional buildings. Indoor Air 9(4):226-252.

Shields, H.C., D.M. Fleischer, and C.J. Weschler. 1996. Comparisons among VOCs measured in three types of U.S. commercial buildings with different occupant densities. Indoor Air 6:2-17.

Spengler, J., H. Burge, T. Dumyahn, M. Muilenberg, and D. Forester. 1997. Environmental Survey on Aircraft and Ground-Based Commercial Transportation Vehicles. Prepared by Department of Environmental Health, Harvard University School of Public Health, Boston, MA, for Commercial Airplane Group, The Boeing Company, Seattle, WA. May 31, 1997.

Wargocki, P., D.P. Wyon, J. Sundell, G. Clausen, and P.O. Fanger. 2000. The effects of outdoor air supply rate in an office on perceived air quality, sick building syndrome (SBS) symptoms and productivity. Indoor Air 10(4):222-236.

Waters, M., T. Bloom, and B. Grajewski. 2001. Cabin Air Quality Exposure Assessment. National Institute for Occupational Safety and Health, Cincinnati, OH. Federal Aviation Administration Civil Aeromedical Institute. Presented to the NRC Committee on Air Quality in Passenger Cabins of Commercial Aircraft, January 3, 2001. National Academy of Science, Washington, DC.

Weschler, C.J. 2000. Ozone in indoor environments: concentration and chemistry. Indoor Air 10(4):269-288.

Wilson, A.L., S.D. Colome, and Y. Tian. 1993. California Residential Indoor Air Study. Vol.1. Methodology and Descriptive Statistics. Integrated Environmental Service, Irvine, CA and Gas Research Institute, Chicago, IL. May 93. 125pp. NTIS PB94-1660551.

Appendix C

Relevant Federal Aviation Regulations

Title 14, Chapter 1 of the Code Federal of Regulations (CFR) pertains to the Federal Aviation Administration.. Within 14 CFR are found the federal aviation regulations (FARs) applicable to the design and operation of commercial aircraft. Section 25 contains air-worthiness standards for transport category airplanes, Section 121 pertains to the operational requirements for air carriers and commercial operators, and Section 125 specifies the certification and operations for aircraft having a seating capacity of 20 or more passengers and the rules that govern people aboard such aircraft.

SEC. 25.831 VENTILATION.

(a) Under normal operating conditions and in the event of any probable failure conditions of any system which would adversely affect the ventilating air, the ventilation system must be designed to provide a sufficient amount of uncontaminated air to enable the crew members to perform their duties without undue discomfort or fatigue and to provide reasonable passenger comfort. For normal operating conditions, the ventilation system must be designed to provide each occupant with an airflow containing at least 0.55 pounds of fresh air per minute.

(b) Crew and passenger compartment air must be free from harmful or hazardous concentrations of gases or vapors. In meeting this requirement, the following apply:

1. Carbon monoxide concentrations in excess of 1 part in 20,000 parts of air are considered hazardous. For test purposes, any acceptable carbon monoxide detection method may be used.

2. Carbon dioxide concentration during flight must be shown not to exceed 0.5 percent by volume (sea level equivalent) in compartments normally occupied by passengers or crew members.

[Doc. No. 5066, 29 FR 18291, Dec. 24, 1964, as amended by Amdt. 25-41, 42 FR 36970, July 18, 1977; Amdt. 25-87, 61 FR 28695, June 5, 1996; Amdt. 25-89, 61 FR 63956, Dec. 2, 1996]

SEC. 25.832 CABIN OZONE CONCENTRATION.

(a) The airplane cabin ozone concentration during flight must be shown not to exceed: (1) 0.25 parts per million by volume, sea level equivalent, at any time above flight level 320, and

(2) 0.1 parts per million by volume, sea level equivalent, time-weighted average during any 3-hour interval above flight level 270.

(b) For the purpose of this section, "sea level equivalent" refers to conditions of 25 deg. C and 760 millimeters of mercury pressure.

(c) Compliance with this section must be shown by analysis or tests based on airplane operational procedures and performance limitations, that demonstrate that either (1) the airplane cannot be operated at an altitude which would result in cabin ozone concentrations exceeding the limits prescribed by paragraph (a) of this section; or (2) the airplane ventilation system, including any ozone control equipment, will maintain cabin ozone concentrations at or below the limits prescribed by paragraph (a) of this section.

[Amdt. 25-50, 45 FR 3883, Jan. 1, 1980, as amended by Amdt. 25-56, 47 FR 58489, Dec. 30, 1982; Amdt. 25-94, 63 FR 8848, Feb. 23, 1998]

SEC. 25.841 PRESSURIZED CABINS.

(a) Pressurized cabins and compartments to be occupied must be equipped to provide a cabin pressure altitude of not more than 8,000 feet at the maximum operating altitude of the airplane under normal operating conditions.

1. If certification for operation above 25,000 feet is requested, the airplane must be designed so that occupants will not be exposed to cabin pressure altitudes in excess of 15,000 feet after any probable failure condition in the pressurization system.

2. The airplane must be designed so that occupants will not be exposed to a cabin pressure altitude that exceeds the following after decompression from any failure condition not shown to be extremely improbable: (i) twenty-five thousand (25,000) feet for more than 2 minutes; or (ii) forty thousand (40,000) feet for any duration.

3. Fuselage structure, engine and system failures are to be considered in evaluating the cabin decompression.

(b) Pressurized cabins must have at least the following valves, controls, and indicators for controlling cabin pressure:

1. Two pressure relief valves to automatically limit the positive pressure differential to a predetermined value at the maximum rate of flow delivered by the pressure source. The combined capacity of the relief valves must be large enough so that the failure of any one valve would not cause an appreciable rise in the pressure differential. The pressure differential is positive when the internal pressure is greater than the external.

2. Two reverse pressure differential relief valves (or their equivalents) to automatically prevent a negative pressure differential that would damage the structure. One valve is enough, however, if it is of a design that reasonably precludes its malfunctioning.

3. A means by which the pressure differential can be rapidly equalized.

4. An automatic or manual regulator for controlling the intake or exhaust airflow, or both, for maintaining the required internal pressures and airflow rates.

5. Instruments at the pilot or flight engineer station to show the pressure differential, the cabin pressure altitude, and the rate of change of the cabin pressure altitude.

6. Warning indication at the pilot or flight engineer station to indicate when the safe or preset pressure differential and cabin pressure altitude limits are exceeded. Appropriate warning markings on the cabin pressure differential indicator meet the warning requirement for pressure differential limits and an aural or visual signal (in addition to cabin altitude indicating means) meets the warning requirement for cabin pressure altitude limits if it warns the flight crew when the cabin pressure altitude exceeds 10,000 feet.

7. A warning placard at the pilot or flight engineer station if the structure is not designed for pressure differentials up to the maximum relief valve setting in combination with landing loads.

8. The pressure sensors necessary to meet the requirements of paragraphs (b)(5) and (b)(6) of this section and Sec. 25.1447(c), must be located and the sensing system designed so that, in the event of loss of cabin pressure in any passenger or crew compartment (including upper and lower lobe galleys), the warning and automatic presentation devices, required by those provisions, will be actuated without any delay that would significantly increase the hazards resulting from decompression.

[Doc. No. 5066, 29 FR 18291, Dec. 24, 1964, as amended by Amdt. 25-38, 41 FR 55466, Dec. 20, 1976; Amdt. 25-87, 61 FR 28696, June 5, 1996]

SEC. 121.219 VENTILATION.

Each passenger or crew compartment must be suitably ventilated. Carbon monoxide concentration may not be more than one part in 20,000 parts of air, and fuel fumes may not be present. In any case where partitions between compartments have louvres or other means allowing air to flow between compartments, there must be a means convenient to the crew for closing the flow of air through the partitions, when necessary.

SEC. 121.578 CABIN OZONE CONCENTRATION.

(a) For the purpose of this section, the following definitions apply:

1. "Flight segment" means scheduled nonstop flight time between two airports.

2. "Sea level equivalent" refers to conditions of 25 deg. C and 760 millimeters of mercury pressure.

(b) Except as provided in paragraphs (d) and (e) of this section, no certificate holder may operate an airplane above the following flight levels unless it is successfully demonstrated to the Administrator that the concentration of ozone inside the cabin will not exceed (1) for flight above flight level 320, 0.25 parts per million by volume, sea level equivalent, at any time above that flight level; and (2) for flight above flight level 270, 0.1 parts per million by volume, sea level equivalent, time-weighted average for each flight segment that exceeds 4 hours and includes flight above that flight level. (For this purpose, the amount of ozone below flight level 180 is considered to be zero.)

(c) Compliance with this section must be shown by analysis or tests, based on either airplane operational procedures and performance limitations or the certificate holder's operations. The analysis or tests must show either of the following:

1. Atmospheric ozone statistics indicate, with a statistical confidence of at least 84%, that at the altitudes and locations at which the airplane will be operated, cabin ozone concentrations will not exceed the limits prescribed by paragraph (b) of this section.

2. The airplane ventilation system including any ozone control equipment, will maintain cabin ozone concentrations at or below the limits prescribed by paragraph (b) of this section.

(d) A certificate holder may obtain an authorization to deviate from the requirements of paragraph (b) of this section, by an amendment to its operations specifications, if (1) it shows that due to circumstances beyond its control or to unreasonable economic burden it cannot comply for a specified period of time; and (2) it has submitted a plan acceptable to the Administrator to effect compliance to the extent possible.

(e) A certificate holder need not comply with the requirements of paragraph (b) of this section for an aircraft (1) when the only persons carried are flight crewmembers and persons listed in Sec. 121.583; (2) if the aircraft is scheduled for retirement before January 1, 1985; or (3) if the aircraft is scheduled for re-engining under the provisions of Subpart E of Part 91, until it is re-engined.

[Doc. No. 121-154, 45 FR 3883, Jan. 21, 1980. Redesignated by Amdt. 121-162, 45 FR 46739, July 10, 1980; Amdt. 121-181, 47 FR 58489, Dec. 30, 1982; Amdt. 121-251, 60 FR 65935, Dec. 20, 1995]

SEC. 125.117 VENTILATION.

Each passenger or crew compartment must be suitably ventilated. CO concentration may not be more than 1 part in 20,000 parts of air, and fuel fumes may not be present. In any case where partitions between compartments have louvres or other means allowing air to flow between compartments, there must be a means convenient to the crew for closing the flow of air through the partitions when necessary.

Appendix D

Additional Monitoring Techniques

In addition to the monitoring methods described in Chapter 7, a host of other techniques exist for monitoring different characteristics of the cabin air or of the occupants themselves. These additional techniques provide more specific information on exposures to cabin air contaminants or on the performance of the environmental control system (ECS) equipment. Some of these techniques have been proposed or used in research investigations of aircraft air quality. These techniques might be considered for subsequent adoption as applications and resources warrant. Several examples of additional monitoring approaches follow.

ELEMENTAL DETECTORS

Phosphorus

Methods exist that are specific for phosphorus-containing compounds such as phosphate ester pesticides and organophosphate esters that are used as additives in certain fluids. Instrumentation is sensitive but expensive. Current instrumentation is not well suited to aircraft environments.

Sulfur

Sulfur is present as an impurity in jet fuel. It may impair the function of an ozone (O_3) scrubber if aspirated into bleed air flow. Real time sulfur dioxide

(SO_2) monitors are available. However, data from such measurements would be of limited utility and do not justify the expense or the use of limited space.

HYDROCARBON DETECTORS

Photoioniziation detectors respond to a large number of organic compounds but are non-specific and have limited sensitivity. In general they are not sensitive enough to detect many of the hydrocarbons at levels that are of potential concern in this setting. Furthermore, these instruments respond with different sensitivity to different compounds, and they are plagued by many of the problems that affect volatile organic compounds (VOC) measurements including drift, surface losses, and marked interference from water vapor. Primarily because of their non-specific response, they are probably of limited additional value unless a measure of total hydrocarbon concentration would be useful.

Catalytic hydrocarbon detectors measure carbon dioxide (CO_2) before and after air is passed over a hydrocarbon oxidation catalyst, which provides a measure of oxidizable carbon species. Such systems must be designed to scrub the majority of CO_2 from the airstream prior to before/after measurement or signal will be swamped by atmospheric CO_2; the same is true of methane and propane. Even taking such precautions, signals will probably still be dominated by ethanol from both breath and beverage service and by acetone from breath. Against this large and varying background, it would be quite difficult to discern changes in CO_2 resulting from a condition such as leaked fluid in the bypass air. Consequently, this approach does not appear to be feasible.

PARTICLE DETECTORS

In addition to the light scattering instrument discussed in Chapter 7, there are direct-reading particle methods based on the behavior of electrically charged particles (Hinds 1999). Instruments using electrical charge include commercial smoke detectors as well as more technically sophisticated electrical aerosol analyzers. Smoke detectors employ an ionizing radiation source to generate electric charge on particles, and the resulting change in electric current is used to sense the presence of particles in air. These devices respond within seconds to relatively high concentrations of fine particles (e.g., combustion aerosols), but may not be suitable for continuous monitoring of lower levels aboard aircraft. Electrical aerosol analyzers have the ability to evaluate particle concentration as a function of particle size, and would thus provide useful information about particle size distribution not obtained from the optical devices described earlier. Such instruments are not now available in compact, portable form, and their cost is also likely to prohibit their use in routine monitoring.

BIOLOGICAL INDICATORS

Some reports of aircraft crew health problems associated with air quality have suggested that exposure to phosphate ester compounds, such as tricresyl phosphate, or their pyrolysis products may have elicited neurologic symptoms. Exposure data relevant to this question are completely lacking because air monitoring equipment has not been in place on the affected aircraft. One possible solution to this problem may lie in the area of biological monitoring for exposure markers. For example, it has been demonstrated recently that alkyl phosphate compounds are present in the urine of workers exposed to organophosphate pesticides (WHO 1996; Lauwerys and Hoet 1993). These pesticides are chemically similar to the phosphate esters commonly used as additives in hydraulic fluids and lubricating fluids employed in aircraft engines and auxiliary power units, and identified as possible causes of neurological problems in cabin crew members (Centers 1992; Craig and Barth 1999; Crane et al. 1983; Daughtrey et al. 1996; Earl and Thompson 1952; Mackerer et al. 1999; Rubey et al. 1996; Wright 1996; Wyman et al. 1993).

The metabolism of these additives in humans produces alkyl- and aryl-substituted phosphates in close analogy to the metabolic fates of the organophosphate pesticides. Therefore, biological indicators of exposure to the phosphate esters may be available. The metabolites are expected to appear in urine within 24-48 h of exposure, but there is evidence that the metabolites continue to be detectable in urine for as long as 14 days after a single exposure (WHO 1996). The analytical method for these metabolites in urine is very sensitive, with lower limits of detection of 0.05 μmole/L of urine or lower having been reported (Nutley and Cocker 1993). Further, studies in agricultural workers have shown that metabolites can be detected in urine samples from workers who display no symptoms of acetyl cholinesterase inhibition (Nutley and Cocker 1993). Thus, the biological indicator of exposure is useful in identifying workers who have been exposed at levels below those associated with acute clinical effects.

Therefore, implementing a biological monitoring program might be possible based on collection of a urine sample either at the end of a flight, or prior to the start of the next flight, after a possible exposure to phosphate esters or their by-products. It may not be necessary to impose this testing as a routine procedure, but biological monitoring could be used whenever a suspected exposure has occurred and complaints are reported. Although this procedure would provide useful objective information regarding the recent exposure history for each individual providing a sample, biological sampling also has certain negative attributes. These negatives include necessary invasion of privacy, need to obtain informed consent, and the additional effort required to keep confidential the data resulting from analyses (Schulte and Sweeney 1995). Very recent work has suggested that in some instances saliva may be substituted for voided urine as an appropriate sampling medium for biological monitoring (Lu et al. 1998). If this were shown to be feasible for phosphate esters, subjects are likely to prefer saliva sampling to urine sampling.

REFERENCES

Craig, P.H., and M.L. Barth. 1999. Evaluation of the hazards of industrial exposure to tricresyl phosphate: a review and interpretation of the literature. J. Toxicol. Environ. Health B Crit. Rev. 2(4):281-300.

Crane, C.R., D.C. Sanders, B.R. Endecott, and J.K. Abbot. 1983. Inhalation Toxicology: III. Evaluation of Thermal Degradation Products From Aircraft and Automobile Engine Oils, Aircraft Hydraulic Fluid, and Mineral Oil. FAA-AM-83-12. Washington, DC: Federal Aviation Administration.

Centers, P. 1992. Potential neurotoxin formation in thermally degraded synthetic ester turbine lubricants. Arch. Toxicol. 66(9):679-680.

Daughtrey, W., R. Biles, B. Jortner, and M. Ehrich. 1996. Subchronic delayed neurotoxicity evaluation of jet engine lubricants containing phosphorus additives. Fundam. Appl. Toxicol. 32(2):244-249.

Earl, C., and R. Thompson. 1952. The inhibitory action of tri-ortho-cresyl phosphate on cholinesterases. Br. J. Pharmacol. 7:261-269.

Hinds, W.C. 1999. Aerosol Technology: Properties, Behavior, and Measurement of Airborne Particles, 2nd Ed. New York: Wiley.

Lauwerys, R., and P. Hoet. 1993. Industrial Chemical Exposure: Guidelines for Biological Monitoring, 2nd. Ed. Boca Raton, FL: Lewis.

Lu, C., L.C. Anderson, M.S. Morgan, and R.A. Fenske. 1998. Salivary concentrations of atrazine reflect free atrazine plasma levels in rats. J. Toxicol. Environ. Health A 53(4):283-292.

Mackerer, C.R., M.L. Barth, A.J. Krueger, B. Chawla, and T.A. Roy. 1999. Comparison of neurotoxic effects and potential risks from oral administration or ingestion of tricresyl phosphate and jet engine oil containing tricresyl phosphate. J. Toxicol. Environ. Health A. 57(5):293-328.

Nutley, B., and J. Cocker. 1993. Biological monitoring of workers occupationally exposed to organophosphorus pesticides. Pestic. Sci. 38(4):315-322.

Rubey, W., R.C. Striebich, J. Bush, P.W. Centers, and R.L. Wright. 1996. Neurotoxin formation from pilot-scale incineration of synthetic ester turbine lubricants with a triaryl phosphate additive. Arch. Toxicol. 70(8):508-509.

Schulte, P.A., and M.H. Sweeney. 1995. Ethical considerations, confidentiality issues, rights of human subjects, and uses of monitoring data in research and regulation. Environ. Health Perspect. 103(suppl.3):69-74.

WHO (World Health Organization). 1996. Selected pesticides. Organophosphorus pesticides. Pp. 237-251 in Biological Monitoring of Chemical Exposure in the Workplace: Guidelines, Vol.1. WHO/HPR/OCH 96.1 Geneva: WHO.

Wright, R. 1996. Formation of the neurotoxin TMPP from TMPE-phosphate formation. Tribology Transactions 39:827-834.

Wyman, J., E. Pitzer, F. Williams, J. Rivera, A. Durkin, J. Gehringer, P. Serve, D. von Minden, and D. Macys. 1993. Evaluation of shipboard formation of a neuro-toxicant (trimethyolpropane phosphate) from thermal decomposition of synthetic aircraft engine lubricant. Am. Ind. Hyg. Assoc. J. 54(10):584-592.

Appendix E

Glossary

ACGIH - American Conference of Governmental Industrial Hygienists.

AFA - Association of Flight Attendants.

AIDS - FAA's Accident/Incident Data System.

Airborne transmission - Exposure to airborne infectious agents in droplet nuclei is considered *airborne transmission*. Droplet nuclei and particles of dust <5 µm can remain suspended in still air longer than larger particles. These particles also can be carried further on air currents and travel deeper into the lungs.

Air-conditioning pack - A cooling device that accepts high temperature and high-pressure bleed air and that expands, cools, and dehumidifies this air to appropriate pressure, temperature, and humidity conditions to be supplied to the aircraft cabin.

Aircraft cabin - The portion of a passenger aircraft intended to be occupied by passengers.

Air exchange rate - The rate that an equivalent volume of air in the cabin is replaced with outside air.

APU - see Auxiliary Power Unit.

ASHRAE - American Society of Heating, Refrigerating, and Air-Conditioning Engineers.

AsMA - Aerospace Medical Association.

ASRS - Aviation Safety Reporting System.

ATA - American Transport Association of America.

Attack rate - The proportion of persons exposed to an infectious agent who become infected.

321

Auxiliary Power Unit (APU) - A turbine engine that is used to power electric generators and provide bleed air for pneumatic and environmental control system use. The APU is normally used during ground operations when the main engines are not operating or are not operating at conditions which allow them to fulfill these needs.

AWS - FAA's Airworthiness Directive.

Bioeffluents - Gases, such as CO_2 in exhaled breath, human body odors, and volatile compounds produced by fungal and bacterial growth, released by humans, animals, microorganisms, or plants.

Bleed air - Compressed air extracted from the compressor section of a turbine engine.

BRE - Building Research Establishment.

BRS - Building Related Symptoms.

Cabin - The section of an aircraft occupied by passengers.

Cabin crew - The flight attendants who are responsible for the safety and comfort of the passengers. Because the majority of exposure and health-effects data has been collected in the cabin and might not be applicable to the cockpit, this report focuses on the cabin crew except where data are explicitly applicable to the cockpit crew. However, many issues that are pertinent to the cabin crew are relevant to the cockpit crew.

Cabin pressure altitude - The distance above sea level at which the atmosphere exerts the same pressure as the actual pressure in the aircraft cabin. Cabin pressure altitude is the static pressure measured within the pressurized fuselage (i.e., cockpit and cabin) that represents the equivalent absolute ambient static pressure at a given altitude for a specific standard day or reference day conditions. The cabin pressure altitude is governed by the pressure schedule as set by the airplane manufacturer. Typical commercial transport airplane pressure schedules top out at a pressure of 10.92 pounds per square inch, which is equivalent to an altitude of 8,000 feet at U.S. standard atmospheric conditions.

Carrier - A person who harbors a specific infectious agent without visible symptoms of the disease; a carrier acts as a potential source of infection to others.

CDC - Centers for Disease Control and Prevention.

CFR - Code of Federal Regulations.

Circulation - Air movement within the aircraft cabin.

CO - Carbon monoxide.

CO_2 - Carbon dioxide.

Cockpit crew - Pilots and flight engineers.

COHb - Carboxyhemoglobin.

Communicable disease - An illness due to a specific infectious agent or its toxic products that arises through transmission of that agent or its products from an infected person, animal, or inanimate reservoir to a susceptible host (synonym: infectious disease).

Communicable period - The time during which an infected person or a carrier can transmit an infectious agent.

Contact - A person or animal that has been in an association with an infected person, animal, or contaminated environment that might provide an opportunity to acquire the infective agent.

Contact transmission -

Direct contact involves the touching of wounds or mucous membranes of one person with contaminated body fluids from another person.

Indirect contact involves the sharing by an infected person and another person of an item that is contaminated with an infectious agent (e.g., soiled tissues, toys, eating utensils, and other items that are touched by hand, nose, or mouth).

Droplet contact involves large particles (>5 μm) that an infected person or a carrier releases during sneezing, coughing, spitting, singing, or talking. Contact occurs when droplets containing infectious agents from an infected person are projected onto the eyes, nose, or mouth of another person, usually within a distance of no more than a few meters.

Contaminants - Any unwanted substance in aircraft cabin air.

COPD - Chronic Obstructive Pulmonary Disease.

Diffuser - A device used to distribute inlet air into the aircraft cabin in the desired manner.

Disinsection - Use of insecticides to exterminate insect pests.

DOT - U.S. Department of Transportation.

Droplets and droplet nuclei - The dried residue of a *droplet* that remains after liquid evaporates is called a *droplet nucleus*. A droplet nucleus would include any microorganisms contained in the original droplet. The important differences between droplets and droplet nuclei are: (1) their size (respectively, greater or less than ~5 μm); (2) the distance that they can travel (respectively, less or more than a few meters); and (3) the site in the respiratory tract at which they deposit (respectively, the airways of the head and the upper respiratory system or the lower lungs). Transfer of an infectious agent in a droplet is considered a form of *contact*, whereas transfer by droplet nuclei is considered an *airborne* transmission.

ECS - see Environmental Control System.

Endotoxin - A class of lipopolysaccharide-protein complexes that are an integral part of the outer membrane of Gram-negative bacteria.

Environmental Control System (ECS) - The combination of equipment and controls used to maintain the environmental conditions in the aircraft cabin.

EPA - U.S. Environmental Protection Agency.

ETS - Environmental Tobacco Smoke.

FAA - U.S. Federal Aviation Administration.

FAR - Federal Aviation Regulation.

FEF - Forced expiratory flow.

FEV_1 - Forced expiratory volume in the first second.

Filtration - Any means of removing contaminants from an air stream, including mechanisms such as mechanical filtration, chemical adsorption, and catalytic reduction.

Flight crew - Pilots and flight engineers employed only on the flight-deck.

Flight deck - Cockpit area of an aircraft.

Fresh Air - Synonymous with outside air. The term fresh air does not imply that it is uncontaminated air.

FVC - Forced vital capacity.

Gasper - An individually controlled air inlet device that a passenger may use to direct a flow of air onto herself or himself.

HEPA - High Efficiency Particulate Filter.

IAM - International Association of Machinists and Aerospace Workers.

Infection - The development or multiplication of a microorganism in the body. (Infection does not always result in a recognizable disease, and individuals sometimes carry pathogens without becoming infected. A person may be infectious (i.e., able to transmit an agent to others) without experiencing symptoms.)

Infectious disease - Clinically apparent or manifest infection with outward signs.

Infectious disease outbreak - Occurrence of two or more cases of infection in a limited time period and geographic region. The first cases that are identified are called primary cases; persons subsequently infected by the primary cases are called secondary cases.

JAA - Joint Aviation Authority

JAR - Joint Aviation Regulation.

MSDS - Material Safety Data Sheets.

MVOC - Microbial volatile organic compound.

NAAQS - National Ambient Air Quality Standard.

Narrow body - Aircraft with fuselages whose diameter is about 12 feet. These aircraft typically have one aisle, with 5 or 6 seats across in the coach section and 4 seats across in the first-class section.

NASA - National Aeronautics and Space Administration.

NIOSH - National Institute for Occupational Safety and Health.

NO_2 - Nitrogen dioxide.

NO_x - Nitrogen oxides, species unspecified.

Notifiable disease - Those diseases for which regular, frequent, and timely information on individual cases is considered necessary for prevention and control; examples include mumps, pertussis, measles, tuberculosis and varicella (chickenpox).

NTSB - National Transportation Safety Board.

O_3 - Ozone.

OSHA - Occupational Safety and Health Adminstration.

Outside air - Air brought into the aircraft cabin from a source outside of the aircraft.

PAH - Polycyclic aromatic hydrocarbon.

Partial pressure - Pressure exerted by a single gas in a mixture of gases; commonly expressed in millimeters of mercury.

Partial pressure of oxygen (PO_2) - The pressure that would be exerted by the oxygen in the air if all other chemical components of air were removed and only the oxygen remained.

PEL - OSHA's permissible exposure limit.

PM - Particulate matter.

$PM_{2.5}$ - Particulate matter less than 2.5 microns in diameter.

PO_2 - see Partial pressure of oxygen.

Pressurization - The increase in pressure of the aircraft cabin air above the ambient outside atmospheric air pressure.

Recirculation - Use of recirculated air in the cabin ventilation system.

Recirculation air - Air that is extracted from the aircraft cabin and then reintroduced to the cabin through the cabin ventilation system.

Reference Concentration (RfC) - RfC is an estimate of a daily exposure to the human population (including sensitive subgroups) that is likely to be without an appreciable risk of deleterious effects during a lifetime.

Relative humidity - The amount of moisture in air compared to the maximal amount the could contain at the same temperature; expressed as a percentage.

Respirable suspended particles (RSP) - Airborne material, including dusts, mists, smoke, and fumes, that is small enough to penetrate the lungs on inhalation (approximately 2.5 µm or less).

RH - see Relative humidity.

RSP - see Respirable suspended particles.

SDRS - FAA's Surveillance Difficulty Reporting System.

SMAC - Spacecraft Maximum Allowable Concentrations.

SO_2 - Sulfur dioxide.

STEL - Short-term exposure level.

TCP - Tri-cresyl phosphate.

TLV - Threshold limit value.

TMPP - Trimethylolpropane phosphate.

TOCP - Tri-ortho-cresyl phosphate.

Tuberculin skin test - An immunological test for tuberculosis in which a purified protein derivative from *Mycobacterium tuberculosis* (called tuberculin) is injected subcutaneously on the lower part of the arm, resulting in a temporary induration (lump) in two to three days if the tested person was previously exposed to the bacterium.

TWA - Time-weighted average.

Ventilation - The process of supplying outside air to the aircraft, distributing this air to the cabin, and providing adequate air motion within the cabin to prevent the air within the cabin from having excessive levels of contamination; may include outside air and recirculated air.

Ventilation rate - The flow rate of outside air supplied to the cabin for ventilation; does not normally include recirculated air even though recirculated air may be used for cabin ventilation.

VOC - Volatile organic compound.

WHO - World Health Organization.

Wide body - Aircraft with fuselages whose diameter is about 20 feet. These aircraft typically have two aisles, with 7-10 seats across in the coach section and 6 seats across in the first-class section.